Tourism and National Parks

In 1872 Yellowstone was established as a national park. The name caught the public's imagination and by the close of the century, other national parks had been declared, not only in the US, but also in Australia, Canada and New Zealand. Yet as it has spread, the concept has evolved and diversified. In the absence of any international controlling body, individual countries have been free to adapt the concept for their own physical, social and economic environments. Some have established national parks to protect scenery, others to protect ecosystems or wildlife. Tourism has also been a fundamental component of the national parks concept from the beginning and predates ecological justifications for national park establishment, though it has been closely related to landscape conservation rationales at the outset.

Approaches to tourism and visitor management have varied. Some have stripped their parks of signs of human settlement, while increasingly, others are blending natural and cultural heritage, and reflecting national identities. This edited volume explores in detail the origins and multiple meanings of national parks and their relationship to tourism in a variety of national contexts. It consists of a series of introductory overview chapters followed by case study chapters from around the world, including insights from the US, Canada, Australia, the UK, Spain, Sweden, Indonesia, China and Southern Africa.

Taking a global comparative approach, this book examines how and why national parks have spread and evolved, how they have been fashioned and used, and the integral role of tourism within national parks. The volume's focus on the long-standing connection between tourism and national parks, and the changing concept of national parks over time and space give the book a distinct niche in the national parks and tourism literature. The volume is expected to contribute, not only to tourism and national park studies at the upper level undergraduate and graduate levels, but also to courses in international and comparative environmental history, conservation studies and outdoor recreation management.

Warwick Frost is Senior Lecturer in Tourism and Heritage at Monash University, Melbourne, Australia. His research interests include environmental history, ecotourism, cultural heritage and the interplay between tourism and the media.

C. Michael Hall is a Professor in the Department of Management, University of Canterbury, New Zealand and Docent in the Department of Geography, Oulu University, Finland. Co-editor of *Current Issues in Tourism* he has published widely in the tourism and environmental history fields, including a long-standing interest in wilderness, national parks and world heritage.

Contemporary geographies of leisure, tourism and mobility
Series editor: C. Michael Hall, Professor at the Department of Management, College of Business & Economics, University of Canterbury, Private Bag 4800, Christchurch, New Zealand

The aim of this series is to explore and communicate the intersections and relationships between leisure, tourism and human mobility within the social sciences.

It will incorporate both traditional and new perspectives on leisure and tourism from contemporary geography, for example, notions of identity, representation and culture, while also providing for perspectives from cognate areas such as anthropology, cultural studies, gastronomy and food studies, marketing, policy studies and political economy, regional and urban planning, and sociology, within the development of an integrated field of leisure and tourism studies.

Also, increasingly, tourism and leisure are regarded as steps in a continuum of human mobility. Inclusion of mobility in the series offers the prospect to examine the relationship between tourism and migration, the sojourner, educational travel, and second home and retirement travel phenomena.

The series comprises two strands:

Contemporary geographies of leisure, tourism and mobility aims to address the needs of students and academics, and the titles will be published in hardback and paperback. Titles include:

The Moralisation of Tourism
Sun, sand . . . and saving the world?
Jim Butcher

The Ethics of Tourism Development
Mick Smith and Rosaleen Duffy

Tourism in the Caribbean
Trends, development, prospects
Edited by David Timothy Duval

Qualitative Research in Tourism
Ontologies, epistemologies and methodologies
Edited by Jenny Phillimore and Lisa Goodson

The Media and the Tourist Imagination
Converging cultures
Edited by David Crouch, Rhona Jackson and Felix Thompson

Tourism and Global Environmental Change
Ecological, social, economic and political interrelationships
Edited by Stefan Gössling and C. Michael Hall

Forthcoming

Understanding and Managing Tourism Impacts
Michael Hall and Alan Lew

Cultural Heritage of Tourism in the Developing World
Dallen J. Timothy and Gyan Nyaupane

Routledge studies in contemporary geographies of leisure, tourism and mobility is a forum for innovative new research intended for research students and academics, and the titles will be available in hardback only. Titles include:

Living with Tourism
Negotiating identities in a Turkish village
Hazel Tucker

Tourism, Diasporas and Space
Edited by Tim Coles and Dallen J. Timothy

Tourism and Postcolonialism
Contested discourses, identities and representations
Edited by C. Michael Hall and Hazel Tucker

Tourism, Religion and Spiritual Journeys
Edited by Dallen J. Timothy and Daniel H. Olsen

China's Outbound Tourism
Wolfgang Georg Arlt

Tourism, Power and Space
Edited by Andrew Church and Tim Coles

Tourism, Ethnic Diversity and the City
Edited by Jan Rath

Ecotourism, NGOs and Development
A critical analysis
Jim Butcher

Tourism and the Consumption of Wildlife
Hunting, shooting and sport fishing
Edited by Brent Lovelock

Tourism, Creativity and Development
Edited by Greg Richards and Julie Wilson

Tourism at the Grassroots
Edited by John Connell and Barbara Rugendyke

Tourism and Innovation
Michael Hall and Allan Williams

World Tourism Cities: developing tourism off the beaten track
Edited by Robert Maitland and Peter Newman

Tourism and National Parks
International perspectives on development, histories and change
Edited by Warwick Frost and C. Michael Hall

Forthcoming

Tourism, Performance and the Everyday: Consuming the Orient
Michael Haldrup and Jonas Larsen

Tourism and National Parks
International perspectives on development, histories and change

Edited by Warwick Frost and
C. Michael Hall

LONDON AND NEW YORK

First published 2009
by Routledge
2 Park Square, Milton Park, Abingdon, Oxon OX14 4RN

Simultaneously published in the USA and Canada
by Routledge
270 Madison Avenue, New York, NY 10016

Reprinted 2010

Routledge is an imprint of the Taylor & Francis Group, an informa business

© 2009 Warwick Frost and C. Michael Hall selection and editorial matter; individual chapters, the contributors

Typeset in Times New Roman by
RefineCatch Limited, Bungay, Suffolk
Printed by the MPG Books Group in the UK

All rights reserved. No part of this book may be reprinted or reproduced or utilized in any form or by any electronic, mechanical, or other means, now known or hereafter invented, including photocopying and recording, or in any information storage or retrieval system, without permission in writing from the publishers.

British Library Cataloguing in Publication Data
A catalogue record for this book is available from the British Library

Library of Congress Cataloging-in-Publication Data
Tourism and national parks: international perspectives on development, histories and change; edited by Warwick Frost and C. Michael Hall.
  p. cm.—(Contemporary geographies of leisure, tourism, and mobility)
Includes bibliographical references and index.
 1. Tourism. 2. National parks and reserves. I. Frost, Warwick.
II. Hall, Colin Michael, 1961–
 G155.A1T5893489 2008
 363.6′8–dc22
                                                                 2008028838

ISBN 10: 0–415–47156–7 (hbk)
ISBN 10: 0–203–88420–5 (ebk)

ISBN 13: 978–0–415–47156–5 (hbk)
ISBN 13: 978–0–203–88420–1 (ebk)

Contents

List of figures and tables x
List of contributors xiii
Acknowledgements xv
Abbreviations xvi

PART I
Introduction 1

1 Introduction: the making of the national parks concept 3
   C. MICHAEL HALL AND WARWICK FROST

2 Reinterpreting the creation myth: Yellowstone
   National Park 16
   WARWICK FROST AND C. MICHAEL HALL

3 American invention to international concept:
   the spread and evolution of national parks 30
   WARWICK FROST AND C. MICHAEL HALL

4 National parks and the 'worthless lands hypothesis'
   revisited 45
   C. MICHAEL HALL AND WARWICK FROST

5 National parks, national identity and tourism 63
   WARWICK FROST AND C. MICHAEL HALL

PART II
New World perspectives 79

6 Framing the view: how American national parks came to be 81
STEPHEN R. MARK

7 John Muir and William Gladstone Steel: activists and the establishment of Yosemite and Crater Lake National Parks 88
STEPHEN R. MARK AND C. MICHAEL HALL

8 Tourism and the Canadian national park system: protection, use and balance 102
STEPHEN W. BOYD AND RICHARD W. BUTLER

9 The Great Barrier Reef Marine Park: natural wonder and World Heritage area 114
LEANNE WHITE AND BRIAN KING

10 'Welcome to Aboriginal Land': the Uluru-Kata Tjuta National Park 128
TAMARA YOUNG

PART III
Old World perspectives 141

11 The national park concept in Spain: patriotism, education, romanticism and tourism 143
JOSE SOMOZA MEDINA

12 The English Lake District – national park or playground? 155
RICHARD SHARPLEY

13 The Peak District National Park UK: contemporary complexities and challenges 167
DAVID CROUCH, DUNCAN MARSON, GEOFF SHIRT, RICHARD TRESIDDER AND PETER WILTSHIER

14 A ticket to national parks? Tourism, railways, and the establishment of national parks in Sweden 184
SANDRA WALL REINIUS

15 'Protect, preserve, present': the role of tourism in
    Swedish national parks   197
    PETER FREDMAN AND KLAS SANDELL

PART IV
Developing world: beyond the eurocentric   209

16 National parks in Indonesia: an alien construct   211
    JANET COCHRANE

17 National parks in transition: Wuyishan Scenic
    Park in China   225
    XU HONGGANG AND ZHANG CHAOZHI

18 'Full of rubberneck waggons and tourists':
    The development of tourism in South Africa's national
    parks and protected areas   238
    JANE CARRUTHERS

PART V
Beyond nature   257

19 National parks as cultural landscapes: indigenous
    peoples, conservation and tourism   259
    HEATHER ZEPPEL

20 National Mall and Memorial Parks: past, present
    and future   282
    MARGARET DANIELS, LAURLYN HARMON, MIN PARK AND
    RUSSELL BRAYLEY

PART VI
Conclusion   299

21 The future of the national park concept   301
    C. MICHAEL HALL AND WARWICK FROST

    Bibliography   311
    Index   353

List of figures and tables

Figures

1.1	Mass tourism development at the beach. Brighton, the first seaside resort.	4
1.2	A contrasting beach illustrates the national park concept – monumental, sublime, easily imagined as untouched. Pacific Rim National Park, Canada.	5
2.1	Hot springs and geysers, scientific curiosities protected at Yellowstone National Park.	20
2.2	The national park at Yellowstone was conceived as a way to manage tourism development. The tourist hub of Mammoth Springs.	20
5.1	The cultural landscape of Castlemaine Diggings National Heritage Park, Australia. Both the ruined buildings and regrown forest are cultural artefacts of the gold rushes.	74
5.2	Red stone markers for fallen Cheyenne warriors at the Battle of Little Bighorn National Monument, USA.	76
5.3	White stone markers for Custer and his men. Battlefields remain contested ground.	77
9.1	The Port Douglas Marina, base for visits to the Great Barrier Reef.	116
9.2	The Low Isles, Great Barrier Reef.	118
10.1	Uluru, the iconic image.	129
10.2	Visitors embarking on the Uluru climb.	134
10.3	Signs at the base of the Uluru climb. The left hand side warns visitors about issues of safety when climbing Uluru.	135
10.4	Continued: The right hand side welcomes visitors to Aboriginal land and explains the Anangu request not to climb Uluru.	136
13.1	The village of Edale, valley farms and high moors, Peak District National Park.	169
13.2	Recreational walkers, Edale, Peak District National Park. Hardly dressed for going beyond made roads.	170

13.3	Grindslow Knoll, High Peak, Peak District National Park. High moors with rough walking tracks.	171
13.4	The 'triangle of recreation' within national parks.	178
16.1	Mountainscape in Java, typical of the monumental scenery which often is included in national parks.	215
16.2	Asian tourists climb to crater rim of Mount Bromo. Note sellers of endangered Javan eidelweiss in foreground. The plant is considered to bring luck, and is illegally picked and sold within the national park.	218
17.1	The current institutional structure of Wuyishan Scenic Park.	229
17.2	Value added as a percentage of GDP for Wuyishan Scenic Park.	233
17.3	A feedback model of protection and utilization of Wuyishan Scenic Park.	236
20.1	Spring visitors to the National Mall.	283

Tables

1.1	Dimensions of the IUCN category of 'National Park'	8
1.2	Number and area of protected areas under IUCN Protected Area Management Categories 2003	12
2.1	Seven principles arising from Yellowstone	28
3.1	Nash's four factors behind America's leadership in national parks	30
3.2	Variables in national parks	31
3.3	The Australian anomaly	36
3.4	Factors behind the late establishment of national parks in Britain, France and Germany	41
8.1	Expansion of the Canadian national park system 1936–1972	107
12.1	Cumbria and the Lake District populations	157
12.2	Day and overnight tourists in the Lake District	159
14.1	Chronological overview of national parks, tourism and railway development in the Lapland Mountains	187
15.1	Estimated visitation to Sweden's nine mountain national parks	203
15.2	Characterization of mountain region national park visitors	203
15.3	Attitudes towards protected areas in general	204
15.4	Factors of importance when nature areas are protected	205
15.5	Most important reasons for protecting new nature reserves and national parks in the mountain region	206
17.1	The budget of the Wuyishan Scenic Park	231
17.2	The tourists and tourism income of the Wuyishan Scenic Park	232

19.1	Biodiversity conservation, indigenous rights and co-managed national parks	262
19.2	Four stages of indigenous co-management of national parks and tourism	265
19.3	Paradigms of protected area management	269
19.4	Values of protected areas	271
19.5	World views on purpose and identity of national parks	273
19.6	Indigenous co-managed national parks in South Africa, South America and Indonesia	274
19.7	Aboriginal joint management of national parks in Australia	279
20.1	Summary of 6 groupings and 18 themes	288

List of contributors

Stephen W. Boyd School of Hospitality and Tourism Management, University of Ulster, Ballywillan Rd, Portrush, Northern Ireland, UK BT56 8JL; sw.boyd@ulster.ac.uk.

Russell Brayley School of Recreation, Health and Tourism, George Mason University, 10900 University Blvd., MS 4E5, Manassas, VA 20110–2203, USA.

Richard W. Butler Department of Hospitality and Tourism Management, University of Strathclyde, 16 Richmond St, Glasgow G1 1XQ, Scotland.

Jane Carruthers Department of History, University of South Africa, PO Box 392, 003 Unisa, South Africa; carruej@unisa.ac.za.

Janet Cochrane International Centre for Responsible Tourism, Leeds Metropolitan University, Civic Quarter, Leeds LS1 34E, England; j.cochrane@leedsmet.ac.uk.

David Crouch Culture, Lifestyle and Landscape Research Group, University of Derby, 1 Devonshire Rd, Buxton SK17 6RY, England; D.C.Crouch@derby.ac.uk.

Margaret Daniels School of Recreation, Health and Tourism, George Mason University, 10900 University Blvd., MS 4E5, Manassas, VA 20110–2203, USA; mdaniels@gmu.edu.

Peter Fredman Department of Tourism, Mid-Sweden University, S – 831 25 Östersund, Sweden; peter.fredman@etour.se.

Warwick Frost Tourism Research Unit, Monash University, PO Box 1071 Narre Warren 3805 Australia; warwick.frost@buseco.monash.edu.au.

C. Michael Hall Department of Management, University of Canterbury, Private Bag 4800, Christchurch, New Zealand; michael.hall @canterbury.ac.nz.

Laurlyn Harmon School of Recreation, Health and Tourism, George Mason University, 10900 University Blvd., MS 4E5, Manassas, VA 20110–2203, USA.

Brian King Centre for Tourism and Services Research, Victoria University, PO Box 14428 Melbourne 8001, Australia; brian.king@vu.edu.au.

Stephen R. Mark US National Parks Service, Crater Lake National Park, Oregon USA; steve-mark@nps.gov.

Duncan Marson Culture, Lifestyle and Landscape Research Group, University of Derby, 1 Devonshire Rd, Buxton SK17 6RY, England.

Jose Somoza Medina University of León, Spain; jose.somoza@unileon.es.

Min Park School of Recreation, Health and Tourism, George Mason University, 10900 University Blvd., MS 4E5, Manassas, VA 20110–2203, USA.

Klas Sandell Department of Human Geography and Tourism, Karlstad University, SE – 651 88, Karlstad, Sweden; klas.sandell@kau.se.

Richard Sharpley Department of Tourism and Leisure Management, University of Central Lancashire, Preston PR1 24E, UK; RAJSharpley@uclan.ac.uk.

Geoff Shirt Culture, Lifestyle and Landscape Research Group, University of Derby, 1 Devonshire Rd, Buxton SK17 6RY, England.

Richard Tresidder Culture, Lifestyle and Landscape Research Group, University of Derby, 1 Devonshire Rd, Buxton SK17 6RY, England.

Sandra Wall Reinius Department of Tourism, Mid-Sweden University, S – 831 25 Östersund, Sweden; Sandra.Wall-Reinius@miun.se.

Leanne White Centre for Tourism and Services Research, Victoria University, PO Box 14428 Melbourne 8001, Australia; leannek.white@vu.edu.au.

Peter Wiltshier Culture, Lifestyle and Landscape Research Group, University of Derby, 1 Devonshire Rd, Buxton SK17 6RY, England.

Xu Honggang The School of Tourism Management, Sun Yat-sen University, China; xuhonggang@yahoo.com.

Tamara Young School of Economics, Politics and Tourism, University of Newcastle, Callaghan 2308, Australia; tamara.young@newcastle.edu.au.

Heather Zeppel Tourism Discipline – School of Business, James Cook University, PO Box 6811, Cairns, Australia; heather.zeppel@jcu.edu.au.

Zhang Chaozhi The School of Tourism Management, Sun Yat-sen University, China.

Acknowledgements

Apart from thanking Warwick for working together on an enjoyable and stimulating project, Michael would like to collectively acknowledge research time spent at Lund University Helsingborg, Umeå University, Oulu University, and Joensuu University at Savonlinna in recent years, which has been extremely helpful in thinking about national parks and protected areas in relation to tourism. There are a number of individuals who have also stimulated thinking and writing in this area at various times, including Tori Amos, Fiona Apple, Patrick Armstrong, Dick Butler, Nick Cave, Bruce Cockburn, Tim Coles, Dave Crag, David Duval, Nicolette Le Cren, Thor Flognfeldt, Ebba Forsberg, Stefan Gössling, Michael James, Simon MacArthur, Steve Mark, Dieter Müller, Stephen Page, Portishead, Margaret Robertson, Dennis Rumley, Jarkko Saarinen, Daniel Scott, Anna Dora Saethorsdottir, Murray Simpson, Penny Spoelder, Nicola van Tiel, Dallen Timothy, Sandra Wall, Sandra Wilson and Brian Wheeler. I must also thank my friends and significant others (JC × 2 and sprog), for their love, support and understanding; hopefully, one day, we will all look back at what we do and why, and laugh. Finally, I would publicly thank and remember my Uncle Harry, one of the true gentlemen of this world, who gave me a copy of *National Geographic* many years ago with an article on John Muir in it, which greatly helped nurture my love of the environment and its history – from little things big things grow.

Similarly, Warwick's interests in national parks can be traced back to childhood. Family holidays nearly always included a trip to a national park, my parents confident they would contain something special. We were rarely disappointed and I still get that tingle of excitement whenever I pass a sign advising I'm about to enter a national park. This project could never have happened without a sabbatical from Monash University and the encouragement and support of the contributors and Andrew Mould and Michael P. Jones at Routledge. Many thanks to Michael for partnering me in this. A debt of gratitude to my Canadian cousins – Susan and Andrew and their families – for hosting my visits to North America. Finally, a big thanks to Sarah, Stephen and Alex for their love and support.

Abbreviations

CEESP	Commission on Environmental, Economic and Social Policy
CPR	Canadian Pacific Railroad
DLS	Department of Lands and Surveys (NZ)
DMP	Destination Management Partnership
FMC	Federated Mountain Clubs
IUCN	International Union for the Conservation of Nature
NPCA	National Parks and Conservation Association
PDNP	Peak District National Park
TILCEPA	Theme on Indigenous and Local Communities, Equity and Protected Areas

Part I
Introduction

1 Introduction

The making of the national parks concept

C. Michael Hall and Warwick Frost

'When I use a word it means just what I choose it to mean', explains Humpty Dumpty to Alice in Lewis Carroll's *Through the Looking-Glass*. When it was published in 1871, a new word – or phrase to be precise – had just been coined. The new expression was 'national park' and it was being used by advocates of a scheme for public preservation of the Yellowstone area of the western US. As the concept took hold and spread around the world, it took on multiple meanings.

For wilderness historian Roderick Nash, national parks were 'the American invention' arising from the combination of the US's 'unique experience with nature in general and wilderness in particular', democracy, affluence and sizeable amounts of land (Nash 1970: 726). However, as the concept of national parks spread, it was not simply duplicated, rather it evolved in numerous ways to adapt to various physical, political and social environments. As more and more countries considered establishing national parks, 'everywhere local culture was as important as foreign example' (Dunlap 1999: 119). In some cases the application of the term may not have been more than the use of an attractive name that could be applied to a piece of land set aside for scenic or recreational purposes, but without understanding its managerial implications. In Australia, the first national park was set aside in 1879 at Port Hacking, south of Sydney. An exhibit organised by the Royal Society of New South Wales in 1878 contained a description of Yellowstone, but it is 'unlikely that Yellowstone National Park provided more than an idea for a name' (Pettigrew and Lyons 1979: 18; also, see, Slade 1985–86, Hall 1992). Turner (1979: 184) similarly notes, 'it is widely known and accepted that The National Park was the first example of this land use designation in Australia, but whether it was either "National" or a "Park" in the sense that Australians now understand the words (either individually or in combination), or as they were used in the United States in the 1870s is more open to question'. In fact the origins of The National Park (now Royal National Park) lie far more in the health-oriented urban parks movement of the Victorian era than they do in the wilds of Wyoming. As the opening remarks of Sir John Robertson at the first meeting of the park's trustees held on 25 September 1879 indicate:

Figure 1.1 Mass tourism development at the beach. Brighton, the first seaside resort.

Having in view the great progress already made, under the Liberal Institutions of this Colony in population etc., the more than probable large increase in population in the Metropolitan District and of the Colony generally within the new few years, and the consequent desirability of securing a suitable area, as to extent, situation etc., for the use and enjoyment of the people of New South Wales; he caused investigation to be made respecting the tract of land bordering Port Hacking etc. A description was then prepared of the boundaries of the fine area, (at least 18,000 acres and probably several thousand additional acres), since dedicated as a National Park and notified in the supplement to the Government Gazette of 26 April, 1879.

(in Stanley 1977)

The recreational focus of Robertson was very much in keeping with the translation in the Australian colonies of the British parks movement. According to Billinge (1996: 450), 'Perhaps the single newest element in the townscape after the general regulation of the street, was the park, and more specifically the recreation ground ... [since] the urban park, as distinct from the garden square, was essentially a nineteenth century phenomenon' and a symbol of civic pride. The acknowledged role of parks as the 'lungs of the city', as a haven from industrialization, was an attempt to recreate notions of community

Introduction: the making of the national parks concept 5

Figure 1.2 A contrasting beach illustrates the national park concept – monumental, sublime, easily imagined as untouched. Pacific Rim National Park, Canada.

well-being, with the main proponents of park development being a middle- to upper-class elite who embodied notions of parks contributing to well-being and reflecting elements of nature that were balanced and inherently good (Young 1996). Billinge (1996: 444) recognized the way in which the Victorians engineered the term 'recreation' 'to perfection, they gave it a role and a geography. Confined by time, defined by place and regulated by content, recreation and the time it occupied ceased to be possessions freely enjoyed and became instead, obligations dutifully discharged.' Such an observation therefore fits well with the origins of the first parks in Australia and is also shown in *An Official Guide to the National Park of New South Wales*, published in 1893, which gives an account of the decision to establish what is now Royal National Park:

> In the early part of the year 1879, several public men, both within and without the walls of Parliament, raised their voices in favour of the Government providing public parks, pleasure grounds, and places of recreation adjacent to all thickly populated centres in New South Wales. A set of resolutions was submitted to the Legislative Assembly, affirming that the health of the people should be the prime consideration of all good Government; and to ensure the sound health and vigour of the

community it was necessary that all cities, towns, and villages should be possessed of pleasure grounds as places of recreation. This necessity was recognised by the leading statesmen of the day, but the resolutions were of a nature so sweeping that their adoption would have imposed an entire change in the policy of the country. At this time Sydney possessed several breathing spaces favoured by nature, but the more densely populated parts of the metropolis and suburbs were destitute of such provision. While the ardour of these well-meaning though impracticable philanthropists was still burning for additional city lungs, the late lamented Sir John Robertson as the acting head of the Government, conceived and developed the idea of bequeathing to the people of this State a national domain for rest and recreation.

(in Stanley 1977)

The recreational dimension was also paramount in the establishment of national parks in England and Wales. The national parks of England and Wales do not generally conform to the International Union for the Conservation of Nature (IUCN) definition of a national park (see Table 1.1 on p. 8) where conservation comes before visitation and are instead cultural landscapes that have been materially altered by human occupation. The purposes of the parks were defined in the landmark legislation that first provided for national parks in England and Wales: 'The provisions of this Act shall have effect for the purpose of preserving and enhancing the natural beauty of the areas specified ... and for the purposes of promoting their enjoyment by the public' (*National Parks and Access to the Countryside Act 1949*, s 5(1)). As the title of the Act suggests, access for countryside recreation was a key component of the mission of English national parks, combined with a desire to retain a 'traditional landscape' and the mixed farming practices that had created it, including the continuation of private land ownership within designated national park areas (MacEwen and MacEwen 1982). Furthermore, the definition of 'national park' in the English and Welsh context has always been different from the American. The landmark Dower report on national parks produced at the end of the Second World War made it clear that they did not mean parks in either the American sense or even with respect to urban parks. To Dower, a national park was 'An extensive area of beautiful and relatively wild country in which, for the nation's benefit and by appropriate national decision and action, (a) the characteristic landscape beauty is strictly preserved, (b) access and facilities for open-air enjoyment are amply provided, (c) wild life and buildings and places of architectural and historic interest are suitably protected while (d) established farming use is effectively maintained' (Dower 1945: 6). His definition was accepted by the subsequent Hobhouse Committee (1947 in MacEwen and MacEwen 1982: 13), which summarised the 'essential requirements' of a national park as 'great natural beauty, a high value for open air recreation and substantial continuous extent'. The sum effect of this approach is evidenced in the 1949 Act, which defined the areas to be designated as 'those extensive tracts of country in England and Wales as to which it appears to the [National Parks] Commission that by reason of

a   their natural beauty, and
b   the opportunities they afford for open-air recreation having regard to their character and to their position in relation to centres of population, it is especially desirable that the necessary measures shall be taken for the purposes mentioned [above] (*National Parks and Access to the Countryside Act 1949*, Part 2).

The importance of local and non-American influences in the development of the national park concept is similarly found in the case of New Zealand where outgoing Premier William Fox left a memorandum to his successor in 1874 in which he suggested the utilisation of hot springs district of the central North Island for "sanitary purposes" '. Fox regarded the American Yellowstone legislation as commendable 'and urged a similar course of action in New Zealand' (Hall 1988b). The efforts of Fox and fear of despoliation of the hot springs and geysers of Rotorua led to the passing of the *Thermal Springs Districts Act* in 1881, with the debate being notable for a comparison with the Yellowstone Park Act made by one of the Government's members (Roche 1987). The focus therefore was on particular dimensions of the Yellowstone reservation and the national park idea reinterpreted through local circumstances and sensibilities.

The spread of national parks may be divided into three broad stages. First, by the end of the nineteenth century, the US had set up a number of large monumental national parks and the idea had spread to the other English-speaking settler societies of Canada, Australia and New Zealand. Second, in the first half of the twentieth century – what Hobsbawm (1994) termed the 'Age of Catastrophe', Europe engaged with the concept. Sweden, Italy, Romania, Greece, Spain, Iceland, Ireland and Switzerland all established national parks. In contrast, the major European powers did not. Britain, and to a lesser extent France and Belgium, focused on the establishment of national parks in its Asian and African colonies. The third stage, after the Second World War was characterized by the global spread of national parks, so that nearly every country boasted them – and indeed one could hardly boast of being a true nation without one – such was the importance of national monuments and nature. This third stage also represents the period at which national parks also came to be perceived from conservation and ecological standpoints in addition to the previous aesthetic, recreation and tourism, and utilitarian perspective.

National parks have spread around the globe. The US provided the original idea and much useful advice, but it imposed no conditions on the use of the term. International organizations, such as the International Bureau on Nature Conservation, the Fauna Preservation Society, the Sierra Club and the Boone and Crockett Club, championed national parks; but they were

Table 1.1 Dimensions of the IUCN Category of 'National Park'

IUCN Category	II
Definition	A national park is a protected area managed mainly for ecosystem protection and recreation: it is a 'natural area of land and/or sea, designated to (a) protect the ecological integrity of one or more ecosystems for present and future generations, (b) exclude exploitation or occupation inimical to the purposes of designation of the area and (c) provide a foundation for spiritual, scientific, educational, recreational and visitor opportunities, all of which must be environmentally and culturally compatible' (IUCN 1994).
Management objectives	<ul><li>to protect natural and scenic areas of national and international significance for spiritual, scientific, educational, recreational or tourist purposes;</li><li>to perpetuate, in as natural a state as possible, representative examples of physiographic regions, biotic communities, genetic resources, and species, to provide ecological stability and diversity;</li><li>to manage visitor use for inspirational, educational, cultural and recreational purposes at a level which will maintain the area in a natural or near natural state;</li><li>to eliminate and thereafter prevent exploitation or occupation inimical to the purposes of designation;</li><li>to maintain respect for the ecological, geomorphologic, sacred or aesthetic attributes which warranted designation; and</li><li>to take into account the needs of indigenous people, including subsistence resource use, in so far as these will not adversely affect the other objectives of management.</li></ul>
Criteria	<ul><li>The area should contain a representative sample of major natural regions, features or scenery, where plant and animal species, habitats and geomorphological sites are of special spiritual, scientific, educational, recreational and tourist significance.</li><li>The area should be large enough to contain one or more entire ecosystems not materially altered by current human occupation or exploitation.</li></ul>
Governance	Ownership and management should normally be by the highest competent authority of the nation having jurisdiction over it. However, national parks may also be vested in another level of government, council of indigenous people, foundation or other legally established body, which has dedicated the area to long-term conservation.

Source: Adapted from IUCN 1994.

lobby groups leaving the management details to individual states. Moreover, much of the original circulation of ideas of the national parks as well as the name took place through often informal networks by which brochures and early park material were sent around the world or in the more formal networks of mountaineering and scientific societies. For example, Myles Dunphy, who has been described as the 'father of conservation in New South Wales' (Australia) (Barnes and Wells 1985: 7, see, also, Meredith 1999), sent copies of American publications on national parks and monuments to 'strategic points in the bureaucracy' (Prineas 1976/1977: 11), even though he had never visited the US parks first-hand. In 1969, the IUCN included national parks as one of its six categories of protected areas, but again, while this was influential, the selection of the category that protected areas were placed in and the reporting process that surrounds them, has depended on the information provided by national governments, 'even though the IUCN 1994 Guidelines for Protected Area Management Categories state:

> It ... follows from the international nature of the system, and from the need for consistent application of the categories, that the final responsibility for determining categories should be taken at the international level. This could be IUCN, as advised by its CNPPA and/or the World Conservation Monitoring Centre (e.g., in the compilation of the UN List) in close collaboration with IUCN.'
>
> (Chape *et al.* 2003: 10–11)

In fact this fluid situation is acknowledged by the *UN List of Protected Areas* when they note 'production of the UN List has been an evolving process since its inception in 1962, moving from the iconic national park "role of honour" concept that characterised the earlier UN Lists to one that reflects the range of protected area objectives and values relevant to the late 20th and early 21st century' (Chape *et al.* 2003). The definitions of IUCN protected area management categories are illustrated in Table 1.2 (on p. 12).

As opposed to Biosphere Reserve or World Heritage listing, there is no international mechanism for accrediting national parks. Each country can be like Lewis Carroll's Humpty Dumpty and apply whatever meaning suits them.

Accordingly, wide variation occurred in the planning and execution of national parks. Advocates could advance a range of different reasons to support establishment – nature protection, tourism, recreation, strengthening national identity. These could conflict, but they could also be used in combinations, resulting in national parks with tourist nodes and less accessible protected scientific areas. Supporters of national parks could call on international precedents, but also compromise their vision to satisfy local perspectives and stakeholders. Indeed, many of the present-day conflicts over exactly what a national park means and how it should balance competing demands for tourism and recreational use as well as conservation, date back to the creation and dissemination of the national park idea. In Australia, and especially New South Wales, the influence of the United States was such that departmental structures, job descriptions and land-use classifications were often 'borrowed' from the American model, as were at times the staff, such as the first head of the NSW Parks and Wildlife Service. For example, the most apparent example of American influences on wilderness preservation was in the choice of the term 'primitive area' in the 1930s to denote wilderness areas within reserves or as potential reserves. However, Marie Byles, one of the most prominent campaigners for the creation of national parks at this time (Butler 1980), wrote:

> When N.S.W. does wake up to the crying need for wild park lands it will be able to benefit by the mistakes made by America. The initial mistake made there was to regard it as sufficient to dedicate land as parks. It was thought that to dedicate them thus would be the same thing as keeping them in their primitive or natural state. But it was soon found that hotels, motor roads and wide advertisement caused tourists to flock there by the thousand. This interfered with the natural ways of wild creatures and few species can survive unless they can live in untouched wilderness.
>
> (Byles 1938: 7)

However, other individuals and jurisdictions were not so critical and aimed to closely follow the North American approach. Until 1952, the national parks of New Zealand were established under a variety of Acts and each park was managed separately. In New Zealand in the 1920s and 1930s bodies such as botanical gardens and mountaineering clubs sought to develop a more systematic approach to park creation and management. Through a comparison of overseas initiatives in park planning, a subcommittee of Federated Mountain Clubs (FMC), the country's leading mountaineering association, decided, 'to put forward suggestions for more systematic general control, based upon the successful and businesslike examples of the United States and Canada' (Thompson 1976: 9). The lobbying of the FMC appeared to have some influence on the New Zealand Government, but the reorganization of the parks had to wait until after the Second World War.

At this time Ron Cooper, Chief Land Administration Officer of the Department of Lands and Surveys (DLS), played a prominent role in the creation of the National Parks Act. Cooper conceived New Zealand's national parks as wilderness to which the general public should have access:

> A national park is ... a wilderness area set apart for preservation in as near as possible its natural state, but made available for and accessible to the general public, who are allowed and encouraged to visit the reserve. In such an area the recreation and enjoyment of the public is a main purpose, but at the same time the natural scenery, flora and fauna are interfered with as little as possible. Such a reserve should contain scenery

of distinctive quality, or some natural features so extraordinary or unique as to be of national interest and importance, and as a rule it should be extensive in area.

(Ron Cooper, 21 January, 1944, in Thompson 1976: 11)

In line with American perspectives, Cooper had a strong anthropocentric perception of wilderness and national parks in which he saw New Zealand's national parks as being recreational in character and did not see them as purely scientific reserves. The recreational importance attached to New Zealand's national parks was demonstrated in the 1952 *National Parks Act* which, following on from North American national park legislation, defined the purpose of the parks as, 'preserving in perpetuity . . . for the benefit and enjoyment of the public, areas of New Zealand that contain scenery of such distinctive quality or natural features so beautiful or unique that their preservation is in the national interest'. The extent of the influence of the North American national park systems on New Zealand is further evidenced by the study tour of these countries by P.H.C. Lucas, Director of National Parks and Reserves, in 1969. The report of the study tour contained a wide account of park management practices and has many sections entitled 'lessons for New Zealand' (Lucas 1970). The report served as one of the major determinants in the direction of New Zealand's national park policies through the 1970s and the early 1980s. Similarly, the review of the administrative structure of national parks and reserves in 1979 also showed a great many American influences (Government Caucus Committee Report 1979) and it was not until the late 1980s and a substantial reorganization of the conservation estate that New Zealand's park system started to develop a more indigenous dimension (Hall 1988b).

This evolutionary process means that there is no single model of national parks. However, the common trap is to assume there is. We tend to look at national park systems in our own countries, which are familiar and close to hand, and see that as the 'correct' model. In this volume a different approach is proposed. If we are to fully understand the development and the future possibilities of national parks, we must take a comparative approach considering the wide range of countries that have embraced and changed the concept. Rather than seeing national parks as a particular model, we must look across a range of jurisdictions and see them as a series of models. With Yellowstone as the common ancestor, there are still clear similarities in these models, but there is also a fascinating range of differences.

This volume therefore aims to chart some of the different approaches to the development of national parks and its long relationship with tourism in both a contemporary and historical perspective. The latter is regarded as extremely important as it helps illustrate the ongoing debate on the role of tourism in national parks and the means by which visitation should be managed. The chapters in this book are divided into five sections. This chapter and the following four provide a contextualization for the social construction of

Table 1.2 Number and area of protected areas under IUCN Protected Area Management Categories 2003

Category	Description	Global no. of categories (2003)	Global no. of categories (2003) (%)	Global area of categories (2003) (km²)	Global area of categories (2003) (%)
Ia Strict Nature Reserve: protected area managed mainly for science	Area of land and/or sea possessing some outstanding or representative ecosystems, geological or physiological features and/or species, available primarily for scientific research and/or environmental monitoring.	4 731	4.6	1 033 888	5.5
Ib Wilderness Area: protected area managed mainly for wilderness protection	Large area of unmodified or slightly modified land, and/or sea, retaining its natural character and influence, without permanent or significant habitation, which is protected and managed so as to preserve its natural condition.	1 302	1.3	1 015 512	5.4
II National Park: protected area managed mainly for ecosystem protection and recreation	Natural area of land and/or sea, designated to (a) protect the ecological integrity of one or more ecosystems for present and future generations, (b) exclude exploitation or occupation inimical to the purposes of designation of the area, and (c) provide a foundation for spiritual, scientific, educational, recreational and visitor opportunities, all of which must be environmentally and culturally compatible.	3 881	3.8	4 413 142	23.5
III Natural Monument: protected area managed mainly for conservation of specific natural features	Area containing one, or more, specific natural or natural/cultural feature that is of outstanding or unique value because of its inherent rarity, representative or aesthetic qualities or cultural significance.	19 833	19.4	275 432	1.5

IV Habitat/Species Management Area: protected area managed mainly for conservation through management intervention	Area of land and/or sea subject to active intervention for management purposes so as to ensure the maintenance of habitats and/or to meet the requirements of specific species.	27 641	27.1	3 022 515	16.1
V Protected Landscape/ Seascape: protected area managed mainly for landscape/seascape conservation and recreation	Area of land, with coast and sea as appropriate, where the interaction of people and nature over time has produced an area of distinct character with significant aesthetic, ecological and/or cultural value, and often with high biological diversity. Safeguarding the integrity of this traditional interaction is vital to the protection, maintenance and evolution of such an area.	6 555	6.4	1 056 088	5.6
VI Managed Resource Protected Area: protected area managed mainly for the sustainable use of natural ecosystems	Area containing predominantly unmodified natural systems, managed to ensure long-term protection and maintenance of biological diversity, while providing at the same time a sustainable flow of natural products and services to meet community needs	4 123	4	4 377 091	23.3
No category		34 036	33.4	3 569 820	19
Total		102 102	100	18 763 407	100

Source: Categories identified in IUCN 1994; Figures derived from Chape *et al.* 2003.

14 *Tourism and National Parks*

the national park concept through the creation myths, diffusion and evolution of the national parks idea, and the relationship of national parks to national identity. The second section looks at the role of 'New World' perspectives on national parks with respect to American and Australian experiences. The third section discusses the role of national parks in Europe with chapters focusing on England, Spain and Sweden. The fourth section seeks to move beyond Eurocentric accounts of parks and provides experiences from China, Indonesia and South Africa. The final section provides two chapters that discuss the explicit cultural dimensions of national park designation.

Some qualifications

Defining national parks is a topic of much debate. The IUCN classification, which has itself been modified over time, provides a definition, but it is conceptual rather than regulatory. Unlike World Heritage status, no country is required to follow this classification. Instead, like Humpty Dumpty, countries have tended to use the term whenever and however they have liked. This results in a number of anomalies. There are some places termed national parks that some would argue are no such thing. Britain's national parks, with an emphasis on recreation and large areas of private land, fall outside the IUCN definition and are sometimes referred to as not 'real' national parks (MacEwen and MacEwen 1982). Nor is this problem confined to Britain; there are many instances of recreational national parks on city fringes and a number of European national parks include private lands. The national park established in Algeria in 1921 (Osborne 1994) is now almost entirely overlooked, as the colonial governor established it. Another 1920s French creation, the national park at Angkor Wat, is no longer given that designation. In contrast, there are many areas that are protected, but which do not have the name national park. Such a situation also makes it difficult at times to chart the development of the national park idea, let alone how many national parks there are in the world or in some jurisdictions. Indeed, it is interesting to note that the IUCN category of national park has actually had a lower rate of growth in terms of numbers of areas than other types of protected areas, which has meant that as a proportion of the world's protected areas it had declined in relative significance, although in 2003 it still accounted for 23.6 per cent of the total protected area of the Earth (Chape *et al.* 2003). Furthermore, it was also more important in relative terms in the Americas and the Caribbean where it was the predominant category of protected region than elsewhere in the world.

It is clear that the world's conservation estate has grown enormously since the first *United Nations List of Protected Areas* was published in 1962 with 9,214 protected areas covering an area of 2.4 million km^2. This had grown to 16,394 in 1972 and 27,794 in 1982. By 1992 there were 43,388 sites listed, while the 2003 edition listed 102,102 sites covering 18.8 million km^2 (7.25 million mi^2). 'This figure is equivalent to 12.65% of the Earth's land surface, or an area greater than the combined land area of China, South Asia and Southeast Asia' (Chape *et al.* 2003: 21). Of the total area protected, it is estimated that 17.1 million km^2 (6.6 million mi^2) constitute terrestrial protected areas, or 11.5 per cent of the global land surface. Nevertheless, biomes, including lake systems and temperate grasslands, remain poorly represented, while marine areas are significantly underrepresented. Marine protected areas occupy an area of approximately 1.64 million km^2 – an estimated 0.5 per cent of the world's oceans, and less than one-tenth of the overall extent of protected areas worldwide. Table 1.2 outlines the different categories of protected areas at the time of the most recent United Nations inventory.

Given these issues, it is tempting to use the term 'protected areas'. However, to do this would jettison the cultural heritage of the term 'national parks'. It would also set ourselves up as arbiters of what places are truly (or not) national parks. In recognizing that countries made deliberate choices to call specific protected areas national parks, we have chosen to include areas with such titles. In accepting self-definition, we are aware that this means considering places that not everybody considers as fully worthy of the title. A further qualification is the dating of establishment of national parks. In some cases there are two to three possible years, including the dates of announcement, legislation and actual opening. Accordingly, there may be some variations between the dates given in this volume and elsewhere.

Finally, we would like to emphasize that one of the key dimensions of national parks and the national park concept is its fluidity. Given that national parks are under greater pressure than ever before to be able to respond to their visitor and conservation values, it is appropriate that the concept be revisited in an international context. Tourism has been a constant presence in national parks and debates over their management since they were first established. But the relative emphasis on tourism and recreation has also shifted with respect to national parks in both space and time. Seen from the ecological perspectives of many contemporary accounts of national parks the concept is often interpreted as always having had such a focus. As the various chapters in the book demonstrate, few things could be further from the truth. The national park concept, like the very landscapes, environments and values that they have come to protect, is constantly changing.

2 Reinterpreting the creation myth
Yellowstone National Park

Warwick Frost and C. Michael Hall

The campfire epiphany

On the evening of 19 September 1870, a group of adventurers relaxed around a campfire at the junction of the Firehole and Gibbon Rivers in the Yellowstone region of the western US. They were members of the Washburn Expedition, a group of businessmen, local officials and journalists accompanied by a small army escort that had come south from Helena in Montana. Their adventure was stimulated by the reports from an earlier smaller expedition and following its directions they had encountered waterfalls, canyons, hot springs and mud pools. Turning for home, they unexpectedly stumbled into the Upper Geyser Basin, which had not been found by their predecessors. This contained, they estimated, nearly a hundred geysers – larger and higher than any they had seen previously (and including what would become known as Old Faithful).

One of the expedition – Nathaniel P. Langford – recorded that the day after finding the Upper Geyser Basin, they had an animated discussion around the campfire:

> The proposition was made by some member that we utilize the result of our exploration by taking up quarter sections of land at the most prominent points of interest, and a general discussion followed. One member of our party suggested that if there could be secured by pre-emption a good title to two or three quarter sections of land opposite the lower fall of the Yellowstone and extending down the river along the cañyon, they would eventually become a source of great profit to the owners. Another member of the party thought that it would be more desirable to take up a quarter section of land at the Upper Geyser Basin, for the reason that that locality could be more easily reached by tourists and pleasure seekers.
> (Langford 1905: 117)

For a while the expeditioners speculated on the potential fortunes to be made by monopolizing the tourist trade. However, Cornelius Hedges, a young Yale-educated lawyer, introduced a quite different idea. Hedges argued:

That he did not approve of any of these plans – that there ought to be no private ownership of any portion of that region, but that the whole of it ought to be set apart as a great National Park, and that each one of us ought to make an effort to have this accomplished.

(Langford 1905: 117–18)

By Langford's account, this suggestion, 'met with an instantaneous and favourable response from all – except one – of the members of the party, and each hour since the matter was first broached, our enthusiasm has increased' (Langford 1905: 118). Returning to civilization, this idea gained momentum and support, culminating in the US Congress establishing Yellowstone Park on 1 March 1872.

Over time, Langford's account of this campfire epiphany has 'evolved into a kind of creation myth' (Sellars 1997: 8). As the national parks concept spread around the world, its advocates harkened back to its campfire origins:

In the wilderness setting and with a backdrop of the vast, dramatic landscape of the western frontier, the origin of the national park idea seemed fitting and noble. Surely the national park concept deserved a 'virgin birth' – under a night sky in the pristine American West, on a riverbank, and around a flaming campfire, as if an evergreen cone had fallen near the fire, then heated and expanded and dropped its seeds to spread around the planet.

(Sellars 1997: 8)

The persuasiveness of the wilderness campfire in converting decision-makers to support national parks was to be demonstrated time and time again. John Muir would take President Theodore Roosevelt on a camping trip into Yosemite and the idea for a national park at Crater Lake would come from a discussion within a tent (see Chapter 7). Another Yellowstone campfire talk involving King Albert of Belgium would lead to one of the first African national parks and the 'second wind' of the national parks movement (Fitter and Scott 1978: 16–17). In the 1970s and 1980s the Australian conservationist, Milo Dunphy, perfected the art of taking key politicians into the bush for 'billy' tea around campfires (Meredith 1999).

Tourism and the establishment of Yellowstone

If we undertake a close reading of the account of this momentous campfire discussion, it is striking how important tourism was in the establishment of the first national park. The spark for the discussion was the realization by the members of the expedition that the scenic wonders they had discovered had commercial possibilities. The members of the expedition were Montana businessmen, attracted to the newly opened territory by its economic opportunities. Under the Homestead Act they could strategically apply for land at key access points. They could then charge admission fees and provide visitor accommodation. As they contemplated the potential, the argument turned to which sights would attract the greater number of visitors. Was it better to control access to the waterfall and canyon or were the geysers likely to be more lucrative?

The members of the Washburn Expedition were fully aware of widespread public interest in natural wonders, how they were being exploited and the potential for overdevelopment. For them, probably the most prominent example of this was Niagara Falls. In the early nineteenth century, Niagara had become popular with tourists. In 1818, William Forsyth built a covered stairway so that paying customers could walk down to and under the falls. In 1822 he built a three-storey hotel that featured a private walkway to the best vantage point. Others followed so that hotels and taverns occupied all the best viewing positions. By the time of the Washburn Expedition, Niagara Falls had become a byword for shoddy and tacky tourism (Healy 2006).

California also provided a model of overdevelopment. In 1852, gold miners had discovered the giant redwoods (*Sequoia Gigantea*). Two years later an entrepreneur cut down the biggest redwood in the Calveras Grove and shipped its bark in segments to England. Here it was reassembled for display at the Crystal Palace and thought by some to be surely a hoax (Runte 1990, Schama 1995). The establishment of Yosemite as a state park in 1864 came with predictions of overdevelopment due to tourism and these were just starting to be realised (Olmsted 1865, Runte 1979).

Those around the campfire were certain that the attractions of Yellowstone would also draw tourists. If they could get in first by lodging strategic land claims, they could either develop profitable tourism operations, or (more likely) sell their property at a substantial mark-up. Hedges, in turn, argued against such developments. In effect, his position was that he did not want Yellowstone to become like Niagara. His idea was that there was an alternative model for managing tourism. That alternative was in public ownership – a *National* park.

The potential for Niagara-style developments was quickly demonstrated. Montana congressman William Clagett recalled:

> In the fall of 1870, soon after the return of the Washburn-Langford party, two printers at Deer Lodge City, Montana, went into the Firehole basin and cut a large number of poles, intending to come back the next summer and fence in the tract of land containing the principal geysers, and hold possession for speculative purposes, as the Hutchins family so long held the Yosemite valley.
> (quoted in Langford 1905: xliv–xlvi)

Following the Washburn Expedition, the US government funded Ferdinand Hayden, Professor of Geology at the University of Pennsylvania, to undertake a scientific survey of Yellowstone. Hayden reported that there were already squatters at Mammoth Hot Springs and argued to Congress that they should not allow entrepreneurs 'to fence in these rare wonders so as to charge visitors a fee, as is now done at Niagara Falls' (quoted in Runte 1979: 45; see, also, Scott 2007). When Yellowstone was debated in Washington, Senator Lyman Trumbull (whose son had been a member of the Washburn Expedition) argued 'some person may go there and plant himself right across the only path that leads to [its] wonders, and charge every man that passes along the gorge of these mountains a fee of a dollar or five dollars' (quoted in Runte 1979: 52).

Once Yellowstone was established, Langford was appointed to the honorary position of superintendant, where he dealt with many proposals for private tourism ventures:

> Soon after the creation of the Park the Secretary of the Interior received many applications for leases to run for a long term of years, of tracts of land in the vicinity of the principal marvels ... These applications were invariably referred to me ... It was apparent from an examination of these applications that the purpose of the applicants was to enclose with fences their holdings, and charge visitors an admission fee. To have permitted this would have defeated the purpose of the act of dedication.
> (Langford 1905: xlix)

While the advocates of Yellowstone did not want what we today would call inappropriate development, they still thought in terms of tourism facilities. Viewing the falls, Langford imagined that they 'could easily be spanned by a bridge directly over and in front of the fall, and fancy led me forward to no distant period when such an effect of airy architecture could be crowded with happy gazers' (Langford 1905: 34). Hedges also wrote a newspaper article envisioning a swing bridge across the falls (Runte 1979: 43).

The railways factor

Yellowstone was a product of the optimism of the time. This was Mark Twain's *Gilded Age* of postwar boosterism. On 10 May 1869, just over a year before the Washburn expedition, the transcontinental railroad had been completed. Built in just six years, it traversed southern Wyoming, only 300 kilometres south of Yellowstone. Other entrepreneurs were planning a second transcontinental railway – the Northern Pacific, which would follow the Yellowstone River to Livingston, only 100 kilometres north of the national park (Lubetkin 2006). From there a trunk line could follow an easy gradient up the Yellowstone River to the national park entrance. The birth of the national parks concept cannot be separated from the contemporary development of railway transport infrastructure (Smith 2004, Runte 2006).

Jay Cooke of the Northern Pacific Railroad Company promoted the Washburn expedition and the National Park campaign as a means to sell

Figure 2.1 Hot springs and geysers, scientific curiosities protected at Yellowstone National Park.

Figure 2.2 The national park at Yellowstone was conceived as a way to manage tourism development. The tourist hub of Mammoth Springs.

bonds for constructing his railway. Langford worked for Cooke. Langford had been the collector of internal revenue for Montana. In 1869, he was put forward for appointment as Governor of Montana (territorial governors were appointed not elected). However, machinations in Washington meant he missed out. Unemployed, he was recruited by Cooke to undertake a lecture tour of the eastern states to promote Montana and the Northern Pacific Railroad. After the expedition Cooke also subsidized sketches and paintings of Yellowstone, which were transported to and displayed in the Capitol in Washington DC (Haines 1972, Runte 1979 and 1984, Sellars 1997, Glick and Alexander 2000, Schullery and Whittlesey 2003, Lubetkin 2006). It is also notable that other members of the Washburn Expedition, including Samuel Thomas Hauser and Walter Trumbull (whose father was a US senator), were heavily involved in railway interests (Haines 1972).

It has been argued that 'Jay Cooke and officials of the Northern Pacific Railroad may actually have suggested the park bill and motivated the interested parties' (Runte 1979: 44). However, the opposite may also be true. Langford, employed by Cooke to promote the region and railway, may have been the one who suggested the national park as a means of achieving this.

Cooke's involvement may be interpreted in two ways. The first (and most commonly advanced) was that Cooke was planning a line to Yellowstone. He did not want the area under the private ownership of numerous smallholders who would charge exorbitant rates (Sellars 1997, Schullery and Whittlesey 2003). On his expedition, Hayden noted that 320 acres had already been pre-empted at Mammoth Hot Springs, specifically on the expectation that the Northern Pacific Railroad would reach there (Runte 1979). Cooke's concerns were clearly apparent in a letter he wrote to one of his railway managers:

> It is important to do something speedily, or squatters and claimants will go in there, and we can probably deal much better with the government in any improvements we may desire to make for the benefit of our pleasure travel than with individuals.
>
> (quoted in Sellars 1997: 7)

It must be understood that the transcontinental railways were very much partnerships between the private sector, local officials and the federal government. To encourage railways, the federal government provided extensive land grants. Once the railways were built, these now accessible lands could be sold, thus paying for the costs of construction. Town, territory and state officials were also involved in lobbying for railway lines in order to develop their local areas. Cooke, Langford and others saw themselves as working together for mutual benefit and broader economic development. For Cooke, it was preferable to continue working with government officials rather than treat with private speculators.

A second interpretation is that Yellowstone could be used to promote the Northern Pacific and a more general destination image of Montana. Building a railroad through sparsely populated Montana was a huge speculative gamble. To succeed, Cooke needed to sell railroad bonds. He had to convince sceptical easterners of the potential of the venture (Lubetkin 2006). The National Park was an endorsement by the US Congress that Montana was a special (even unique) place. This could, in effect, be used as a brand, attracting media coverage and attracting potential investors and customers (note that while Yellowstone was actually in Wyoming, access was via Montana). As the Yellowstone proposal was debated, the *Helena Herald* (a Montana newspaper) argued it was:

> A means of centering upon Montana the attention of thousands heretofore comparatively uninformed of a Territory abounding in such resources of mines and agriculture and of wonderland as we can boast, spread everywhere about us.
> (quoted in Glick and Alexander 2000: 184–5)

Origins of the term National Park

In recent years there has been a growing dissatisfaction with the simplicity of Langford's account. He did not publish his diary until 1905, 35 years after the events described and it has been suggested that he may have edited them with the benefit of hindsight. Runte argued 'that the explorers used the term *national park* at this time is more than open to question', citing that none of their published accounts of the expedition used the term (Runte 1979: 42). Furthermore, Runte noted that the Act establishing Yellowstone refers to it as a public park and does not use the term National Park at all. His view was that Yellowstone was intended to be like the state park at Yosemite, but as Wyoming was still a territory, it came under federal control, essentially becoming national by default (Runte 1979: 42). Finally, Schullery and Whittlesey (2003) found that none of the magazine and newspaper articles of the time, nor any of the diaries of the expedition members, tell of the campfire discussion.

However, some care needs to be taken here. In his 1905 account, Langford went to some pains to explain that it was based almost exactly on his diary, with only some personal expressions deleted and 'here and there [I] added an explanatory annotation or illustration' (Langford 1905: lxi). Throughout the published journal, Langford used footnotes to clearly distinguish his later observations. Furthermore, in 1905, none of the surviving expedition members challenged Langford's recollections. Langford may have embroidered the details, but clearly in 1870 some sort of discussion about the desirability of public control had taken place (Schullery and Whittlesey 2003).

While the 1864 declaration of Yosemite as a state park was influential, what was proposed for Yellowstone was intended to be something more. During the campaign for Yellowstone, its advocates positioned it as a place of national rather than territorial significance.

What is intriguing is where the concept of a national park came from. Did the idea just suddenly occur to Hedges sitting around the campfire or Cooke and Langford in an office, or did it go back earlier? In 1865, Thomas Meagher, Acting Territorial Governor of Montana, had suggested that Yellowstone be reserved and it has been argued that Hedges was present at the time (Haines 1972, Sellars 1997, Schullery and Whittlesey 2003). This may have implanted a seed, though at this time there had been no expeditions to the region. Another possibility is that David E. Folsom made a similar comment to Washburn before the party left on its expedition. Folsom had led the earlier small expedition and was a friend of Langford's (Haines 1972, Schullery and Whittlesey 2003). However, Langford noted that Folsom, fearful of being ridiculed, was shy about publicizing what he had found and that his conversation with Washburn was unknown to Hedges at the time (Langford 1905).

The term 'national park' had already been applied to Yosemite, albeit as a descriptor rather than a formal title. In 1868 the state of California had published the *Yosemite Book*, written by state geologist Josiah Whitney and intended to promote the park. In this, Whitney described Yosemite as a 'National public park' (quoted in Runte 1990: 33).

There were even earlier antecedents. In 1832, the artist George Catlin had visited the Great Plains and in his subsequent book proposed a 'Nation's Park' be established (quoted in Nash 1970: 728–30, Boyd and Butler 2000: 14–15, Hall 2000: 58). However, Sellars (1997) argued that there is little evidence that Catlin's proposal had any direct influence on the Washburn expedition. Another possible influence is the poet William Wordsworth, who had concluded his popular guide to England's Lake District with:

> Persons of pure taste throughout the whole island, who, by their visits (often repeated) to the Lakes in the North of England, testify that they deem the district a sort of national property, in which every man has a right and interest who has an eye to perceive and a heart to enjoy.
> (Wordsworth 1810: 92)

Greater consideration should be given to three general influences that were at work, both on the proposers of the national park concept and those they sought to convince. These influences were the spread of the romantic viewing of nature, the development of public parks and the rise of national institutions. By 1870, these three concepts were sufficiently well developed for them to be combined and crystallized in an idea like a national park.

Reinterpreting nature

The century or so before the Washburn Expedition had seen a revolutionary reinterpretation of how people viewed nature. Previously, people saw the wilderness as savage and hostile. In America, the pioneer 'lived too close to wilderness for appreciation . . . the *conquest* of wilderness was his major concern' (Nash 1967: 24). However, in the eighteenth and nineteenth centuries the spread of romanticism altered perceptions, wilderness became 'associated with the beauty and godliness that previously had defined it by their absence' (Nash 1967: 46; see, also, Thacker 1983).

The nature of beauty was reordered by works such as Edmund Burke's *Philosophical enquiry into the origin of our ideas of the sublime and beautiful* (1757). Wild places, particularly mountains, evoked sublime feelings of awe, exultation, mortality and even terror. These were feelings so profound that visitors were almost unable to describe them (Nash 1967). Indeed, writers aimed to outdo each other in conveying their inadequacies.

Following these conventions, Langford described the Yellowstone canyon as 'a scene composed of so many of the elements of grandeur and sublimity ... calculated to fill the observer with feelings of mingled awe and terror' (Langford 1905: 30). And when he looked down into the canyon 'I realized my own littleness, my helplessness, my dread exposure to destruction, my inability to cope with or even comprehend the mighty architecture of nature' (Langford 1905: 32).

Romantics such as Jean-Jacques Rousseau and Lord Byron in Europe and later, James Fenimore Cooper, Henry David Thoreau and John Muir championed ideas that the wilderness allowed people to be free. Visiting wild places became particularly popular with urban dwellers. This was essentially what the Washburn Expedition was about – a group of professional and business men taking a break from their normal lives. Educated men, they were aware of the ideas and language of the romantics. When the Washburn Expedition was slowed by heavy scrub, Langford jokingly reminded them of their purpose by reciting from Byron's *Pilgrimage of Childe Harold*. He particularly emphasized the line 'There is pleasure in the pathless woods' (Langford 1905: 69). For Nash, Byron's work was a manifesto for a generation and 'climaxed European Romanticism's century-long achievement of creating an intellectual framework in which it [wilderness] could be favourably portrayed' (1967: 50). It was significant that Langford and his colleagues shared these values.

An increasingly romantic view of nature went hand in hand with greater scientific interests. The nineteenth century became the grand age of the amateur naturalist and collector. Theodore Roosevelt provides an instructive example of such interests. His father was one of the founders of the Museum of Natural History and young Theodore was an enthusiastic naturalist who became a skilled taxidermist (Cutright 1985). For Yellowstone to be protected it was necessary for the claims of the Washburn party to be substantiated by Hayden's scientific expedition.

The development of public parks

The term 'park' was originally applied to the private estates of royalty, nobles and wealthy landowners. Most began as hunting reserves, but by the eighteenth century fashions had changed and they were recreated as romanticized landscapes. In London, growing population pressure led to a number of royal hunting reserves becoming public parks. In 1637, Hyde Park was opened to the public and in the early eighteenth century it was landscaped as a more formal park. In the early nineteenth century St James and Green Park were made accessible to the public and Marylebone Park recreated as Regent's Park. Other instances of royal hunting reserves becoming public urban parks included Phoenix Park in Dublin (opened 1747), Holyrood (or Queen's) Park in Edinburgh, Bois de Boulogne in Paris and the Tiergarten in Berlin.

Increasing industrialization resulted in growing concerns about the housing, health and general welfare of the working classes. Urban parks and outdoor recreation were seen as a counter to the growth of slums, immorality, disease and despair. Accordingly, nature became part of the reform agenda for political radicals and religious dissenters. In 1841, Thomas Cook began organizing excursion trains to take workers out of the cities and into the countryside for temperance meetings. In 1843, work began on Birkenhead Park in Liverpool. The park was the centrepiece of a model housing estate catering for workers from the nearby docks. Often referred to as the first 'public park', Birkenhead was important in being developed by local authorities and made permanently accessible to everybody, rather than a converted royal reserve, which was only open at the discretion of the monarch.

The radical passion for parks spilled over into Britain's settler colonies. When Adelaide in South Australia was surveyed in 1836, its formulaic grid pattern was enveloped in a green belt of parks and reserves. In 1844, the Melbourne town council petitioned the governor for parks to be reserved. In language that could have been directly imported from Britain, they reasoned:

> It is of vital importance to the health of the inhabitants that there should be parks within distance of the town where they could conveniently take recreation therein after their daily labour ... experience in the mother country proves that where such public places of resort are in the vicinity of large towns the effect produced on the minds of all classes is of the most gratifying character.
>
> (quoted in Wright 1989: 34)

The US was initially slow to follow. A crucial connection came with the 1850 visit to England of New York farmer Frederick Law Olmsted. After his ship docked in Liverpool, he took a few days to take in the sights. A baker in Birkenhead suggested he had to visit the new park. Olmsted was impressed, writing that he 'was ready to admit that in democratic America there was nothing to be thought of as comparable with this People's Garden' (Olmsted 1852: 52). He further enthused:

> And all this magnificent pleasure ground is entirely, unreservedly, and for ever, the people's own. The poorest British peasant is as free to enjoy it in

all its parts as the British queen. More than that, the baker of Birkenhead has the pride of an OWNER in it. Is it not a grand, good thing?

(Olmsted 1852: 54)

Returning to the US, Olmsted and Calvert Vaux were the winners of a competition to design Central Park in New York. Inspired by Birkenhead, it was intended to also be a 'People's Park'. Emphasizing its national significance, Vaux described it as 'the big art work of the Republic' (Vaux 1865: 385). Olmsted also saw it as linking together democracy and art:

> It is of great importance as the first real park made in this country – a democratic development of the highest significance and on the success of which, in my opinion, much of the progress of art and esthetic culture in this country is dependent.
>
> (Olmsted 1858: 201)

In their artistic approach to designing Central Park, Olmsted and Vaux aimed for a 'picturesque' effect. They wished to create a landscape, which evoked a romanticized wilderness, contrasting starkly with the formality of aristocratic European gardens and the straight lines of the surrounding American urban grid-plan. Accordingly, for Central Park:

> Its informality was democratic, and it offered the right synthesis of the beautiful and the sublime: an overall composition of smoothness, harmony, serenity, and order, with an occasional reminder of the awesome grandeur of a mountain, a deep crevasse, long waterfall, or steep crag.
>
> (Cranz 1982: 24)

As such, at the time of the Washburn Expedition, the concept of public parks was highly topical. A radical, democratic idea, it had been successfully transplanted to New York and was now spreading across the country. Indeed, it is worth noting that at the time of the Washburn Expedition, San Francisco had just commenced to develop the Golden Gate Park as its version of Central Park. Furthermore, in the case of Yosemite (where Olmsted was the first head of the commission of management), California also presented a model of how the concept of an urban park could be transposed to a non-urban setting.

National institutions

The Age of Enlightenment also heralded the dawn of the Age of Nationalism (Anderson 1983: 11). It became far more common for people to identify themselves with a country and nation rather than a local region or town. In settler societies such as the US, notions of 'nation-ness' and national institutions were much more advanced than in most of Europe (Anderson 1983: 50).

The pace of this development quickened dramatically in the years before Yellowstone. The US had just been through the Civil War, which had established the paramountcy of the nation (the *union*) over individual states. The Washburn Expedition were all Yankees (some had fought in the Civil War) and supporters of the US's Manifest Destiny to spread across the Americas. They were also aware of recent developments in Europe – the unification of Italy and Germany, the decline of the Austrian Empire and the reversion of France to republicanism – the ongoing Franco-Prussian War was also a campfire topic (Langford 1905: 13 and 120). These changes were reflected in the fashion for prefixing new institutions as 'National'. France's National Museum of Natural History provided a template to follow.

When Yosemite was declared a park during the Civil War, it made little sense to call it national. The decision to protect Yosemite came during the build-up to a presidential election and the result of the Civil War was still in doubt. California contained many Confederate supporters. In a delicate political situation, on the one hand establishing Yosemite demonstrated the Union's commitment to California, but on the other emphasizing it as symbolic of the nation may have been incendiary. However, by the 1870s, attitudes had changed, and the designation of Yellowstone as national suggested reconstruction and unity.

The Yellowstone model

Yellowstone created a template for future national parks, both in the US and elsewhere. It established seven main principles (see Table 2.1, overleaf). These were highly influential on later national parks, though not set in stone. As will be seen in the next chapter, the evolution of national parks would include many variations from this initial model.

US national parks after Yellowstone

Yellowstone imprinted a strong pattern on future national park developments in the US. However, initially there was a substantial pause. In 1875, Mackinac National Park was established, taking in a small island of about 1,000 acres on Lake Huron in Michigan. Mackinac gained its national status as the US Army maintained a small post on the island; when that closed in 1895, it reverted to a state park. In hindsight, Mackinac was a strange anomaly and is often overlooked in discussions of early national parks (Sellars 1997).

Though they were not created as national parks, the influence of Yellowstone may be seen in the establishment of the Niagara Falls Reservation and Adirondack Forest Preserve (both in New York and both in 1885). While there was some discussion of Adirondack becoming a national park, it was eventually preserved as a state entity (Nash 1967). Meanwhile, as the situation remained static in the US, Australia, New Zealand and Canada enthusiastically embraced the national parks concept (see Chapters 1 and 3).

Table 2.1 Seven principles arising from Yellowstone

1. The term national park is applied for the first time. While used for a specific place, it is general enough to be easily transferred elsewhere in the US and other countries. Most importantly it captures the public imagination.
2. Yellowstone is reserved for its natural wonders and monumental scenery. This establishes influential precedents for what type of natural environments should be preserved elsewhere. The monumentalism enshrined by Yellowstone will dominate the national park concept for decades and is still highly influential today.
3. The natural wonders and monumental scenery are preserved for visitors to enjoy. The national park is established to manage these tourism flows and prevent the inappropriate tourism developments that have occurred elsewhere.
4. National park status is conferred by the national government. It is the highest level of park status, decided upon by the highest level of government.
5. The national park is a permanently protected area. This is confirmed by the defeat in 1886 of a proposal to allow a railway to cross the park (Nash 1967: 114–15).
6. National park status comes after scientific investigation (the Hayden Expedition) and public debate.
7. There is little consideration of indigenous people. Even though the Washburn Expedition was wary of encountering Crow Indians, they socially constructed Yellowstone as a wilderness without people through the argument that the Indians regarded it as taboo (Langford 1905: 58 and 97).

It was nearly 20 years after the establishment of Yellowstone that the US began to extend the concept elsewhere. Major new national parks included Yosemite, Sequoia and General Grant (1890), Mount Rainier (1899), Crater Lake (1902), Glacier (1910), and Rocky Mountain (1915). As Runte argued, 'when the national park idea enjoyed a true resurgence, the areas set aside were unmistakably in the image of Yellowstone and Yosemite' (1979: 55–6). The tendency was to the monumental appealing to tourists. These new national parks were spectacular mountain landscapes sculpted by volcanoes or glaciers and surrounded by temperate forest. Other environments, including coastlines, deserts, grasslands and wetlands, were conspicuously absent. It would not be to the Everglades (1934) that a national park would be established to protect wetlands. There were also some Mackinac-style anomalies – Sully's Hill in North Dakota (1904) and Platt in Oklahoma – small parks created through influential local politicians (Sellars 1997).

As these national parks were established, it became increasingly obvious that the initial management structure of Yellowstone was inadequate. Congress had gladly allocated land for Yellowstone, but it was less forthcoming in funding. While a national institution, the intention was that its management would be through a local commission. This model was then applied to subsequent national parks, and unfortunately its deficiencies were repeated.

Such were the problems of adequately managing Yellowstone that in 1886 the army was brought in to control it (Hampton 1971, Scott 2005).

The military had previous experience with Mackinac and had resources and expertise, particularly in constructing infrastructure. However, this too was not a satisfactory long-term solution. Accordingly, the situation at Yellowstone and the other major national parks stimulated a campaign to create a specialized national parks service. When this was achieved in 1916, it marked the end of the first pioneering phase of the national park concept.

3 American invention to international concept

The spread and evolution of national parks

Warwick Frost and C. Michael Hall

The 'American Invention'

In 1970, Roderick Nash described the idea of national parks as an 'American Invention'. As he argued, 'the concept of a national park reflects some of the central values and experiences in American culture' (Nash 1970: 726). Not only had the US conceived of the idea, it had also taken the lead in promoting and exporting the concept around the world (Nash 1970: 726). Nash's sentiment would be echoed by Alfred Runte, who would declare that 'the United States, recognized for its Declaration of Independence and the Constitution, has also bequeathed to the world its most stunning example of landscape democracy – the national park idea' (1984: 5).

Nash's phrase (and Runte's affirmation of it) raises interesting questions regarding national parks today. How did the concept of national parks spread from the US to the rest of the world? What factors aided its diffusion and which worked against it? Were national parks directly copied from the US or were they modified for local conditions?

The last question is important, given that Nash saw the American invention of national parks as primarily a product of American history (see Table 3.1). This American experience was so important that Nash concluded that 'had any of these factors been missing, national parks could not have arisen in the United States in the form they did' (1970: 726).

This suggests a major paradox. If national parks arose from uniquely American factors, how could the concept spread to other countries where these conditions were not present and, in some cases, arguably even antithetical?

Table 3.1 Nash's four factors behind America's leadership in national parks

- America's unique experience with wilderness created an appreciation of nature.
- Democratic ideology ensured that national parks be public not private.
- The US had significant amounts of undeveloped land available for national parks.
- The US was sufficiently affluent to afford the luxury of national parks.

Source: Nash 1970.

This *internationalization* of the national parks concept needs to be thoroughly examined.

The characterization of national parks as a specifically American invention invites comparisons with other noted American products and concepts. McDonald's, Coca-Cola, westerns, jazz and rock 'n' roll come to mind as originating in the US and then spreading around the world. Indeed, it is not too difficult to imagine national parks as a brand, both as a brand name and an image. While there are often claims for various American products to be the world's most recognized brands, the concept of a national park is at a similar level, being universally known and valued.

Thinking of national parks in this way highlights the paradox of conceptualizing them as the American Invention. In one sense McDonald's and Coca-Cola provide good analogies for the national parks concept (as distasteful as this may be for some readers!). As McDonald's and Coca-Cola have spread around the world, they have done so with what appears at first glance to be a uniform, centrally controlled product. Coca-Cola, for example, is made to a single formula for its syrup concentrate, which is then mixed with filtered water and sugar/sweeteners and carbonated by its franchises. This means that although Coca-Cola is marketed as a global brand there are differences that cater to local tastes, while in many markets there are also of course a selection of other colas ranging from Pepsi to Mecca Cola. Often portrayed as a globally monocultural product, McDonald's has increasingly adapted to meet local tastes through the availability of different meat and vegetable products. Indeed, there is significant variation in McDonald's offerings even in the US.

As with other aspects of American culture such as Coca-Cola and McDonald's, when the national parks concept was transferred outside of the US, it evolved to be highly variable. While Yellowstone was copied, it was not duplicated. Both within the US and around the world, national parks have therefore followed many varied models. Some of the variables that have developed over time are set out in Table 3.2.

These variations reflect the influence of local natural, social and political environments. A reconsideration of Nash's (1970) four factors is instructive. They are specifically a set of factors operating within and exclusive to the US in the late nineteenth century as reinterpreted in the late twentieth century.

Table 3.2 Variables in national parks

- Size
- Ecosystems
- Protection focus (eg landforms, wildlife, cultural heritage)
- Visitor focus and infrastructure
- Level of government
- Land ownership
- Indigenous involvement

Such factors do not necessarily apply in other countries or at other time periods. Accordingly, as circumstances varied, so too did the application of the national parks concept. Different countries adapted the concept to suit their needs.

Of crucial importance was that the national parks concept spread without any central institutional control. The US did not establish a National Parks Service until 1916. Up until then each national park had been established under individual legislation. Furthermore, 10 foreign jurisdictions had established their own series of national parks (Canada, New Zealand, the six Australian states, Sweden and Switzerland) by this stage, several of which were arguably more administratively developed than the American model (Nash 1970). While greatly influenced by the American model, these governments did not seek American permission, nor did the US seek to impose any conditions on the use of the name. This institutional freedom continued after the establishment of the US National Parks Service and continues today.

The lack of any world controlling body reinforces the international variability of the national parks concept. To return to earlier analogies, it is as if Coca-Cola or McDonald's did not enforce any standards with their franchisees and subsidiaries. Potentially, the controlling body could have been the World Conservation Union (International Union for the Conservation of Nature and Natural Resources), founded in 1948 as part of UNESCO. It could have operated in a similar way to the United Nations' later arrangements for World Heritage listings. However, this did not occur, as too many countries had already established their own systems and organizations.

In the international spread of the national parks concept, it is possible to distinguish four patterns (or perhaps more accurately variations). These patterns can only be broad and we accept that they are open to criticism. However, given the large number of countries that have established national parks, it is important to distinguish some trends in the evolution and spread of the concept. Accordingly, we propose four main models of development based on loose political–geographic associations. These we term as: the New World, the Developing Countries of Africa and Asia, Europe (the Old World) and Totalitarian States.

The New World

National parks were a New World invention and it was not surprising that they rapidly spread from the US to similar settler societies. Taking their cue from America's innovation at Yellowstone, Australia established its first national park in 1879 and Canada and New Zealand theirs in 1887. That the idea was taken up so quickly reflected the shared heritage of these four societies. All were rapidly expanding frontier societies. They shared a common language, had high levels of literacy and drew on common cultural values, particularly in literature, art and philosophy (Johns et al. 1998, Tyrrell 1999, Horne 2005). Canada shared a long border and many geographic features with the US.

Though further distant, Australia and New Zealand were linked to the US through the Pacific economy and the 'Pacific Exchange' of migrants, ideas, biota and technology (Tyrrell 1999). Newly settled by Europeans, they had a much stronger sense of their 'nation-ness' than most Old World countries (Anderson 1983). Creating national parks was, in part, a way of reinforcing their national identity (Hall and Shultis 1991; see, also, Chapter 5).

Canada

Canada and New Zealand tended to follow the Yellowstone model. Canada's first national park was created in 1885, only 13 years after the establishment of Yellowstone (Nicol 1969). An Order in Council established an area of 25.6 km^2 at Banff in the Rocky Mountains, as a Hot Springs Reserve for the 'great sanitary advantage' of the public (in Marsh 1983: 275–6). The influence of American national park philosophy was indicated as both Yellowstone and the Arkansas Hot Springs were used as models when Canada's first national park was officially established around the Banff Hot Springs Reserve in 1887 (and initially called the Rocky Mountain National Park). As in the American case, the first Canadian park was inspired by the dominant utilitarian desire to develop otherwise low-value land via tourism.

Banff was not reserved as a wilderness park. The reserve was created in the tradition of the upper-class European spa and received much support from the Canadian Pacific Railway (CPR) (Hart 1983, Marsh 1985). As the 1887 debates of the Canadian Senate indicated, the reservation was to become a park of designed, ordered nature in the manner of the British aristocracy, 'in order to make a park of this tract of land, of course, it becomes necessary to improve it to a certain extent' (quoted in Brown 1969: 99). The scenery of the area was recognized as being one of the main attractions of the region, but it was not to be accepted on its own terms. The core of the park, the Banff townsite, represented tamed nature set within British aesthetic tradition and landscape design.

As in the United States the initial purpose of Canadian national parks was to promote economically beneficial tourism. The Canadian federal government saw national parks as a means, along with extractive industries such as lumbering and mining, to develop the western lands of the nation. Similarly, railroad companies were a vital factor in the creation of Canadian parks. The CPR played a major role in expanding the boundaries of the Rocky Mountain National Park in 1902 in order to include Lake Louise. Similarly, the creation of Glacier National Park in the Selkirk Mountains of British Columbia, and of Yoho National Park in the Rockies, were also the result of the CPR's desire to encourage tourists to the region. The dominion government's willingness to use tourism as a tool for regional development also led to the establishment of Waterton Lakes National Park in 1895 (Marsh 1983).

The Dominion Forest Reserves and Parks Act 1911 and the consequent establishment of a Canadian park service marked a new stage in park administration but, in keeping with the utilitarian attitudes of the time, the legislation imposed no preservation function on the parks and did not distinguish them from the commercially oriented forest reserves. The Act also saw the installation of J.B. Harkin as Commissioner of the Canadian National Parks (Nicol 1969). In a similar fashion to Stephen Mather in the United States National Park Service, Harkin was faced with an economic problem in the parks – how to obtain the funds to conserve, manage and develop the parks? Tourist traffic became one of the major sources of revenue and Harkin actively encouraged the development of roads, townsites and automobile tourism (Alderson and Marsh 1979), 'realising that to persuade politicians to create national parks in all the provinces would require that their value be clearly demonstrated by increased visitation and economic return' (Marsh 1983: 278). As Harkin expressed in *Conservation is the New Patriotism* (1922), 'we . . . stand very closely by the economic view in order to secure the whole-hearted interest of the people of Canada in the conservation of forests and the wild life . . . and we may have to show that the movement will pay for the efforts many times over' (quoted in Nash 1969: 77).

New Zealand

The first parks and reserves established in New Zealand were often on Maori land. As in Canada and the United States, the tourism potential of hot springs as spas and resorts served as a catalyst for the creation of national parks (Hall and Shultis 1991, Star and Lochhead 2002; see, also, Chapter 1). The nucleus of present-day Tongariro National Park was gifted to the New Zealand Government in 1887, with the park finally legally designated in 1894. The considerable delay between the deeding of the land of the Maori Chief Te Heuheu Tukino to the Crown, and the actual establishment of the park, reflected the Government's concern that only 'worthless' land would be incorporated in the park. 'There had to be absolute certainty that land being added to the park had no economic value' (Harris 1974). In speaking to Parliament on the proposed park, the Hon. John Ballance (New Zealand 1887: 399) stated: 'I may say that this land is particularly suited for a national park. It has all the appearance of a park in itself, and many persons, looking at it, would imagine it had been laid out artificially, and created at enormous expense for the purpose of a park'. Ballance's comments bear witness to the aesthetic sensibilities of the colonial 'aristocracy'. The desire for parkland replete with British fauna and flora was a testament to the recreational sentiments of Victorian New Zealand and dominated the calls for the preservation of 'untouched' wild nature. However, the 'worthless lands' view of national parks, so characteristic of early attitudes towards parks in Australia, was also dominant in New Zealand. For example, in discussing Tongariro National Park, the Hon. John McKenzie, Minister for Lands, was reported as telling Parliament that, 'anyone who had seen the portion of the country . . . which he might say was almost useless so far as grazing was concerned, would admit that it should be set apart as a national park for New Zealand' (New Zealand 1894: 579).

In a similar fashion to Canada and the United States, the New Zealand government saw national parks as a means to economically develop areas through tourism (Star and Lochhead 2002), the aesthetic values of regions being the attraction to the tourist. To quote Ballance again on his proposal for a Tongariro National Park: 'I think that this will be a great gift to the colony: I believe it will be a source of attraction to tourists from all parts of the world and that in time this will be one of the most famous parks in existence' (New Zealand 1887: 399).

Australia

In contrast to New Zealand and Canada, Australia rapidly diverged from the standard US Yellowstone model. Its first national park was on the outskirts of Sydney. Simply called The National Park, it was later renamed the Royal National Park following a visit of Queen Elizabeth in 1955. A large area of pleasant coastal bush, it was hardly monumental. Instead, it was intended for the recreational use of the city's population and the acclimatization of exotic animals (Hutton and Connors 1999). At the time, Australia was still a number of individual self-governing colonies and the New South Wales government created this national park.

So different and so varied was the Australian experience with national parks that it is appropriate to term it the 'Australian Anomaly', although some parallels do exist with that of other federal systems such as Canada (Hall and Shultis 1991). Apart from the wide range of natural features protected, this anomaly was distinguished by massive variations in size, purpose and governance. Some of the main features are set out in Table 3.3 overleaf.

Latin America

The history of conservation and national parks in Latin America has tended to be overwhelmed by the focus on its northern neighbours (Coates 2004). Accordingly, our comments are brief, though we stress there is much potential for further research. The greatest attention has tended to be on Mexico (Salcido 1995, Simonian 1995, Nelson 2000, Coates 2004). In Mexico, national parks began to be created in the 1930s under the administration of Lazaro Cardenas, who also established the Department of Forestry, Fish and Game, which supervised the designation of 40 parks, based on scenic beauty, recreational potential and ecological value. Initial designations in the Mexican park system emphasized the highland forests and significant attention was given to their tourism potential. Similarly, national parks in Costa Rica have focused on ecotourism in mountainous rainforests and been greatly influenced by US developments and tourism markets (Evans 1999, Honey 1999).

Table 3.3 The Australian anomaly

1 **Multiple agencies.** In Australia, national parks were primarily established and operated by the six individual states that had previously been separate colonies, not the national government. Upon Federation in 1901, the states retained control of lands, which were envisaged as a source of government revenue. Accordingly, the states also retained the power to convert these lands into national parks. After the Second World War, each of the states created their own national parks agencies, which continue to operate quite separately, although policy coordination meetings are held at a ministerial and administrative level. A seventh agency developed when the Federal government began to establish national parks and reserves in the Northern Territory and other Australian Federal Territories, such as Christmas Island. Even though the Northern Territory has now achieved self-government, the national parks that were originally established by the Commonwealth (Kakadu National Park and Uluru-Kata Tjuta National Park) are still administered by the Federal government, while the Territory has also established its own national parks. Similarly, the Australian Capital Territory (ACT) also has its own national park – Namadgi National Park – which lies under ACT jurisdiction.

2 **Numbers of national parks.** The Australian state governments have tended to create large numbers of national parks (see Chapter 4 for a more detailed discussion). Establishing national parks, particularly in recent times, was often popular with some stakeholder groups and voters and this was especially stimulated by the growth of a number of environmental interest groups (Hutton and Connors 1999, Anderson 2000). This has meant that some national parks were created which would probably not have been given that status in other countries. For example, the state of Victoria is approximately the same size as the US states of Minnesota, Wyoming or Utah. Yet it contains over 50 national parks compared to one in Minnesota, two in Wyoming and five in Utah. Indeed there are only a handful of countries in the world with more national parks than Victoria. However, the issues of designation perhaps become a little more apparent when it is realised that Minnesota has 72 state parks, Wyoming 12, and Utah 40. To make matters even more confusing the different Australian park systems also have a range of protected area reservation categories. For example, New South Wales has 11 state parks that are maintained by the Department of Lands as opposed to the national parks that are maintained by the Parks and Wildlife Service. Victoria also has national parks and state parks, but both come under the jurisdiction of Parks Victoria.

3 **Emphasis on recreation.** The early national parks tended to be along the coast, near population centres and with a strong emphasis on recreation (see Chapter 4 for a more detailed discussion). The National Park catered for nearby Sydney. In South Australia, Belair (1891) was a 1,990 acres former government farm just out of Adelaide. It too was conceived in terms of urban recreation. Very small national parks were declared at rainforest gullies and waterfalls, really little more than popular picnic grounds at picturesque locations (in New Zealand as well as in some cases in Australia such sites were often made scenic reserves). These micro-national parks included 557 acres at Ferntree Gully (reserved 1881, but not officially proclaimed until 1928), 750 acres at Tarra Valley (1909), 324 acres at Witch's Falls (1907) and a tiny 49 acres at Bulga (1904). Advocates of mountainous national parks, such as Mount Buffalo (1898) and Lamington (1915), argued that these would function as 'sanitoriums' for jaded city-dwellers (Webb and Adams 1998, Frost 2004). This emphasis on health parallelled the proposals for Yellowstone and Banff. In contrast, national parks came much later to remote, often more arid,

areas. For example, Uluṟu was not 'discovered' by Europeans until 1873, was virtually unknown to tourists until the 1960s and was only declared a national park in 1977 (Hall 1992; see, also, Chapter 10).

4   **Variety.** Whereas the early national parks of the US, Canada and New Zealand tended to be spectacular mountains and alpine areas, Australia incorporated a far wider range of landforms. Three in particular are worth noting as early examples of diverging considerably from the US model. First were coastal national parks, including the Royal (1879), Kuring-Gai Chase (1894), Wilson's Promontory (1898) and Mallacoota (1909). Second was the protection of rainforests, starting with small waterfall gullies, but proceeding to the creation of Lamington (1915). At 47,000 acres, this was the world's first large rainforest national park. Third was the establishment of Wyperfeld (1909). Protected for its semiarid vegetation and wildlife, dry lakes and ancient sand dunes, with no mountains or spectacular geological attractions, this was a very different model of a national park to anything that had come before. Goldstein (1979b: 120) argued that the reasons for the reservation of such areas 'could be traced to battles fought by particular organizations or individuals, or the foresight of particular Lands Department surveyors, rather than a general awareness by the public'. However, some parallels may possibly be made to the use of national monument designation in the US under the Antiquities Act of 1906, which provided for the reservation of sites of scientific interest.

5   **The Queensland 'Ultra-Anomaly'.** It was in Queensland that the greatest anomalies occurred. Less developed than the southern states, its governments were often stridently pro-economic development. It was then curious that in 1907, this was the first place in the world to establish a framework where national parks could be declared by administrative processes rather than legislation (similar to the declaration of national monuments in the US). As national parks could be easily and quickly established, they proliferated. On Tamborine Mountain, farmers donated patches of rainforest and each was designated a separate national park (Curtis 1988). By 1948, this one state had 225 national parks (Groom 1949: 204). At that time, this was well over half the number of national parks in the world.

Developing countries

After the First World War, national parks began to be established in Africa and Asia. Most of these were primarily established to protect wildlife rather than for scenery. They were the product of European colonial administrations and the impetus for them often came from European-based institutions. Highly influential was Britain's Society for the Preservation of the Wild Fauna of the Empire (later shortened to Fauna Preservation Society). Established in 1903, its members were a 'star-studded caste' (Fitter and Scott 1978: 8). Drawn from the British upper class, they included the Secretary of State for the Colonies, colonial governors and members of the British parliament. It formed a strong coalition with US decision-makers, particularly the similarly elite and influential Crockett and Boone Club, and boasted Theodore Roosevelt as an honorary member. Nicknamed the 'Penitent Butchers', its members were strongly interested in the preservation of game animals. They took the lead in a number of campaigns for national parks within the British Empire and also influenced other European powers (Fitter and Scott 1978, MacKenzie 1988, Jepson and Whittaker 2002).

Just as important as the Anglo-American hunting lobby were a range of continental European groups pushing for protected areas in their colonies. These tended to take an international approach, be more science-based and less interested in hunting (Jepson and Whittaker 2002). They often operated at both the domestic and colonial levels. For example, the Dutch had the Society for the Preservation of Monuments of Nature in the Netherlands active within the home country and the Netherlands Indies Society for the Protection of Nature operating in what would become Indonesia (Boomgaard 1999).

These European groups adopted an international approach, developing similar structures and aiming to work cooperatively. In 1928, the French, Belgians and Dutch formed the International Bureau on Nature Conservation. After the Second World War this would become the International Union for Nature Protection (ultimately, the International Union for the Conservation of Nature). Working together, the International Bureau on Nature Conservation, the Fauna Preservation Society and the Boone and Crockett Club championed the cause of National Parks in Africa and Asia. In 1933, they staged the London Conference on African Wildlife (held in the House of Lords under the patronage of the British Prime Minister), which achieved general agreement that national parks and sanctuaries were the best mechanism for protecting wildlife (Jepson and Whittaker 2002).

After the Second World War, the governments of newly independent countries also took a strong interest in national parks. While they could have abandoned them as colonial relics, they tended to embrace them. This was partly as a means of promoting their new identities internationally and partly to generate revenue from tourism (MacKenzie 1988).

Africa

Although legislation had been enacted in several colonies, including Natal, as well as the Orange Free State and the South African Republic in the nineteenth century, such regulations had little effect on preventing the loss of wildlife (Grove 1995). In the late nineteenth century, the colonial powers were increasingly concerned with over-hunting in Africa, particularly the devastating effects of the ivory trade. In 1900, the world's first international environmental agreement, the Convention for the Preservation of Animals, Birds and Fish in Africa – was signed by Britain, France, Germany, Italy and Portugal. The Convention aimed to protect game animals and was designed to control trade as well as suggest conservation measures such as closed seasons for hunting, although the convention was never put into effect. The British initially tried to combat over-hunting through transplanting measures developed in Britain, such as licences and closed seasons. In contrast, the Germans introduced game reserves. These were defined protected areas where shooting was forbidden. Game reserves captured the public's attention as a much more effective solution and the British began to adopt them (MacKenzie 1988, Jepson and Whittaker 2002). The first wildlife conservation body in southern Africa, the Natal Game Protection Association was founded in 1883 and President Paul Kruger of the Transvaal Republic supported the establishment of the Pongola Game Reserve in 1894. Arguably these reserves are the foundations for the present-day national park and conservation system in southern and eastern Africa (see Chapter 18). Whether or not Germany would have eventually converted its game reserves to national parks is a moot point, for after the First World War it lost all of its colonies.

Instead, it was Belgium that was the first colonial power to make the breakthrough. In 1919, King Albert toured the US with the Director of the Brussels Museum of Natural History. Members of the Boone and Crockett Club escorted them to Yellowstone, where the issue of Belgium establishing national parks was enthusiastically discussed. This ultimately led to the 1925 establishment of the Albert National Park in the Belgian Congo. This was specifically to protect mountain gorillas and was initially not open to the public (MacKenzie 1988, Jepson and Whittaker 2002).

In 1926, South Africa established the Kruger National Park. In many respects this was the exception to the other colonial national parks discussed in this section. South Africa was self-governing; the decision was not made by colonial authorities (though it suited the imperial agenda). Local tourism was an important factor in Kruger's establishment; there was a sufficient affluent urban market to generate considerable demand. In 1928, Kruger was visited by only 800 people in 240 cars, but by 1938 it received 38,000 visitors in 10,000 cars and was starting to attract international tourists (MacKenzie 1988; see, also, Chapter 18).

In the early discussions regarding Kruger the example of Yellowstone was invoked (Carruthers 1995a). However, a different model was also developing, one which conceptualized it as a game sanctuary. Indeed the terms national park and sanctuary were often used interchangeably (Carruthers 1995a). What evolved at Kruger was very different to Yellowstone. It was a vast area, not of geological wonders or monumental scenery, but of charismatic African fauna. Initially its administrators (coming from the hunting tradition) thought in terms of preserving game animals. To achieve this they considered eradicating the main predators (as had occurred at Yellowstone). However, in the 1930s they came to realize that the increasingly more important tourists demanded to see lions (MacKenzie 1988, Carruthers 1995a).

In 1930 and 1931 the Fauna Preservation Society funded assessments of East and West Africa. While these stimulated a general push for national parks, there was a further hiatus in Africa. After the Second World War, Kenya, Tanganyika and Uganda quickly established national parks (MacKenzie 1988). In the case of Kenya the British government, as a result of pressure from British conservationists, appointed a game committee in 1939 to study and make recommendations regarding the establishment of game parks in the East Africa Protectorate and other African colonies. The committee mainly comprised British naturalists, aristocrats, explorers and senior administrative officials. The committee was to plan the location, extension, constitution, control and management of game parks and the forms of recreational activities that should be permitted in the parks. The game committee made recommendations that were approved by the colonial legislature in 1945 and which led to the creation of the pioneer national parks in Kenya, which included Nairobi (1946), Amboseli (1947), Tsavo (1948) and Mt. Kenya (1949) (Akama 1996).

Asia

Up until the 1930s, Asia trailed behind Africa for two reasons. First, Asian wildlife generally did not have the same cachet as Africa's among game hunters. Second, the nature of colonialism was different in Asia. Europeans were more ready to accept and value Asian institutions and customs, often leaving in place local rulers and recognizing land rights. Growing nationalist movements made the Europeans wary of upsetting local communities. There were also no European settler societies to lobby for game protection, as in South Africa and Kenya. On the other hand there were a broader range of colonial powers and countries (such as China, Thailand and Japan), which maintained their independence.

In 1934, the United Provinces (later Uttar Pradesh) in India established national parks legislation. India had a strong forest conservation system, which was often viewed as the ideal model for others to follow. The strength of the forestry lobby may have mitigated against the rapid spread of national parks. In 1938, British colonial administrations in Malaya and Ceylon established national parks (MacKenzie 1988, Kathirithamby-Wells, 2005). In the Dutch East Indies, 17 wildlife sanctuaries were established between 1932 and 1940. At the time the Dutch decided these were equivalent to national parks elsewhere, but chose not to use that term. These sanctuaries would become national parks in the 1980s (Jepson and Whittaker 2002; see, also, Chapter 16).

While it is generally argued that Africa was the leader and Asia followed (MacKenzie 1988), this was not always the case. Before the Second World War there were two Asian instances that followed very different patterns. The first was the establishment by the French colonial authorities in Cambodia of the Angkor National Park in 1924 (Cresswell and MacLaren 2000). The French had been on the periphery of the European national parks movement, apart from linking it with colonial acclimatization schemes; they did not have a strong record of interest in the concept (Osborne 1994). What they created differed quite markedly from all models elsewhere as this national park was for the protection of an archaeological site rather than natural features or wildlife.

The second variation occurred in Japan, where 12 national parks were established between 1934 and 1936 (Tamura 1957). These were mainly in mountainous volcanic areas. The other side of the Pacific's Rim of Fire, they mirrored Yellowstone monumentalism. Europeans had not colonized Japan and these developments were quite separate from the colonial push within Africa and Asia. There was a long cultural history of the appreciation of nature and Japan's cold and mountainous environment was often evoked as creating its particular culture and national identity (Kalland and Asquith 1997). By this period, Japan was becoming increasingly militarist and expansionary, acquiring its own colonies. Its national parks were symbolic of growing nationalism and were not intended as game reserves.

Europe

The major European powers were late to establish national parks. Britain did not establish national parks until 1949, France 1963 and Germany 1969. This lateness was due to a combination of cultural and environmental factors (see Table 3.4).

The first national parks in Europe were established by lesser powers. Sweden created the first (1909), followed by Switzerland (1914) and Spain (1918). It is notable that these were essentially countries without colonies and which were trying to remain neutral. As such, their national parks were intended to generate great symbolic power. After the First World War, they were joined by Italy (1922), Iceland (1928), Poland (1932), Romania (1935) and Greece (1938). All of these countries had a history of being dominated by other European powers, and again their establishment of national parks was linked to affirming their national identity (see Chapter 5).

Table 3.4 Factors behind the late establishment of national parks in Britain, France and Germany

1 **Comparisons with the US.** Decision-makers were aware of Yellowstone and other national parks being renowned for their spectacular monumental scenery. Their view was that there were no comparable rugged mountain areas worthy of being national parks within their countries (Mair and Delafons 2001, Uekoetter 2006). Belgium and the Netherlands also took a similar view.
2 **National parks were for colonies.** As such, energy was diverted towards the colonies and imperial organizations.
3 **Lack of public land.** As most land was in private hands, the cost of purchasing it for national parks was prohibitive.
4 **Cultural confidence.** As major imperial powers, their national identity and tourist appeal was defined in terms of cultural heritage and historic sites.
5 **Other priorities.** The two World Wars, the Great Depression and other political issues diverted attention. For example, in Britain the Addison National Park Committee (1929–1931) recommended the establishment of two national authorities – one for England and Wales and one for Scotland. Though initially approved, the plan was abandoned due to government cost-cutting in response to the Great Depression (Mair and Delafons 2001).

These early European national parks tended to follow a model of protecting monumental mountain scenery. While in some cases they incorporated former royal hunting reserves, the protection of wildlife was a secondary concern. Given the concentration of different nations within Europe, it was not surprising that there were often different views as to what should constitute a national park. An example of such conflict occurred when Germany conquered Poland. German conservationists proposed that Poland's Ludwigshöhe National Park be downgraded as 'the park does not match our ideas of a national park' (quoted in Uekoetter 2006: 155).

While continental national parks tended to follow the American model, what developed in Britain was another major anomaly. In the 1920s Lord Bledisloe mounted a campaign for the Forest of Dean to be declared a national park. This led to a government inquiry under Christopher Addison. Significantly, Addison had previously established the Ministry of Health. Much of the discussion of his committee concerned the health and recreational benefits of national parks. Accordingly proximity to urban centres was more a priority than natural values (Mair and Delafons 2001).

The Great Depression and the Second World War delayed progress. When national parks were revisited after the War, it was against a background of massive social change, including the introduction of a welfare state and a programme of nationalization. Drawing heavily on the Addison Committee, national parks were seen in terms of domestic recreation. Ten national parks were established under the 1949 legislation. All were in England and Wales. Curiously, none were established in Scotland (which would wait until 2000 for a national park). This is illustrative of the logic employed at the time – national parks were for urban recreation, tourists were a lower priority. While landscapes like the Scottish Highlands were attractive to international tourists, they were perceived as too distant from the major cities and accordingly were not protected (Parker and Ravenscroft 2000).

The other great variation that occurred in Britain concerned land ownership. In the US national park establishment was usually confined to government-owned land. The federal government was loath to resume private land. In some cases it would purchase land for national parks, often with the assistance of private benefactors, but it was conscious that this was a politically sensitive issue. In creating the Great Smoky Mountains National Park (1934), for example, 6,600 separate tracts of land had to be purchased and transferred to the federal domain. These included the land of 18 timber companies, 1,200 farms and 5,000 second homes. This was only achievable through private donations, with John D. Rockefeller Jr contributing approximately 40 per cent of the funding (Young 2002).

In Britain, nearly all of the land under consideration was privately owned and the government could not countenance its resumption or purchase. Instead, it simply included the private land – including farms and whole towns – within the national parks. In effect, national parks were a sort of rural planning scheme, allowing freeholders to continue to operate, albeit with conditions on further development and the guarantee of access to recreational walkers (Parker and Ravenscroft 2000).

Totalitarian countries

As an American invention, national parks are often closely linked with American ideals of democracy. How then, was such an American idea received by regimes that were clearly hostile to America and/or anti-democratic? How did countries like the USSR, Nazi Germany, Fascist Italy and Imperial Japan adapt the concept of the national park?

Russia was initially greatly influenced by the American development of national parks. In 1913, the first Russian Conservation Fair included an exhibit on the US system. In 1916, the Tsarist government of Russia established a 200,000 hectares *zapovednik* (nature reserve) on the shores of Lake Baikal. Following the Revolution, further *zapovedniki* were established, contacts were made with the Parks Services of the US and Canada and the magazine *Okhrana prirody* (Conservation) published features on Yellowstone (Weiner 1988).

However, problems arose in the late 1920s. Two rival bodies established systems of *zapovedniki*. *Narkompros* (People's Commissariat of Education) viewed the reserves in terms of scientific research, whereas *Narkomzem* (People's Commissariat of Agriculture) saw them in terms of economic utility. Collectivization increased the pressure to resume reserves for agriculture and logging or to use them for acclimatization experiments. Under Stalin, pro-conservation scientists and officials were purged and the use of the US as a model faded (Weiner 1999). In effect, the USSR had a system of national parks, but Russian terminology was used to separate it from the rest of the world. It would not be until 1983 that *zapovedniki* would begin to be designated as national parks under the policy of *Perestroika* (Weiner 1999).

Nazi Germany also did not establish any national parks. Instead, having inherited what was widely regarded as the best system of nature protection in Europe, they tended to continue focusing on nature reserves. An example of this might be seen in the establishment in 1937 of a nature reserve at Schorfheide near Berlin. Championed by Hermann Göring, it comprised 141,000 acres of forest and lakes. Göring stated that he wished to create a reserve 'similar in design to the national parks in North America' (quoted in Uekoetter 2006: 100). Nonetheless, the area was designated as a nature reserve and not a national park.

In contrast, Fascist Italy and Imperial Japan were enthusiastic creators of national parks. They appear to have had few problems in using a concept invented in America. Indeed, Mussolini was quite keen to imitate America – where there was a large Italian diaspora – and be seen as a modernist (he was also an enthusiastic supporter of creating an Italian film industry modelled on Hollywood). Accordingly, in the 1920s and 1930s he approved Gran Paradiso, Abruzzo and Circeo national parks (Sievert 2000).

The Japanese drew heavily on Yellowstone as a model, sending staff to the US on investigative missions. In 1930, a National Parks Investigation Committee was formed, resulting in the establishment of 12 National Parks between 1934 and 1936. The enthusiasm for national parks was due to the areas protected combining strong natural, national and religious values (Tamura 1957, Kalland and Asquith 1997).

The development of national parks and conservation in these totalitarian states raises major questions, which require more research. Some writers are perplexed by the possible ramifications of enthusiasm for nature protections in such repressive regimes. Uekoetter (2006) argued that we should not read too much into such connections. In support, he cited Israeli novelist Ephraim Kishon. Replying to a question that one might easily share similar views to the Nazis on some topics, Kishon observed that it would be a mistake to see such things as intrinsic to Nazism, quipping, 'I will not start smoking because Adolf Hitler hated cigarettes' (quoted in Uekoetter 2006: 3).

Another paradox for the totalitarian states arose from their rapid building of war economies. The establishment of national parks would seemingly have been at odds with a centrally controlled state focused on industrialization and autarky. Certainly the conservation movement in the USSR was stymied by the introduction of collectivization. Yet some of these states did divert resources to conservation. In some cases it was to provide prestigious rewards (essentially private game reserves) for the elites (Uekoetter 2006). In other cases, national parks diverted attention from other developments. For example, in draining the Pontine Marshes for agriculture, Mussolini set aside 8,000 acres for the Circeo National Park (Sievert 2000). Most importantly, national parks were enthusiastically embraced as tangible symbols of national identity and pride (see Chapter 5).

Summary

While the idea of a national park was an 'American invention', what was adopted around the world was not a direct copy of the American model. Yellowstone and subsequent developments served as a powerful inspiration, but as countries took up the idea, they quickly modified it substantially for their own purposes and conditions. National parks therefore became a curious concept, an international brand – recognised and respected around the world – but varying greatly depending on local physical, political and economic environments. The implications of the varieties of protected areas that describe themselves as national parks – whether or not they meet the IUCN categories – have left a legacy that many governments, as well as conservationist and tourism stakeholders are grappling with to the present. However, in nearly all jurisdictions tourism and recreation has had a vital role in justifying the establishment of national parks, along with other more local emphasis with respect to conservation, governance and other forms of development.

4 National parks and the 'worthless lands hypothesis' revisited

C. Michael Hall and Warwick Frost

'Nothing dollarable is safe, however guarded'

John Muir (1910: 263)

INTRODUCTION

One of the most fascinating controversies in the study of national park history has been centred around the proposition that national parks were, and to a certain extent still are, only created because they were regarded as 'worthless lands' without alternative economic uses. The concept was initially applied to Yellowstone and the early national parks in the United States by historian Alfred Runte (1972, 1973, 1977, 1979, 1983, 1987, 1997a). As Schullery and Whittlesey (2003: 73) commented:

> Runte pointed out that Congress, before designating public lands as national parks, had to be persuaded that those lands were 'worthless' for other purposes. Thus the willingness to behave altruistically was contingent upon the impossibility of commerce. Historians have quarreled over this 'worthless lands' hypothesis ever since Runte offered it, and we remain uncertain just how significant a factor it was in congressional debates over Yellowstone. But merely pointing out that discussion of the point arose whenever a park was proposed, Runte injected some much-needed skepticism into the study of park origination.

The 'worthless lands' hypothesis has prompted substantial debate in the United States (e.g. Hampton 1981, Cox 1983, Sellars 1983, Utley 1983, Winks 1983, *Journal of Forest History* 1984, Grant 1994, Spence 1996, Wright and Mattson 1996, Strittholt and Dellasala 2001, Schullery and Whittlesey 2003). This debate has been replayed in the settler societies of Australia (e.g. Hall 1989, 1992, 2007, Common and Norton 1992, Pressey 1992, Pressey and Tully 1994, Pouliquen-Young 1997, Wilson and Bryant 1997, Lowry 1998, Crowley 1999, Dunlap 1999, McCarthy and Lindenmayer 1999, Mendel 2002, Mendel and Kirkpatrick 2002, Cary *et al.* 2003, Kelly *et al.* 2003, Frost

2004, Lindenmayer and Burgman 2005, Lindenmayer and Fischer 2006) and New Zealand (Hall and Shultis 1991, Hall and Higham 2000). More recently, discussions of competing economic interests and the need for advocates to present national parks as having no other worth have extended to other countries, including Russia (Weiner 1999), Italy (Sievert 2000), South Africa (Carruthers 1995a), Malaysia (Kathirithamby-Wells 2005), the Solomons Islands (Bennett 2000) and Sweden (Reinius Wall, this volume).

The purpose of this chapter is to revisit the value of the 'worthless lands' hypothesis to understanding the establishment of national parks. The chapter is divided into three sections. The first is a review of the 'worthless lands hypothesis'. The second is an examination of the application of the 'hypothesis' to the creation of national parks in Australia. The third considers the contribution of the hypothesis to understanding the nature of conservation reserves systems.

The 'worthless lands' hypthesis

The most pervasive form of opposition to the creation of national parks is that based upon economic or material values. As Hampton observed with respect to the establishment of national parks in the US:

> In the nineteenth century a sceptical Congress had to be assured that proposed parks contained nothing of exploitable value. In the twentieth century materialistic opponents have operated most efficiently in blocking extensions of older parks and modifying boundaries and provisions of ingress in the new ones.
>
> (1981: 45)

The strength of the economic opposition to national parks stands in clear opposition to the supposed idealism and altruism of the park ideal encapsulated in the myths surrounding the creation of Yellowstone (Schullery and Whittlesey 2003). Indeed, it has come to be argued that the dominance of material over aesthetic and ecological values was such that national parks were only designated on land that was regarded as worthless. The clearest expression of the notion that national parks were created because they were regarded as 'worthless' lands comes from the work of Alfred Runte. According to Runte:

> An abundance of public land that seemed worthless – not environmental concern or aesthetic appreciation – made possible the establishment of most national parks in the United States. Nothing else can explain how aesthetic conservationists, who in the past have represented only a small minority of Americans, were able to achieve some success in a nation dominated by a firm commitment to industrial achievement and the exploitation of resources. A surplus of marginal public land enabled the

United States to 'afford' aesthetic conservation; national parks protected only such areas as were considered valueless for profitable lumbering, mining, grazing, or agriculture. Indeed, throughout the history of the national parks, the concept of 'useless' scenery has virtually determined which areas the nation would protect and how it would protect them.

(1973: 5)

Runte's 'worthless lands' argument arose from the very first speech in Congress which contained elements of the national park idea. In 1864, Senator John Conness of California on introducing a bill to cede Yosemite to the State of California as a park noted, somewhat paradoxically, that the location in the Sierra Nevada mountains was 'for all public purposes *worthless* [our emphasis], but which constitute, perhaps, some of the greatest wonders in the world' (quoted in Runte 1979: 48–9). The speech reflected the dominant utilitarian attitude of the time: 'The wording reassured Conness's colleagues that no universally recognized alternative to preservation had been detected in the Yosemites' and, hence, they 'certainly could afford to recognize the valley for its substantial "intrinsic" worth' (Runte 1983: 135).

The first national parks were national monuments, regarded as natural expressions of American independence from the Old World (Sax 1976). An abundance of monumental scenery was the most important criterion for the establishment of a park. For example, in *Picturesque America*, Bunce reported that with respect to Yellowstone:

> This remarkable area has recently been set apart by Congress for a great national park. It certainly possesses striking characteristics to which it has been devoted, exhibiting the grand and magnificent in its snow-capped mountains and dark canons, the picturesque in its splendid water-falls and strangely-formed rocks, the beautiful in the sylvan shores of its noble lake, and the phenomenal in its geysers, hot springs, and mountains of sulphur. It may be claimed that in no other portion of the globe are there united so many surprising features – none where the conditions of beauty and contrast are so calculated as to delight the artist, or where the phenomena are so abundant for the entertainment and instruction of the student.
>
> (Bunce 1872: 293–4)

However, an abundance of scenery had to be matched by an absence of exploitable material wealth in the form of minerals, lumber or land suitable for agriculture. For instance, the world's first national park, Yellowstone, was only preserved after Dr Ferdinand V. Hayden, head of the United States Geological and Geographical Survey of the Territories, reported to the House Committee on Public Lands that the proposed park took 'nothing from the value of the public domain'. In particular, Hayden stressed that settlement was 'problematical' unless there were 'valuable mines to attract people', but it was highly improbable 'that any mine or minerals of value' would be found in the area (quoted in Runte 1973: 5). Interestingly, Bunce (1872) also reports:

> In the report to Congress by the Committee on Public Lands we learn that 'the entire area comprised within the limits of the reservation is not susceptible of cultivation with any degree of certainty, and the winters would be too severe for stock-raising . . . These statements make it evident that, in setting apart this area 'as a great national park and pleasure-ground for the benefit and enjoyment of the people,' no injury has been done to other interests. The land did not need to be purchased, but simply withdrawn from 'settlement, occupancy or sale;' and hence, by timely action, a great public benefit was secured, which in a few years time would have been impracticable, or at least attainable only with great difficulty. The time is not distant, in the opinion of the Congressional committee, when this region will be a place of 'resort for all classes of people from all portions of the globe.' The Northern Pacific Railroad, now rapidly advancing toward completion, will render the park easily accessible; and, this once accomplished, the marvels of the strange domain will tempt the curious in great numbers to visit it. As a place of resort for invalids, the Yellowstone Valley, on account of its pure and exhilarating atmosphere, is believed to be unexcelled by any portion of the globe; and, if this anticipation prove true, there will be additional reason to be gratified at the wise forethought which secured it for public uses forever.
>
> (Bunce 1872: 316)

Yellowstone was therefore preserved not so much for what it was, but for what it was not. Similarly, the parks that were established by Congress between 1899 and 1919, including Mount Rainier (1899), Crater Lake (1902) (see Chapter 7), Glacier (1910), Rocky Mountain (1910) and the Grand Canyon (1919) were all in areas considered worthless for material economic use. The new parks contained little of material economic value to the nation, and were regarded as containing no significant timber, mineral, grazing, or agricultural resources (Runte 1973, 1977, 1979). Indeed, as Runte (1973: 8) pointed out in relation to individual park legislation: 'no vested economic interests were affected, nor was any federal agency prevented from developing parkland that met the economic demands of businessmen, agriculturalists, or mining concerns'.

Virtually all national parks created in America up to the 1930s were examples of rugged scenery – 'Mountain top parks' comprising but 'a fringe around a mountain peak', a 'patch on one slope of a mountain extending to its crest', or 'but portions of a slope' (Wright et al. 1933: 37, 39). Yet such mountain top parks were of some economic value. The Romantic appreciation of picturesque landscape as well had led to the development of a tourist industry associated with national parks. Railroad companies were major advocates of the concept of national parks in the United States. In Canada, national parks were also created because of the desire of government and the railroad companies to develop the western provinces through tourism (Hall and Shultis 1991). However, according to Runte (1983: 1940–1941), 'tourism does not contradict the worthless lands hypothesis – it supports it. In the chess game of scenic preservation, ecology was the pawn – only economics could checkmate economics'. Similarly, as Runte states in the preface to the second edition of *National Parks: The American Experience*:

> I stand by my original interpretations. Among them none has been more debated than my observation that Congress allowed only those lands considered worthless from a natural resources standpoint to be set aside permanently as national parks ... Perceptions of what Congress itself considered 'worthless' varied with both the time and place, particularly after the turn of the century, when the 'See America First' campaign provided the national parks with a unique commercial foundation of their own through tourism. This observation itself is not intended to refute their ecological and scenic significance. More to the point, it merely underscores the persuasiveness of economic arguments in determining precisely which scenery the nation felt it could afford to protect in perpetuity. As I originally explained, the term 'worthless' grew out of the congressional debates. The word consistently referred only to the absence of natural resources of known commercial value, not to scenery, watersheds, or wildlife with obvious inspirational or biological – if not direct monetary – worth. ... The point is that Congress, at least with respect to the western parks, did not use the term 'worthless' to describe real estate. Rather it was meant to assure prospective miners, loggers, farmers, and ranchers that national parks to be carved from the public domain were unsuitable for sustaining the traditional economic pursuits of the American frontier.
> (Runte 1997b)

Tourism provided the national parks with a defence mechanism. The economic value of tourism proved a valuable weapon for preservationists in the development of more parks and in the protection of others. Tourism gave parks a material value. Aesthetics were important, but only so far as they enabled the procurement of the tourist dollar. Materialistic considerations were still the dominant factor in the establishment of national parks. For Runte (1997b) 'The commercialization of Yellowstone and its counterparts invites historians, both now and in the future, to inquire again whether Americans truly value the protection of wilderness and wildlife, or whether most people simply prefer (or at least accept) that the parks be resorts ensconced in a more pristine setting'.

The rise of ecology gave another aspect to the reasons for park creation. However, according to Runte (1983: 138), 'ecological needs have come a poor second because the nation has been extremely reluctant to forego any reasonable opportunity, either present or future, to develop the national parks for their natural resources'. National parks are still not defined or established on the basis of abundance of ecological criteria. For Runte (1973: 11) the American government 'protects primarily what the business ethic of the nation allows it to protect'. In the search for compromise and 'balanced' land use, ecology is no match for material gain.

However, Runte was criticized for understating the level and range of competing economic interests for lands proposed as national parks. With such demand, it was pointed out, they could not be described as 'worthless'. Sellars argued, 'the American experience with national parks plainly shows the public's determination to preserve park lands in the face of sometimes immense economic values' (1983: 134). Wright and Mattson noted that national park proposals were, 'met with varying degrees of opposition because lands that could be used for consumptive economic purposes were being withheld from development' (1996: 9). Cox, a third critic, stated that:

> Many a park came into being only after a long struggle against those who feared a 'locking up' of resources. Such objections would have made no sense if the land involved had been worthless, yet the history of the parks movement is replete with examples of parks that were established only after extensive debates over the propriety of doing so.
> (1983: 145)

Nevertheless, it may be wondered to what extent such criticism has arisen because Runte is, in essence, debunking one of the great myths of American society, namely the supposed idealism and altruism inherent in the American creation of national parks. As Runte (1997b) himself notes, 'The evidence for this interpretation is abundant; it is simply not always popular to accept. To reemphasize, Americans prefer to think of their national park system as an unqualified example of their statesmanship and philanthropy.' In opposition to Runte, one commentator suggested that the purpose of those who shaped the American national park system 'Was, and is, to find the compromise between altruism and materialism that best captures the public interest' (Utley 1983: 142). However, the question needs to be asked as to what interest for which public? As Runte (1983: 141) concluded:

> The 'worthless lands' hypothesis does not deny the achievements of preservation; it merely asks why the United States still seem to weigh economic issues more seriously than ecological ones. Perhaps one day Congress will establish a national park without even asking about its other potential uses. Perhaps – but that day is not history yet.

The creation of national parks in Australia

Hall (1989, 1992) examined how the perception of national parks as worthless lands applies to Australasia and concluded that the first national parks in

Australia were created for reasons of tourism, scenic beauty and a lack of intrinsically valuable resources. In Australasia, the preservation of flora and fauna for scientific reasons initially received only limited attention while parks were often regarded as wastelands that could be made productive through the creation of tourism opportunities.

As in the US, nineteenth-century Australia was dominated by utilitarian ideals. In 1856, Jevons noted, 'while money is to be made, trees will never in Australia be spared for mere ornament' (quoted in Collison Black and Kinekamp 1972: 214). By 1892, some 95 million hectares of New South Wales forests had disappeared through clearing or ringbarking, 'a little more than a quarter of the total area which was under forest when the white man came a century earlier' (Bolton 1981: 45). However, despite the warnings of forest experts, including George Goyder and the South Australian Forest Board, grazing and agricultural interests maintained the need to clear the land of trees for economic reasons. As W.E. Abbott expressed the situation in 1880 at a meeting of the Royal Society of New South Wales, 'The very rapid spread of the ringbarking in spite of the opposition of all the lovers of fine scenery, and of so many scientific men . . . proves that there must be a clear gain to the graziers in getting rid of the timber' (quoted in Bolton 1981: 44).

In Australia in the latter half of the nineteenth century a debate on forest conservation occurred strikingly similar to that in the United States. In part this was due to the same intellectual impetus – the publication of George Perkins Marsh's (1864) book *Man and Nature* – within a year of its publication it was being used as the main justification for the creation of state forests. Utilitarian conservation became the dominant theme in public policy, especially in Victoria, the most densely populated of the colonies. A series of Royal Commissions and government inquiries bemoaned rapid forest clearance and argued for the creation of forest reserves (Powell 1976, Frost 1997, Hutton and Connors 1999, Tyrrell 1999). One of the main advocates for a utilitarian approach to forest conservation was Ferdinand von Mueller, Victorian Government Botanist. In 1871, he wrote of the need to conserve forests for the continued economic well-being of future generations:

> But this formation of dense and at the same time also thriving settlements, how is it to be carried out, unless indeed we place not merely our soil at the disposal of our coming brethren, but off with this soil also the indispensable requisites of vigorous industrial life, among which requisites the easy and inexpensive access to a sufficiency of wood stands wellnigh foremost.
>
> (quoted in Powell 1976: 70)

The dominant utilitarian conservationist perspective combined with the influence of British imperial attitudes also saw forest lands being concerned with single rather than multiple-purpose forestry (Carron 1985). This provides a significant divergence to US conservation in the twentieth century.

Nevertheless, the much-celebrated Australian bush ethos offered very little real protection for the quickly diminishing wilderness. The first nature reserves were not created to protect entire landscapes, but, as in Canada, New Zealand, and the United States, were established to provide a degree of protection for scenic sites of significance for commercial tourism.

New South Wales

The first reserve in Australia, which may claim some association with the national park concept, was the reservation of an area of 5,000 acres (2,025 hectares) in the Fish River (Jenolan) Caves district in the Blue Mountains in October 1866. The caves, which had previously been a refuge for Aboriginals and bushrangers (Havard 1934), were covered by legislation that was intended to protect 'a source of delight and instruction to succeeding generations and excite the admiration of tourists from all parts of the world' (Powell 1976: 114). The protection of natural monuments with tourism potential parallels the first US federal reservations at the Arkansas Hot Springs, Yosemite and Yellowstone. In 1870, the head of Jamieson Creek in the Blue Mountains was reserved, while the Bungonia Lookdown was reserved in 1872. Both areas were 'beauty spots', which provided views of spectacular gorges (Goldstein 1979e, Prineas and Gold 1983).

In 1879, seven years after the creation of Yellowstone, 18,000 acres (7,284 hectares) of land were set aside as The National Park at Port Hacking, south of Sydney. This area was increased to 14,000 hectares the following year (Black and Breckwoldt 1977: 191). The creation of the National Park (later Royal National Park) was inspired more by a desire to ensure the health of Sydney's working population than to provide a wilderness experience. According to a member of the New South Wales Legislative Assembly, John Lucas, the park was created 'to ensure a healthy and consequently vigorous and intelligent community . . . all cities, towns and villages should possess places of public recreation', while Sir Henry Parkes commented, 'The Honourable Member says it is a wilderness and that years must elapse before it can be of any use, but is it to remain a wilderness? . . . certainly it ought not to remain a wilderness with no effort whatever to improve it' (quoted in Mosley 1978: 27). The most likely model for the park was the large 'common' parks of urban Britain, and it represented an antipodean version of the then popular views of the negative effects of the city on health and morality (see also chapters 1, 2 and 3 of this volume). The National Park was to be 'a sanctuary for the pale-faced Sydneyites – fleeing the pollution – physical, mental and social, of the densely packed city', not an escape to wild, untamed nature (quoted in Pettigrew and Lyons 1979: 15, 18).

Bearing in mind the purpose of the government of the day 'to bequeath to the people . . . a national domain for rest and recreation', some of the provisions of the deed of grant of the park make clear the utilitarian and recreational goals:

And we do hereby empower the Trustees of the National Park in their discretion to set apart and use such portions of the said Park as they may from time to time think necessary for the purpose following, that is to say, first, ornamental plantations, lawns, and gardens; second, zoological gardens; third, racecourses; fourth, cricket, or any other lawful game; fifth, rifle butt or artillery range; sixth, exercise or encampment of Military or Naval forces; seventh, bathing places; eighth, for any public amusement or purpose which the Governor for the time being may from time to time by notification in the Government Gazette, declare to be an amusement or purpose for which the said National Park, or any portion or portions thereof, may be used.

And we hereby also declare that it shall be lawful for the Trustees of the National Park to grant licenses to mine upon and under the said land for and to take away and dispose of, as the licensees may think fit, all coal, lime, stone, clay, brick, earth or other mineral (excepting gold or silver) that may be found in the said lands for such period and upon such conditions and upon such terms of payment for such licenses whether by payments at fixed period or by royalty, paying to the Trustees of National Park in addition to any payment or royalty payable to us, our heirs and successors under any mining act now in force in our said Colony for the purpose aforesaid such sum as may be agreed upon, and to sink, erect, make, maintain, and use such pits, drives, roads, railways, tramways, engine houses or buildings as the Trustees of the National Park may from time to time think expedient.

(Stanley 1977: 3)

According to Pettigrew and Lyons (1979: 17), the area reserved, was available as 'a consequence of the poor quality of much of it and of the Georges River between it and the expanding Sydney'. Nevertheless, as Australian authors have proudly pointed out, New South Wales was the first colony or country to actually include the term 'national park' in legislation (Moseley 1978) (the 1872 legislation for Yellowstone lacked the words 'national park'). As Robin (2000) comments, 'it was a great relief to the "red faced" planners of the American Centennial that the Australian park managers "courteously" did not press a claim for priority in the National Parks Idea in 1970' when the centenary of Yellowstone was being celebrated.

The reasons for establishment of Australia's first national park therefore resemble those operating in North America. First, there was no cost to the government in the reservation of land as the Crown already held it. Second, the land was regarded as worthless with no value for agriculture, although timber cutting and grazing were allowed to continue in the park until well into the twentieth century. Third, railway development was integral to the park's development. However, in contrast to the American situation, the park was established to provide for mass recreation rather than the elite commercial recreation that characterized the early days of Yellowstone. In addition, the landscape value of the national park was related to the coast and rivers rather than mountain scenery or spas as in the United States and Canada. As in the US, the area was 'improved' with suitable types of development such as military parade grounds, picnic areas, bandstands, and zoological displays. Despite these 'improvements', the park has become one of the major peri-urban components of the national park system of New South Wales.

Western Australia

In 1872, an area of 175 hectares (432 acres) on Mount Eliza, near Perth, Western Australia, was established as Perth Park and gazetted as a public park and recreation ground. The park, enlarged to 397 hectares in 1890 and renamed King's Park in 1901 on the accession of King Edward VII to the throne, contained areas of natural bushland as well as botanic gardens, lawns and reservations for Perth's water supply. Black and Breckwoldt (1977) suggest that King's Park could be regarded as Western Australia's first national park, but although the park may have similarities with early national parks in other states, the term was never applied to it, quite possibly because it pre-existed Yellowstone. As in other Australian states, natural history and scientific societies were active in advocating the creation of national parks and nature reserves. But the efforts of many of the early champions of national parks 'were thwarted by the developmental urge and particularly by the exceptions made for mining interests and pioneer settlement in most of the reservations' (Powell 1976: 115).

A native flora and fauna reserve of 64,777 hectares, with apparently little potential for agriculture, was established in south-eastern Western Australia in the Murray River area between Pinjarra, North Dandalup and the Bannister River, in 1894. Despite the efforts of the Western Australian Natural History Society, the reserve was altered to that of 'Timber – government Requirements' on 7 April 1911 (Australian Academy of Science Committee on National Parks (Western Australian Sub-Committee) 1963: 17–18). The change in designation of the Murray River Reserve resembles attempts to revoke national park designation for some areas of the early American parks for purposes of mining. In these cases, the central issue for legislators was not the protection of scenery or flora and fauna but whether the land could contribute to economic development.

The first reserve in Western Australia to which the term 'national park' was applied was an area of 43 hectares at Greenmount near Perth in 1895. Enlarged by a further 1,423 hectares in 1900 it was named John Forrest National Park, an extremely apt dedication given Forrest's interest in natural history. The focus of the park was the attractive waterfalls of Jane Brook, as memoranda between Forrest and the Lands Department indicated:

> I should like a National Park reserved above the tunnel on the Eastern Railway, a beauty spot. I should be glad to advise the Surveyor-General.
> (Memo the Premier John Forrest to the Minister for Lands, 26 August 1900)

> Please note Premier, I think we have anticipated this request by making reserves to include the waterfalls in this locality.
> (Memo Minister for Lands to Under Secretary for Lands, 28 August 1900, *Lands and Surveys Department File No. 10617/99*)

A railway line also ran through the park and a station was established at the main public picnic grounds. The railways were critical in ensuring the prosperity of the park as a major recreation destination.

South Australia

In 1891, the government of South Australia passed the *National Park Act*, which set aside the Old Government Farm at Belair, an area of 796 hectares as a reserve. The Act was designed to 'establish a national recreation and pleasure ground as a place for the amusements, recreation and convenience of the Province of South Australia'. Despite attempts by the Field Naturalists Section of the Royal Society of South Australia to allow the park to be retained in its natural state, the Playford government insisted that the park be organized along the lines of Sydney's recently established national park (Black and Breckwoldt 1977). 'The area was developed with tennis courts, ovals, pavilions and walking trails through the bushland. Stands of ornamental trees were planted along curving drives through the park and it became a favoured picnic area for the people of the Adelaide region' (Goldstein 1979a: 215). The general perception of national parks in South Australia at that time was well summed up in a letter that appeared in the *Register* in October, 1884:

> National parks will be useful, not only as preserves for indigenous plants and animals, but also as recreation grounds for the people. It is well to consider how comparatively few and small are the areas of this description which will be permanently available for the residents of the Adelaide Plains ... there must come at a time when these plains will be thickly populated from hills to sea, and then, if not now, the need for more breathing space will be recognised. The Mt. Lofty Range is gradually passing more and more into private hands, and before many years have elapsed it will be difficult to find a place where one may enjoy the beauties of nature without fear of trespassing.
> (quoted in Nance 1986: 215)

Despite continued pressure from the field naturalists for the creation of a comprehensive system of fauna and flora reserves, the South Australian government emphasized the protection of popular recreation interests (Powell 1976). For instance, the South Australian *National Park Act* of 1891 gave the Commissioners power to 'set apart such portions as they think fit for the conservation of water, for the purposes of sports and games, for landscape gardening, for temporary platforms along the railway line, for enclosures of birds and animals and any other purpose for public enjoyment they think fit . . .' (Black and Breckwoldt 1977: 192). The emphasis continued in the South Australian *National Pleasure Resorts Act* of 1914. Again, railways played a prominent role in the establishment of an Australian state's national park system reflecting government intent to provide recreational and tourism opportunities for the general population.

Victoria

In 1866, Tower Hill, an area of 597 hectares near Warrnambool in southern Victoria, was set aside as a public park in order to preserve its dormant volcanic cone. In 1882, Fern Tree Gully, a small temperate rainforest area of the Dandenong Ranges near Melbourne was reserved for recreation in its natural state (Anderson 2000; Bonyhady 2000). In 1892, the Victorian parliament passed the *Tower Hill National Park Act*, thereby creating Victoria's first national park. The Tower Hill Park was designed mainly for recreational purposes and the Crown reserved the right 'to all gold, silver and coal, within the park, its right to prospect or mine there, and its liberty to occupy any parts of the park which might be required for the roads, railways, water courses, reservoirs, drains, sewers, etc.' (quoted in Black and Breckwoldt 1977: 191). Some of its local advocates linked its reservation to an unsuccessful campaign for the establishment of the national capital in the area. Once that failed, the national park tended to be neglected and used for grazing and quarrying. Following increasing criticism of its mismanagement, in 1960 it was converted into a state game reserve (Anderson 2000; Bonyhady 2000).

From 1898 onwards a number of areas were set aside as temporary reserves under the *Land Act* as 'sites for national parks' (Goldstein 1979b). These included a temporary reserve of 36,842 hectares at Wilsons Promontory, which became Victoria's second national park in 1905. This campaign was stimulated by a number of proposals to develop Wilsons Promontory for settlement, including one to attract immigrant Scottish crofters (Anderson 2000).

Despite the provision of temporary reserves for parks, the creation of national parks in Victoria was not part of an organized conservation movement, rather it was a result of significant individuals and associations, which were strictly local in their initiation and subsequent control. At Bulga National Park, for example, management was vested with the local council and it was the shire engineer's idea to shift a surplus swing bridge and place it over a rainforest gully, thus creating the world's first temperate rainforest viewing structure (Frost 2002).

Queensland

Significant individuals also played a major role in the establishment of national parks in Queensland. In 1878, a pastoralist from the McPherson Ranges near the Queensland–New South Wales border – Robert Collins – was impressed by Yosemite while on a trip to the United States (Goldstein 1979c, Frost 2004). After his election to the Queensland parliament, Collins argued that the rainforested Lamington Plateau be declared a national park. In his view it had great potential as a health and tourism resort:

> Who does not long to get away from the moist heat of Brisbane ... in summer-time ... It is no wonder that people really needing rest and change in a cooler and more invigorating climate go to New Zealand and Tasmania, or to the mountainous regions of Victoria or New South Wales ... [Yet] within sight of Brisbane is a fine area of habitable country, and with a climate more equable than perhaps any New Zealand town enjoys – volcanic soils of surpassing richness, shady forests and scrubs, cool running streams, and splendid, bold mountain scenery.
> (Collins 1897: 21)

In 1906, Queensland passed its State Forests and National Parks Act. This allowed the Secretary for Public Lands to declare national parks by regulation rather than through individual legislation enacted by Parliament and was the first legislation in the world concerning the procedures to be followed in the systematic establishment of national parks.

The bill was introduced in the lower house by Joshua Bell, Secretary for Public Lands. He saw National Parks being created in:

> Areas which, either from a climatic or scientific point of view – or from both – are localities that are likely to become popular resorts as the population grows larger – places to which those who desire to take a holiday may like to go from time to time, and know that they will get pure air, good scenery and country life.
> (Queensland 1906: 1541)

In the upper house, the legislation was introduced by Andrew Barlow, Secretary for Public Instruction, who acknowledged the influence of other jurisdictions:

> This measure simply follows in the wake of the southern States [of Australia] and the United States of America in attempting to create a national park. We all know what the great national park of the United

States is – where the gigantic trees are. It is very desirable that a small bit of nature should be reserved before it is all destroyed and cultivated and broken up.

(Queensland 1906: 1930)

The first national park created under the Act was an area of 131 hectares at Witches Falls on Tambourine Mountain. The national park was only created after the Queensland Inspector of Forests judged it as 'unfit for any other purpose' (Powell 1976: 114). However, Frost (2004) notes that many of the Queensland areas that were nominated for national park status in the early part of the twentieth century were under threat of development from agriculture, with their rainforests attracting potential dairy farmers. The Queensland case therefore has some differences to that of the early American experience although some similarities, in terms of utilitarian justifications, including tourism, health, flood protection and water supply remain. For example, the justifications for declaring Lamington National Park in Queensland (Frost 2004) closely resemble the reasons given for the establishment of Egmont National Park in New Zealand in 1900 as that country's second national park. Nevertheless, it is noticeable that Lamington Park was declared not long after a similar battle for Hetch-Hetchy in the United States was lost by conservationists (Runte 1990).

Tasmania

Tasmania was the last Australian state to establish a national park. In 1915, an area of 11,000 hectares at Mount Field, 70 kilometres to the west of Hobart, was set aside as a national park although scientific associations such as the Royal Society of Tasmania had been unsuccessfully urging the creation of flora and fauna reserves since the late 1840s.

Under the *Waste Lands Act* of 1863 and the subsequent *Crown Lands Act* a variety of reserves were established for their scenic value. A list published in 1899 included twelve reserves comprising six scenery reserves, three cave reserves, two falls reserves and a fernery reserve. One reserve not listed included an important area of 120 hectares at Russell Falls proclaimed in 1885 (in Mosley 1963: 211).

As in other Australian states, institutions such as the Tasmanian Museum, scientific and natural history organizations, and certain committed individuals played a major role in creating national parks. Initial attention focused on creating a national park in the Mount Field–Russell Falls area. A National Park Association was formed, which successfully agitated for the establishment of a reserve of some 11,000 hectares for the purpose of a national park. The campaign was a blend of the elements that occurred in campaigns for national parks in the United States: aesthetics, science and tourism. The proposed Tasmanian reserves were described as 'living museums' for the preservation of Tasmania's flora and fauna; the moral overtones of scenic preservation are witnessed in the description of hunters and timber interests as 'spoilers and utilitarians' (quoted in Mosley 1963: 215), a statement which could equally have been made by the American wilderness preservationists of the time.

The Tasmanian government favourably viewed the tourism aspects of national parks. The *Mercury* spoke of the Mount Field area as 'a public resort of quite first class value ... and more especially for the citizens of Hobart and the increasing number of tourists'. The recently established government Tourist Bureau took an active role in scenic preservation, while the advantages of increased revenue on the Derwent Valley Railway, due to increased usage, were also noted (quoted in Mosley 1963: 213–14, 215).

The *Scenery Preservation Act* of 1915, under which the Mount Field national park was created, was 'advanced legislation for its time' and the Scenery Preservation Board, which the Act created 'was the first special authority in [Australia] for the creation and management of parks and reserves' (Goldstein 1979d: 142) and one of the first in the world. However, the approach of the Scenery Preservation Board was to protect scenic 'nature's monuments', as evidenced by the Surveyor-General's announcement at the first board meeting on 16 July 1916, that he had requested his surveyors to take special note of, 'waterfalls, forest clad outcrops, attractive and commanding viewpoints, or other places of historical or scientific interest and natural beauty suitable for reservation' (quoted in Mosley 1963: 217).

National parks as conservation systems

The early days of national park and wilderness preservation in Australasia closely parallel the North American situation. Emphasis was placed upon scenic values, the preservation of nature's monuments, health, water supply, and, to a small extent, the preservation of fauna and flora for their scientific values. Nevertheless, national parks were generally regarded as relatively worthless lands that could only be made productive through the creation of tourism opportunities, especially in relation to the railway system. The preservation of aesthetic or ecological values was only a minor force in the creation of the first Australian national parks, even though there was some scientific recognition of the value of conserving fauna and flora. Even as late as 1974, the groundbreaking Committee of Inquiry into the National Estate (1974: 77) reported, 'National parks and other large reserves have generally been made only in areas unwanted for any other purpose. Sectional pressures have ensured that other areas, whether their potential is for agriculture, grazing, mining, forestry, water storage or settlement, have largely remained unreserved'. However, how does this fit with current preservation systems?

In the case of Tasmania, for example, it has long been recognized that following the initial pattern of establishment as well as subsequent revocations the Tasmanian reserve system is strongly biased towards the protection of alpine regions and buttongrass plains – areas with minimal value for agriculture, forestry and mineral extraction. In a restating of the 'wasteland hypothesis', Bell and Sanders (1980: 59) provided a 'residual' explanation of the pattern of Tasmanian reserves. That is, they are lands left over from the needs of economic interests, or as Mercer and Petersen (1986: 139) expressed it: 'areas are reserved only if they are seen to have no commercial value'. More recently, Mendel and Kirkpatrick (2002) undertook a study of the reservation status of plant communities in Tasmania in 1937, 1970 and 1992. They found that before the 1970s the representation of plant communities in the Tasmanian National Park system was strongly biased toward the reservation of communities that are not economically valuable – usually those occurring in scenic alpine and subalpine areas. In 1970, less than 3 per cent of economically valuable communities were reserved above 15 per cent of their pre-European area, compared with over 17 per cent of communities without economic value. According to Mendel and Kirkpatrick (2002), these statistics support the 'worthless lands' hypothesis. Between 1970 and 1992 there was a dramatic rise in the representation of communities in the reserve system in response to a change in motivations for reserve establishment and the availability of gap analyses for many plant communities and species. In 1992, one-third of the plant communities in Tasmania had 15 per cent of their pre-European area represented in the reserve system, with over 20 per cent of economically valuable communities and 58 per cent of non-economically valuable communities having over 15 per cent of their pre-European area reserved (Mendel and Kirkpatrick 2002).

Similar findings have been made with respect to other Australian states. Pouliquen-Young (1997) concluded that the system of reserves had been biased in Western Australia towards a very large number of very small reserves, particularly in the southwest of the state, dominated by agriculture and forestry, and a few very large reserves, in the arid remainder.

> Because of the belated response of governments and their 'worthless lands' approach to the concept of conservation through reserves, the process of implementing reserves has been largely opportunistic in Australia, relying on acquiring 'worthless land' to rapidly increase the size of the system in areas of low land-use competition, and on the affordability of land and pastoral leases in areas of more intense land use.
> (Pouliquen-Young 1997: 178)

In South Australia nearly two-thirds of the reserves established prior to 1960 were for public amenity. 'As a result much of the land reserved by the early 1960s was selected by the government because it was considered unsuitable for development . . . ecosystems characteristic of highly prized farmland were rarely considered for reserve selection despite being the most vulnerable to clearance and most in need of protection' (Bryan 2002: 201). Similarly, Harris (1974: 141) stated: 'Before the 1960s there was little aim evident in the acquiring of parks and reserves . . . many areas, the large flora and fauna reserves especially, were created not because they conserved unique landforms or rare flora or fauna, but because they were regarded at the time as useless wastelands'. According to Bryan (2002), this has resulted in a reserve system with a similar pattern to that noted in Western Australia, a vast network of reserves in the arid regions and many fragmented and small reserves in the agricultural regions. Indeed, with respect to biodiversity representation through national park establishment, Prato and Fagre (2005: 177) argue that although 'worthless lands' is a 'poor choice of terms', evidence does indicate striking gaps in the conservation of biodiversity and biases in reserve systems toward those areas historically perceived to have no significant economic value.

Conclusions

As Runte (1983: 138) observed, 'everyone would prefer to attribute the national park idea to idealism and altruism'. However, the truth of the matter is that the early creation of national parks in Australia and the United States has been as much dependent on the absence of material wealth as it has been on the weight of aesthetic and ecological/scientific arguments. The notion that national parks were, and to an extent still are, created because they can be regarded as 'worthless lands' serves as a timely reminder of the contestation between interests involved in the establishment and maintenance of national parks. Runte (1997b) argues, 'The worthless-lands speeches were not "rhetorical ploys." They were, in fact, serious assessments of national park lands based substantially on the findings of government resource scientists' and, in reference to his original works, 'If I were writing them today, I would add only a few more examples and quotations to support my initial discussions of monumentalism and the worthless-lands thesis'. Indeed, as will be noted in the next chapter, Runte (1987, 1997a) stressed that monumentalism was more than a metaphor to help the average American visualize nature's wonders, but was part of America's attempt to overcome cultural anxiety.

The creation or continued protection of a national park is not a rational process. It is a political battle, a process that involves the values of interests in the struggle for power relative to government decisions. As Runte (1983: 141) observed: 'Over the long term, preservationists never "win" environmental battles; they can only hope to minimize their losses . . . the essence of any political struggle is compromise. Compromise in the ecological context, regrettably, is simply another definition for loss.' Some readers will no doubt disagree with Runte's statement, in the same way that his original 'worthless' argument also provoked substantial reaction from many. Yet Runte's arguments are important as they focus on the way in which the national parks concept, along with associated ideas such as conservation, are socially constructed. As such, the worthless lands hypothesis is of importance, not only as a significant concept of national park history, but also as a way of explaining the changing pattern of biodiversity conservation – what Hinds (1979) described as the 'cesspool hypothesis' – whereby much of the national park and reserve system land was historically the 'crap' nobody wanted. McDonald (1987) similarly noted such a distribution along with the location of Aboriginal reserves in Australia. However, what is undoubtedly required is greater analysis of the validity of the concept in a non-New World setting. Nevertheless, the 'worthless' thesis is significant for emphasizing that the creation of national parks is not purely the result of aesthetic and ecological considerations but was, and to an extent still is, strongly determined by the materialist utilitarian perception that national parks are, in essence, 'worthless lands'.

5 National parks, national identity and tourism

Warwick Frost and C. Michael Hall

Discussions of national identity and tourism have mainly been concerned with museums and cultural heritage (McLean 1998, Howard 2003, Timothy and Boyd 2003). However, increasingly, researchers are extending their investigations of this complex relationship to other fields of tourism. These include destination marketing (McCrone *et al.* 1995, Morgan and Pritchard 1999, Pritchard and Morgan 2001); the marketing of national heroes and icons (Frost 2006, Light 2007); festivals (Picard and Robinson 2006) and attractions, including theme parks, zoos, monuments, battlefields and other historic sites (Stanley 2002, Pretes 2003, Chronis 2005, Ryan 2007, Frost and Roehl 2008). That natural attractions, including national parks, often function as symbols of national identity, pride and achievement, is also beginning to receive greater attention from researchers (Wall 1982, Hall 1985, 2002a, Carruthers 1995a, Zimmer 1998, Sievert 2000, Kathirithamby-Wells 2005, Knudsen and Greer 2008).

Identity is closely linked with the concept of heritage – those things (tangible or intangible) that we value as special and worth preserving. Identity arises from a shared heritage:

> Whether we are discussing the family photograph album or the national park, a major outcome of conserving and interpreting heritage, whether intended or not, is to provide identity to that family or that nation. There may be other purposes as well, such as legitimation, cultural capital and sheer monetary value, but the common purpose is to make some people feel better, more rooted and more secure.
>
> (Howard 2003: 147)

This heritage and identity may be tangibly represented in landscapes:

> Every mature nation has its symbolic landscapes. They are part of the iconography of nationhood, part of the shared set of ideas and memories and feelings which bind a people together.
>
> (Meinig 1979: 164)

National identity was a product of the eighteenth-century Age of Enlightenment in Western Europe. Prior to this people tended to think in terms of themselves as belonging to local or regional identities or in terms of their religious denominations or personal loyalties to particular rulers. The growth of nation states, democratic revolutions and settler colonies contributed to a change in how people saw themselves (Anderson 1983). The nineteenth and early twentieth centuries witnessed rising nationalism reflected in the unification of small states with common languages (e.g. Germany and Italy) and the creation of new countries (e.g. Belgium, Greece, Romania and Czechoslovakia). Furthermore, the break-up of European empires saw the establishment of independent nations throughout Asia, Africa and the Americas.

These new nations were 'imagined communities' (Anderson 1983). Unlike local communities, it is not possible for all members of a nation to meet and interact personally. They accordingly need to 'imagine' their connections and oneness (Anderson 1983). Furthermore, nations might be formed from the amalgamation of quite disparate ethnic and language groups with national political boundaries being formed by colonial convenience rather than any relationships with cultural and language affiliation. For these nations to bond, there was a need for institutions to be developed and promoted, which could be shared, identified with and 'glue' the nation together. Such institutions could include national heroes, myths, icons, commemorations and monuments, art (literature, paintings, cinema), cuisine, customs and traditions.

While national identities may represent a shared heritage, there is also much potential for dissonance and conflicts – both within the nation and with other nations. A shared national heritage may exclude and marginalize. While, 'heritage benefits someone . . . [it also] disadvantages someone else' (Howard 2003: 4). While making, 'some people feel better, more rooted and more secure . . . [it] simultaneously makes another group feel less important, less welcome and less secure' (Howard 2003: 147). Sites of national importance are open to multiple, often contested, interpretations. Battlefields – by definition sites of struggle between competing interests – may be commemorated in different ways by different groups (Buchholtz 2005, Chronis 2005, Ryan 2007). Differing interests may seek to monopolize control and 'ownership' of what they perceive to be their national heritage and identity. Accordingly, 'heritage battles are not just against vandals, but also those who would also claim the same heritage' (Lowenthal 1998: 230).

Tourism was important in this construction of national identity. People were encouraged to visit sites of national importance, utilizing their holidays to reinforce their national pride and devotion. Thomas Cook rose to prominence when he was invited to organize tours for working people to visit Britain's Royal Exhibition. These tours were intended to bring people to the capital and instil pride in the nation's achievements. Today, there are strong flows of domestic tourism to national capitals, a civic duty to experience national institutions such as museums, parliaments and memorials (Hall 2002b).

Governments and other stakeholders also often look to international tourists to validate their national importance. Such international tourists are often motivated by the desire to experience a country's particular culture, history and atmosphere. Attractions are developed to project images of national identity. These range from formal representations, such as museums, monuments and historic sites, to more popular attractions such as theme parks, festivals and historic/cultural precincts. International tourist interest in such attractions affirms that the national identity projected is both important and worthwhile.

Of course there is still the potential for dissonance, as representations of national identity (especially official ones) differ from the expectations of tourists. Two examples demonstrate this. First, whereas the real Vlad Dracul is regarded as a national hero in Romania, tourists expect something closer to the Hollywood image of Dracula (Light 2007). Second, Scotland's National Wallace Monument was experiencing declining visitation until the release of the film *Braveheart*. Catering to increasing numbers of international tourists, a new statue of William Wallace was deliberately modelled on actor Mel Gibson, who played the Scottish patriot in the film (Beeton 2005).

These links between national identity and tourism suggest a different way of looking at national parks. While they are often seen as 'national' in terms of public ownership and management, they may also be viewed as major contributors to the construction, promotion and acceptance of national identity (Wall 1982). That is, in creating national parks, nations are clearly making a statement about who they are and how they wish to be seen by both their citizens and the international community.

This chapter aims to examine some of these linkages between national parks, tourism and national identity. It is divided into five sections. The first considers how the early national parks in the US filled a major gap in the national psyche. The second explores the role of national parks in the nation-building process for a number of European states. The third extends that discussion to Asian nations. The fourth section outlines how the desire to protect and promote national heritage has led to the creation of special types of national parks. The final section focuses on how the naming (and renaming) of national parks reflects national interests.

Creating the US's national identity

In the century or so after declaring independence, Americans were acutely conscious of shortcomings in their national identity. Despite substantial material progress, they were deficient in some of the qualities that they saw as defining a country as a major power. Western Europe was seen as the font of all modern culture. It contained the artistic and literary heritage, the ancient monuments and ruins, which the US as a new country lacked. In comparison to a culture stretching back thousands of years, the US had only 200 years of European settlement (Runte 1979, Eagles and McCool 2002). Like many newly independent countries, it had severed ties with old cultural traditions (e.g. royalty, aristocracy) and not yet developed its own new ones.

The poverty of this comparison was particularly apparent in the landscape of the Hudson Valley. Longer settled than most American regions, it boasted historic battlefields and a pleasing romantic combination of farms, villages, forests and mountains. The Hudson River School celebrated this landscape in art. American writers such as James Fenimore Cooper and Washington Irving used it as a setting, with the latter retelling European folktales in this new landscape (Runte 1979, Sears 1989). However, it had neither the antiquity to compare with Europe nor the grand scenery to distinguish it from that of Europe. Only Niagara Falls had that quality, but it was blighted by inappropriate development (Chapter 2).

Westward expansion revealed natural wonders that exceeded those of Europe and the popularization of its natural wonders awaited what would be labelled the Rocky Mountains School of Landscape Painting (Runte 1979). In 1852, Californian miners discovered the giant redwoods. No European trees could compare to them in terms of size or age. For religious writers, 'the pious notion that the Big Trees were somehow contemporaries of Christ became a standard refrain in their hymns of praise' (Schama 1995: 190). Horace Greeley went back further, writing in the *New York Tribune* that they 'were of substantial size when David danced before the ark, when Solomon laid the foundations of the Temple, when Theseus ruled in Athens, when Aeneas fled from the burning wreck of vanquished Troy' (quoted in Runte 1979: 22). What Europe still retained from 2,000 years ago was often in ruins, but these were living and thriving, along with nearby Yosemite a symbol of divine recognition of the US's special status. Something that was also encapsulated in the Rocky Mountains School:

> One distinction was the compulsion of artists in the West to cut their canvas by the yard instead of by the foot. Others sacrificed realism, as if to suggest that the mountains of the region were even higher, its canyons far deeper, and its colors more vivid than in real life. Still, while exaggeration was out of place in the Hudson River School, its practice in the West was in keeping with pronouncements that the region was in fact America's repository of cultural identity through landscape.
>
> (Runte 1979: 23)

That the redwoods distinguished the US was apparent in the controversy over naming the new discovery. British botanists saw the opportunity to honour *their* national hero, naming it *Wellingtonia Gigantea* (and indeed that name is still used in some historic gardens, such as England's Biddulph Grange with its recently restored Wellingtonia Avenue). To the Americans' relief, a French botanist ruled that it was related to the previously named *Sequoia Sempiverens* and renamed it *Sequoia Gigantea* (Schama 1995).

The national symbolism of Yosemite and the giant redwoods is reflected by

Abraham Lincoln signing the order for their preservation in 1864, in the midst of the Civil War. Schama argues that this only makes sense if Lincoln accepted that 'Yosemite and the Big Trees constituted an overpowering revelation of the uniqueness of the American Republic' (1995: 191). Less romantically, it may have been a smart political move in the run-up to the presidential elections to demonstrate to faraway California that Washington was still thinking of them.

Yosemite was proclaimed as comparable with the best mountain scenery of Switzerland (Olmsted 1865). Indeed, it only took a short time for boosters to argue it was far better (Runte 1979). Its preservation and promotion, coming strategically at the end of the Civil War, provided a new view of the US's national identity. This was reinforced with the establishment of a National Park at Yellowstone. The US was now entering its Gilded Age, the Civil War behind it, the future marked by growing prosperity and confidence. As the US expanded, 'the natural marvels of the West compensated for America's lack of old cities, aristocratic traditions, and similar reminders of Old World accomplishments' (Runte 1979: 22).

The second wave of American national parks was also linked to concerns about national identity. As Runte noted,

> ... the national park idea evolved to fulfil cultural rather than environmental needs. The search for a distinct national identity, more than what have come to be called 'the rights of the rocks,' was the initial impetus behind scenic preservation. Nor did the United States overrule economic considerations in the selection of the areas to be included in the national parks. Even today the reserves are not allowed to interfere with the material progress of the nation.
>
> (1997a: xxii)

The 1890 declaration that the frontier was now closed highlighted a growing social crisis. In the increasingly urban US, commentators began to see that the lack of pioneering opportunities might be leading to physical and moral degeneration. This provoked a search for a new national identity:

> Eastern men began to romanticize aggressive male endeavour and to appropriate its symbols, its rhetoric and its psychological rewards. They were increasingly drawn to all-male leisure activities such as sport and hunting ... the cult of the cowboy soldier arose in art, drama, and fiction.
>
> (Watts 2003: 7)

National parks provided a substitute for the frontier, allowing both opportunities for recreational adventure and the psychological comfort that there were wild places that could not be developed. Theodore Roosevelt symbolized this new identity. A New Yorker who idealized the West, the self-proclaimed 'Rough Rider' channelled his aggressive energies towards expansionary war against Spain, but also towards creating national parks. In founding the Boone and Crockett Club he co-opted the legends of two highly popular frontier heroes to conservation.

While it is a topic that requires much further research, it seems likely that national parks were important to national identity in the other settler societies of Canada, Australia and New Zealand (Hall 1985), although the association of nature to national identity came much later in the colonial societies than in the US.

Kline (1970) in her comparison of American and Canadian views of nature noted that although stemming from a similar cultural heritage the two views were quite dissimilar. According to Kline (1970: 42) Canadian writing was 'marked by a profound fear of the natural world', as opposed to the American embracement of wilderness virtues. The best nature according to early Canadian writers was tamed, impotent nature – the tidy hedgerows and gardens of Old England.

Early American writers also demonstrated the same attitude of terror towards the wilderness, but it did not evolve into the fear that characterized the Canadian attitude, despite being faced with a similar landscape. As Atwood (1972: 54) wrote, 'nature seen as dead, or alive but indifferent, or alive but actively hostile towards man is a common image in Canadian literature'. Whereas with American writers, 'the terror never took over – not in their minds, and not in the characteristic expression of the first half of the nineteenth century' (Kline 1970: 52).

By the 1840s the United States had 'nationalised' nature (Miller 1967). Nature's monuments such as the Niagara Falls were acclaimed as manifestations of America's independence from Europe. However, they were not endowed with the same symbolism on the Canadian side of the border. Canada rejected the notion that revolution was the affirmation of natural rights. Instead, rights were derived from 'the toil of previous generations', and from civilization (Kline 1970). Nevertheless, artistic renditions of the Canadian drive to the west were depicted in Romantic and naturalistic 'images of the noble adversary, the strong and resistant land itself' (Francis 1982: 8). As in the United States and Australia (for example in the work of Frederick McCubbin, Arthur Streeton and others of the Heidelberg School), the Romantic depiction of the landscape played a major role in developing positive national attitudes towards nature and, following that, the creation of national parks and scenic reserves. Furthermore, as in the US, the Canadian Romantic vision was also successfully blended with utilitarian goals by the railroads (Hart 1983):

> In the mid-eighties this romantic involvement with the land found concrete manifestation in the ribbon of steel that sought to physically bind the nation together. The promoters of the CPR [Canadian Pacific Railway] understood the force of images, and they encouraged the association artists made between their road and the picturesque wonders

it opened. If they saw themselves as nation builders, they were also eager to be seen as enriching the cultural life of the nation ... The remarkable success of the CPR programme in promoting interest in artistic views of the Rockies and the West Coast represents the first significant instance of a widespread acceptance in Canada of the myth of the land as the basis of a national art.

(Reid 1979: 68)

The linkage of nature to national identity came later to Canada, Australia and New Zealand, but the promotion of natural icons was an integral part of tourism promotion by the end of the nineteenth century and national parks became tied to such scenic sights in a number of locations. However, in many cases the national parks followed the identification of scenic locations. For example, Federation Peak in Tasmania, so named in 1901 in honour of the federation of the Australian colonies into the Commonwealth of Australia, yet predated the creation of a national park in the area by over 50 years. In the contemporary context national parks have very much become part of the identity of New World countries, particularly in terms of their branding, but the historical account of the relationship between identity and national parks needs considerably more research.

National identity for European states

Just as the US saw national parks as providing the cultural credibility it lacked in comparison to Europe, in the early twentieth century many European countries utilized national parks to declare their national integrity in the face of competing claims from stronger neighbours. Indeed, it is extraordinary that European national parks before the Second World War were all declared by nations that had recently come into being and/or were struggling to justify their continued existence. In contrast, the major powers lagged behind; Britain waiting until 1949, France until 1963, Germany until 1969 and Russia not till 1983.

In 1909, Sweden was the first European country to declare a national park. A major power from the sixteenth to eighteenth centuries, Sweden's adoption of national parks must be seen in the context of its decline in influence and need to make a new place for itself in twentieth-century Europe. Of particular significance was Norway's declaration of independence from Sweden. This occurred in 1905, four years before the first national park.

Switzerland inaugurated the Swiss National Park on 1 August 1914. That date was specifically chosen, as it is the Swiss national holiday, commemorating the creation of the Swiss Confederation in 1291. By utilizing that date the Swiss connected two major national institutions and reaffirmed their nationhood (Walter 1989, Zimmer 1998). Furthermore, on the same day that Switzerland created its national park, France and Germany mobilized their troops and Germany began its invasion of neutral Belgium. While Switzerland was to preserve its neutrality, as a country surrounded by combatants with a population consisting of German, French and Italian speakers, its national integrity was under threat.

Spain, another neutral country, established its legislative process and its first national park during the First World War. Having lost its last colonies to the US during the 1898 war, the new national park served the function of promoting national morale and encouraging the Spanish to look forward rather than backwards (see Chapter 10).

Italy achieved unification in 1870. Historically, the peninsula comprised a series of small competing states and had for centuries been a battlefield for French, Spanish and Austrian interests. Though politically unified, the new nation still suffered from parochial views and interests. Organizations like the Touring Club Italiano saw encouraging Italians to travel as a way of stimulating an underdeveloped national identity. The declaration of the Swiss National Park on the Swiss–Italian border stimulated ideas that Italy too should have national parks (Sievert 2000). As the movement gained momentum, it received a boost from the election of the Fascist Party to government:

> Mussolini wanted to counter the perceived inaction of previous governments with the appearance of dynamism for his new regime. The upshot for nature protection was rapid government approval of both the Abruzzo park and Gran Paradiso National Park in northwest Italy.
>
> (Sievert 2000: 173)

However, Mussolini's grasp of the propaganda benefits of national parks proved to be at odds with those of conservationists. In the 1930s he commenced the draining of the Pontine Marshes as a showcase of Fascism's innovation and dynamism. This involved the destruction of major wetlands and clearance of over 20,000 hectares of ancient forest. Mussolini also decided to retain a small section untransformed, some 7,000 hectares, declaring it the Circeo National Park in 1934 (Sievert 2000). Not for the last time would the declaration of a small national park be used to divert criticism from major environmentally damaging projects.

New nations created out of the wreckage of old Europe were among the first to create national parks. Romania and Greece had achieved independence from the Ottoman Empire in the nineteenth century and had expanded further in the aftermath of the First World War. Romania, created from the amalgamation of Wallachia, Moldavia and Transylvania, established a national park in 1935. Symbolically, it was located in the Transylvanian Alps, an area that it had gained in 1919 and was subject to competing territorial claims from Hungary. Greece, drawing on its ancient Hellenic culture, created the Mount Olympus and Parnassus National Parks in 1938. Poland, recreated as a buffer state between Germany and Russia, established two national parks in 1932. Iceland, granted independence in 1918 (though in a union with Denmark), created a national park in 1928. The Irish Free

State, established in 1921, opened a national park in 1932. Yugoslavia created Triglav as an Alpine Conservation Park in 1924 (though it was not officially a national park until 1961). Under Tito, Yugoslavia would establish national parks in 1949. Czechoslovakia would also create its first national park in 1949.

These forces are still apparent in the new Europe in which new states continue to be created. This is most clearly demonstrated in the case of Scotland, granted its own parliament and possibly heading towards independence. When Britain decided to create national parks in 1949, these were established in England and Wales, though curiously not in Scotland (Chapter 2). Half a century later, the new Scotland has acquired the institutional trappings indispensable to a new nation, including national parks at Loch Lomond and the Trossachs (2000) and Cairngorms (2003).

National identity in colonial and post-colonial countries

During colonial times, national parks were decided upon by imperial officials located in London, Paris and Brussels (Chapter 3). Such national parks could be read as affirming the imperial identity, making a statement that these areas were part of a greater British, French or Belgian empire. Indeed, the authorities did not intend these national parks to foment notions of local identity, for their resources were increasingly being devoted to discouraging independence movements.

However, in some places the situation was more complex than this. In Malaya, the 1930s campaign for a national park became intermeshed with issues of local identity and the Pan-Malay concept. Malaya was composed of a range of small states. These included some directly ruled by Britain and some with local rulers under British direction. The push for a national park was firmly based on game protection, being led by local British officials and imperial organizations such as the Society for the Preservation of the Fauna of the Empire (see Chapter 3). In the lobbying for a national park, the idea became linked to another concept of a federation of the Malay states. Indeed, it is notable that the eventual national park spanned three states. A Wild Life Commission was set up and from 1930 to 1931 investigated the issue, recommending a national park. Its hearings generated substantial newspaper discussions of both concepts and were also notable for gathering evidence from local Malay and Chinese people in favour of the proposals (Kathirithamby-Wells 2005).

After independence, many countries suffered from a 'post-colonial crisis of identity' (Hall and Tucker 2004: 12). National parks might be variously seen as relics of colonialism, centrally imposed burdens on marginalized local peoples, opportunities for the making of new national identities or sources of much needed tourist revenue. This ambivalence towards national parks is particularly well illustrated in the case of Indonesia (see Chapters 3 and 16).

National parks and cultural heritage

While national parks are often thought of as protecting natural heritage, they may also include (and protect) sites of cultural heritage. Indeed, this is the common situation, as there are very few parts of the world that have not been disturbed by human behaviour. Accordingly, national parks may contain landscapes that are mainly the result of human activity. Though often imagined as completely natural, these are essentially human artefacts requiring careful management to maintain.

Two examples demonstrate how national parks may contain highly modified landscapes. The first is the English moors – treeless uplands popularly seen as primeval wilderness. Such is the appeal of the moor landscape, that it is the dominant landscape in seven of England's ten original national parks (Shoard 1982). Yet, while moors have gripped the public's imagination, most are:

> relatively recent landscapes, created at most 4000 years ago through the destruction of forest to provide wood, charcoal or sheep runs. What is more, most moors rely on Man's activities – burning and the grazing of animals – for their continued existence: left to itself, heather moorland reverts to scrub or woodland within about 60 years.
>
> (Shoard 1982: 58)

The second example is Denmark's Thy National Park. Comprising coastal dunes, forest and heath, it (like England's moors) is popularly imagined as wild and ancient. However, its existence is due to extensive plantings and landscaping in the nineteenth century. Its current management plan includes converting 3,000 hectares of forest to heather through a regular cycle of controlled burns, tree removal and restrictions on grazing (Knudsen and Greer 2008).

In many countries – particularly the US, New Zealand and Australia – there have been attempts to remove all evidence of human occupation from national parks. This has included the buying out of farmers and second-home owners (Young 2002); the removal of indigenous peoples (see Chapter 19); the demolition of buildings and fences and landscaping works to make national parks look more natural (Griffiths 1996). As one senior park official in Australia was reported as saying in the 1980s, 'there's a very conscious policy not to acknowledge history' (quoted in Griffiths 1996: 255). When historic sites have been preserved, the tendency has been to interpret them as from the earliest pioneering days, with no mention of a continuing cultural tradition (Griffiths 1996, Young 2002).

However, it should be recognized that many other countries have accepted that cultural heritage is an integral part of the landscapes of national parks and have valued and protected it equally with natural features. In Japan, many national parks contain ancient shrines and temples, demonstrating a long-standing reverence for nature. The Wicklow Mountains National Park in Ireland contains a glaciated valley and lake (Glendalough). On the shores of the lake are the extensive remains of a monastic settlement established by St Kevin in the sixth century. Like the Book of Kells, this demonstrates Ireland's role in preserving western culture during the Dark Ages.

In Iceland, Þingvellir National Park is noted for its glaciated lake and volcanic formations. It is equally important for its ancient and continuing linkages to Icelandic national identity. In 930, the Alþingi – one of the world's first parliaments – was established here (the national park was created in 1928 to commemorate the 1000th anniversary of this). This was also the place where Christianity was officially adopted by Iceland. Finally, this national park was chosen to be the site of the declaration of the Republic of Iceland in 1944.

Another approach has been to create protected areas specifically for cultural heritage, which have a status equivalent to that of national parks. In 1906, the US enacted legislation to allow the president to protect antiquities – defined as of historic, cultural or scientific interest – as National Monuments. The catalyst for this was the uncontrolled private collecting of Native American artefacts from the southwest. However, President Theodore Roosevelt grabbed the opportunity to declare a wide range of places as national monuments, including the Grand Canyon, the Devils Tower and Mount Olympus. In time, the Grand Canyon and Mount Olympus would become national parks, though the Devils Tower remains a national monument (Runte 1979, Rothman 1989). Some national monuments would gradually expand, becoming very much like national parks, albeit without the name. Dinosaur National Monument in Utah, for example, was established in 1915. It protected 80 acres of fossil beds being excavated by the Carnegie Museum. Over time it has grown to 210,844 acres of canyons and mountains, though only the original 80 acres carries any dinosaur fossils.

When the US National Parks Service was created in 1916, it was also given responsibility for national monuments. In time it would also acquire national battlefields, national preserves, national seashores, national wildlife reserves, national rivers and national recreation areas. The logical extension of these concepts is the creation of national parks specifically based on cultural heritage. Accordingly, a number of countries have opted for historical or cultural heritage national parks. Turkey, for example, has established the Ancient Troya National Park and the Gallipoli Peninsula Historical National Park. Both were important sites in antiquity, with the latter also notable for its World War One campaign. Gallipoli is also represented as the birthplace of the modern Turkish state, for it was here that Kemal Ataturk came to prominence.

Managing historic national parks raises special issues. In 2002, the Castlemaine Diggings National Heritage Park was established in Australia. This national park protects the site of a major gold rush of the 1850s. It comprises 7,500 hectares, which was previously classified as an historic reserve and state forests. Now regrown forest, it contains extensive ruined mine workings and abandoned settlements. In managing and interpreting the national parks, there is a conscious effort to recognize the combination of cultural and natural values and the many layers of history. An example of this may be seen at Eureka Gully, which was worked with various methods from the 1850s to the 1980s and features the ruins of miners' stone huts clustered around Aboriginal rock wells (Frost 2005).

Naming and renaming national parks

Giving a name to a geographical feature or place demonstrates power and ownership. It makes a statement about who is in charge. This may often be seen in national parks, which have been consciously given names that represent certain perspectives. In some cases, changes in political regimes and/or public opinion had led to changes in the names of national parks.

National parks in European colonies were often named after royalty or important government officials. These names demonstrated the sovereignty and power of the colonial authorities. In 1925, when the Belgians established a national park in Africa, they named it after their king – Albert. In 1938, the first national park in Malaya was called after the British monarch – King

Figure 5.1 The cultural landscape of Castlemaine Diggings National Heritage Park, Australia. Both the ruined buildings and regrown forest are cultural artefacts of the gold rushes.

George V – and there was a conscious decision to make the national park Malaya's contribution to the twenty-fifth anniversary of his rule. In 1952, Uganda established a national park, which was then renamed after the incoming monarch – Queen Elizabeth. However, once these countries gained independence, these names were no longer appropriate and were changed. Albert became Virunga, King George V became Taman Negara and Queen Elizabeth became Ruwenzori (though after a regime change it reverted to its original name) (MacKenzie 1988, Kathirithamby-Wells 2005). Australia's colonial ties and arguably, lack of cultural independence, are perhaps evidenced in the renaming of The National Park to the Royal National Park in 1955 when Queen Elizabeth II drove through the park on the way to Wollongong.

South Africa provided an intriguing example of the political nature of naming national parks. Immediately after the First World War there was a strong push for creating a national park. Most of its advocates were of British background and favoured the title of South Africa National Park. However, there was little support from the Boers. The political compromise suggested by some advocates was to link the new national park to the growing mythologization of the Boer Voortrekkers and promotion of the Afrikaans language. By suggesting that the name be Kruger National Park they were able to gain support for their proposal (Carruthers 1995a).

In Australia, most national parks carry indigenous names, usually the names of localities or natural features. However, there have been contests over whether to use indigenous or European names. An early example was the Dandenong National Park created in 1941. It was quickly renamed the Churchill National Park to honour Britain's Second World War leader.

A greater contest has continued to rage over Uluru. In 1873, the explorer William Gosse named it Ayers Rock after Henry Ayers, the then premier of the colony of South Australia (and hardly a major figure in Australian history). Over a century later, the 1985 handover of the area to its traditional indigenous owners was marked by changing its name back to Uluru. However, many tourism operators have refused to accept the change, perhaps reflecting their distaste for the handover. Accordingly, the nearby resort and airport are branded as Ayers Rock and even official roadsigns do not use Uluru.

A comparable dispute over name changes occurred in the US in 1991. What had been the Custer Battlefield National Monument (and often referred to as Custer's Last Stand) was renamed the Little Bighorn Battlefield National Monument. The decision recognized that a neutral geographically based name was more appropriate than one which just referenced one side in the battle. Nevertheless, it generated great animosity among some stakeholders (Buchholtz 2005).

Figure 5.2 Red stone markers for fallen Cheyenne warriors at the Battle of Little Bighorn National Monument, USA.

Conclusion

National parks are human artefacts with multiple purposes and meanings. Primarily, we think of them in terms of protecting nature and satisfying tourists. However, they have other purposes, which, depending on circumstances, may be just as important. As explored in this chapter, national parks may be utilized to promote and extend national identity.

Governments and other stakeholders often see national parks as important vehicles for spreading messages about national identity. The audience for these messages may be both internal and external. Furthermore, the interest shown by international tourists is often important in affirming national identity.

National parks have been portrayed as symbolizing national characteristics and culture. They have been used as tangible foci for national pride. They have demonstrated that countries are modern and internationally significant, one of a group of countries special enough to have national parks. Through such international comparisons, national parks affirm and symbolize nationhood. National parks have also symbolized empires. They have marked regime changes – used by incoming powerbrokers to demonstrate their legitimacy and international recognition. They have honoured national heroes

Figure 5.3 White stone markers for Custer and his men. Battlefields remain contested ground.

and royalty. They have marked disputed territories as belonging to a particular country. Advocates of national parks have played the national identity card in order to succeed and political opportunists have used national parks for their own ends.

These political uses of national parks demonstrate the importance of the title 'national'. These are not just grand parks or wilderness reserves or any one of a number of other possible titles. These are consciously established as national parks because of their significance to national identity.

Part II
New World perspectives

6 Framing the view
How American national parks came to be

Stephen R. Mark

There is a widely persistent myth that men from the Washburn-Langford-Doane Expedition in 1870 gathered around a campfire one night to discuss what should be done with the wonders they had visited in what is now the northwestern corner of Wyoming (see, also, Chapter 2). After one proposed that the expedition's members might claim land at the most prominent points of interest, another declared that the whole of the Yellowstone region be set aside as one great national park and remain open to all. This sentiment met with a favourable response from all but one of those present, and supposedly launched an eventually successful campaign to establish the world's first national park. However, stirring it might be, abundant evidence has been gathered to debunk the campfire story. Instead, officials from the Northern Pacific Railroad (who stood to gain from handling future tourist traffic) suggested the park bill and then motivated the principal movers with their lobbying efforts to get it passed (Runte 1979).

Nevertheless, the idea of establishing national parks by whatever means did not drop from the sky when part of the Yellowstone Plateau was so designated by the United States Congress in 1872. As an institution, they are not the product of effort by any one person or group, although such activism is extremely important (see Chapter 7). Luminaries such as John Muir in the United States and Paddy Pallin in Australia (Hall 1992) have played key roles in securing legislative action to establish individual parks, but the 'idea' originates from a cultural response to nature organized around what came to be known as the *sublime* during the eighteenth century (Nicolson 1959, Weiskel 1976, Novak 1980). However, even more fundamental is seeing nature as a picture, something that coalesced around the Italian Renaissance (Hunt 1986). Not only did the physical remnants of ancient Greece and Rome inspire painters (who began to add scenic backdrops to religious paintings) and other artists, but classical myths also provided ways to organize the very few large and private ornamental gardens of the sixteenth and seventeenth centuries, which were the early precursors of the public parks movement (Hadfield 1985).

Renaissance gardens also incorporated natural features like caves, since they were associated with contemplation and creativity in the classical world of Greece and Rome. As places where knowledge or poetic inspiration followed from encounters with the underworld, it is little wonder that some of these portals engendered their muses, oracles, and nymphs to offset the prevailing ambience of disorientation and mystery (Ross 2000). Other naturally occurring features like waterfalls, pools, springs, rock outcrops (especially if these furnished a vista), and groves of trees have also possessed intellectual and mythological significance (Hunt 1986). The development of 'taste' among the rich and powerful took a new turn in England of the eighteenth century, where the rigid lines, order and formality that dominated Italian and French estates were cast aside for more informal landscapes imitative of wild nature (Hadfield 1985, Short 1991). Some estate owners created their own grottos, pools and groves in this new template, embellishing them with a greater diversity of plants than what Britain could offer – many being American species only recently described by botanists. They chose serpentine paths and carriage roads as the main circulation devices since the gardens had now fused with whole parks, the latter being formerly uncultivated and used only to enclose deer or other game (Shepard 1967, Cosgrove 1998).

In being designed to yield individualized 'scenes' alluding to classical literature or pastoral perfection, the landscape gardens became so popular among the elite that they conditioned a collective response to nature. As 'pleasure grounds' such scenery furnished models for the eventual appearance of public parks during the nineteenth century, at a time when industrial capitalism increasingly pushed its workforce into rapidly growing cities (Conway 1991, Lasdun 1992). Designers of the new public spaces wanted to bring the country into city life, just as the early Romans and subsequent *literati* desired, believing that 'rustic' facilities in 'naturalistic' landscapes best allowed the populace to enjoy fresh air, exercise and moral improvement. The first city parks in the United States followed European precedents (Taylor 1999), but the idea of establishing public pleasuring grounds beyond city limits can be largely attributed to the opportunities presented by closing the once uncontrolled frontier. It also signalled acceptance of the idea that one should travel away from the distractions of towns and cities to experience nature, but as a cultural construction perceived as more primitive and egalitarian than urban life, and idealized in some places to be an Edenic garden.

Tourism initially developed with the rise of informal landscape gardens in the eighteenth century, since the former followed from the wealth used to create the latter. It did not take English tourists (as well as wealthy western Europeans and even Americans) long to go beyond art galleries and the pastoral landscapes of Italy to add experiencing the 'sublime' as part of their itinerary. Many believed that knowledge and perhaps inspiration could be found in jagged mountains, a stormy sea, or wherever the scenery aroused feelings of awe and reverence. The sublime could even include caves, provided they were large and interesting enough to attract the few who undertook the rigours of travel. The Mammoth Cave in Kentucky, for example, could fit into any nineteenth-century 'grand tour' of the American sublime – one that first included the Hudson River Valley and Niagara Falls, but subsequently expanded across the continent to embrace wilder (yet also Edenic) places like Yosemite, Yellowstone and the Grand Canyon (Sears 1989). Through a kind of perceptual lens attuned to seeing scenery as art in nature, pursuit of the sublime eventually helped link nascent regional identity with these seemingly wild places. Their sheer scale made them symbols of a 'new' nation, where landscape painters and photographers marketed their views to the upper classes but also to a slowly emerging middle class (Runte 1979).

Widespread interest among the educated in natural history also fuelled scenic tourism's fascination with the sublime. The eighteenth-century Enlightenment brought with it a belief that the average person possessed the potential to understand and appreciate nature, and even the universe. Creation was thought to be ordered from evidence gathered by newly established sciences, so natural history sought to arrange all life on earth into a single system such as the Linnaean, one that might illuminate relationships among organic forms. Natural history also spurred a search for beauty and significance in sublime landscapes, so that nature writing became popular – impressing readers with the marvels of life and landscape in an accurate, but often teleological, way.

Artists and writers also stood at the forefront of imbuing sublime landscapes with spiritual and national traits. Connected with them is the desire for a golden age – one of the most persistent, oldest, and most basic myths in all of Western thought. It is where human endeavours attained such heights that they were considered worthy of emulation, or at least remembrance, in a present blemished by imperfection (Diamond 1986). The mountainous and picturesque district of Arcadia furnished the mythic setting in classical Greece for such an age, as an abode of a simple, pastoral people dwelling in rural happiness. Landscape gardens of the eighteenth and nineteenth centuries therefore sought to express both the idyllic and heroic, though they were just as much an expression of power over nature and people. The importance of these gardens to public parks of the twentieth century is as a form where evolving perceptions of nature manifested themselves in a number of persistent design principles. Although the historical development of landscape gardens is a complex subject and one open to multiple interpretations, these places could be both didactic and scenic, yet they also made an overt connection between nature perfected and personal liberty in the minds of those who created them. The best creations evoked what some called the 'genius of the place', by bringing forth the emblematic or 'iconic' qualities associated with each site (Loukaki 1997). By responding to what nature dictated, landscapes derived from an Arcadian vision might achieve unity and harmony with circulation systems, structures and other features built from native materials at the appropriate scale.

Visitors to parks, whether public or private, quickly picked up cues in those designed and developed this way. As part of expecting to enter a special place, the so-called 'rustic architecture' constituted a more subtle way of selling the idea of travelling to see the sublime in what could become a national park. Subordinating facilities by integrating them into the larger scene enhanced the park experience in a curious way, sold the parks just as the scenic photographers had. It became part of the tourist experience in the largest and most prestigious national parks, particularly around main hotels or administrative headquarters (Runte 1979).

Unlike the private estates that included landscape gardens, designation of the earliest national parks did not involve purchase of private lands because Congress would not buy land for them. Created from pieces of a larger 'public domain' acquired by the federal government from its aboriginal occupants and other nations, parks like Yellowstone, Yosemite, Sequoia, Mount Rainier and Crater Lake had been part of a 'frontier'. The frontier, especially as this concept pertains to the United States, is sometimes defined as a shifting or advancing zone that marks the successive limits of Euro-American settlement. As a cultural construction, this idea resonated with a society that saw its expansion as a political unit in the form of margins in settled or developed territory (Turner 1920). National parks could not be established until the land was effectively under control of the federal government, yet also had to be free of private holdings in order for Congress to even consider the designation.

Parks also had to be part of the larger industrial economy in order to attract tourists. Access to national and international markets was impeded in the nineteenth-century American West (where the candidates for national park status were located) by the lack of a transportation system capable of profitably carrying heavy loads of agricultural or other commodities over long distances. By better linking western products to a larger market economy, railroads received federal subsidies in the form of land grants to put an end to the farmer's isolation. Their tracks also prompted investors to expand agricultural production in addition to other types of extraction such as minerals, fish and logs. This fuelled population growth, but all of these outcomes were predicated on producers being close enough to the tracks so that additional transport costs (in the form of tolls and freight charges on the relatively few wagon roads suitable for hauling goods in quantity) could be minimized (White 1991).

Commodity production had to come before tourists, due to the latter being relatively few in number, but national parks and other sublime wonders quickly became part of promoting western development (Runte 1979). Regular stagecoach service helped augment railroad lines as a means to reach customarily remote destinations like national parks. Often, rough wagon roads and trails also provided access to the relatively few, but visitors to the parks seemed to beget more visitors, especially once automobile stages and motorcars became favoured and feasible modes of transport after 1910 or so. Only when public subsidies for automobile roads and associated infrastructure became available could the relatively few national parks and areas with status as equivalent reserves (such as national monuments) seem important enough to merit a bureau created specifically to manage them. The National Park Service was born in 1916 (the same year that the first federal aid for state highways became available) to administer 17 national parks and 22 national monuments. Those numbers expanded dramatically in less than two decades, as tourist infrastructure responded to public investments in roads.

Even before highway construction accelerated during the 1920s, American conservationists had succeeded in securing large reservations from the public domain intended for permanent use as public forests and parks (Runte 1979). This represented a departure from the time when individuals could hunt, fish, gather food, use forage, or cut timber without the need to recognize formal ownership boundaries and other types of controls needed to prevent conflicts among users tied to national or global markets. In short, the 'frontier commons' could persist where competition for resources was slight. A shift, however, began during the 1890s in the federal government's role to a more active agent whose willingness to regulate (rather than dispose of) the public domain it still controlled. The most widespread change came in the form of reserving forests, to be accompanied by permits for grazing and prohibitions on activities such as setting fires to improve forage. Impetus for imposing what initially were called 'forest reserves' on what the government retained as public domain came from a well-placed few who recognized that eastern forests had been reduced rapidly through land clearance and logging. Ownership of forested land in the United States increased after 1850, for example, to the point where four-fifths of all standing timber had fallen into private hands by 1886 – more than doubling the figure alienated from the public domain just 30 years before (Runte 1991).

Statutory loopholes and abuses in selling public land led to concentrated private ownership of forests in the east, as well as fraudulent acquisition of timber by syndicates in the western states. Fears of denuded watersheds and a host of uncontrollable impacts following from such an ownership pattern in the 1890s prodded Congress to pass legislation that allowed the President to proclaim forest reserves in unallocated public land containing timber. As a sort of safety valve that might ensure more far-sighted use of natural resources, the forest reserves also sprang from the realization by some that much of the mountainous west remained unsuitable for settlement, but might provide an all-important water supply and a resource base that could be managed in perpetuity (Runte 1991). Some of the early national parks and monuments were created out of the reserves, since these lands contained many millions more acres than any public park aside from Yellowstone. Not that getting bills establishing national parks through Congress was easy; only six made it over the span of 30 years since the act establishing Yellowstone National Park in 1872 had passed (Runte 1979).

Squeezing enough money out of Congress to operate the parks and forests at even the most minimal level proved almost equally difficult. Appropriations eventually increased, but only with the support of three presidents (Theodore

Roosevelt, William Howard Taft and Woodrow Wilson) and other progressives who regulated the use of resources on federal lands as part of a larger agenda of reform. This 'conservation movement' was bipartisan and astute enough in its public relations to use how Americans viewed themselves to advantage through recasting the frontier in their favour.

Despite its limitations, a great number of Americans have believed that the frontier produced a distinct national character – one based on self-reliance, love of liberty, aptitude for innovation and belief in progress (Eutlain 1999). Whether this is true or not, such thinking prevailed at the beginning of the twentieth century, at a time when scientists and government officials sought to regulate use on what had formerly been (at least since 1850 or so) a frontier commons. Popular fiction during the first couple of decades after 1900 generally portrayed forest and park rangers as pioneers, who in facing an untamed land alone, constantly faced the threat posed by wildfires and unscrupulous forest users who stole timber, set fires, or otherwise harmed the public interest through self-serving and expedient actions that worked against the long-term vision of conserving natural resources (LaLande 2003). The early park rangers enjoyed a similar image and performed most of the same duties as their counterparts in the national forests, except that they also began to take on an educational role to help visitors better understand an area's natural history.

Even with the constant reminder of the menace that wildfire posed to western communities (1910 was a particularly bad year), federal subsidy in the form of appropriations for developing infrastructure on the national forests and national parks was slow in coming. Private land produced far more timber, and most investments made in the national forests served a relatively small number of constituents (as was true in the early national parks), so Congress understandably proved reluctant to provide much in the way of funding. More lavish subsidies for developing the west (which began with grants Congress gave to railroads during the Civil War) were tied to faster returns (Pyne 2004).

More than any other single factor, access in the form of new roads ensured the long-term viability of national parks as an institution (Runte 1979). By 1920, more roads opened the possibility of new uses like recreation tied to the expanding ownership of automobiles, and thus more constituent support. The Good Roads Movement also adopted language of the frontier as it pressed for 'new roads over old trails' (Cole 1996). Even if far from unified, this coalition consisted of farmers, country doctors, travelling salesmen, manufacturers and urban professionals. Beyond creating new possibilities for touring, the movement also sought to broadcast its message that more and better roads could aid progress by eradicating rural isolation. As a result, road advocates adopted promotional strategies first developed by the railroads to emphasize how a system of both long-distance and local roads served to unite the nation (Shaffer 2001).

As automobile ownership in America began to surge, Congress passed the first federal highway act in 1916. The legislation also carried three key stipulations. First, no state could receive federal funds without setting up a highway department. This acted as the impetus for California and other states to begin implementing the plans laid years earlier for a system of trunk roads. The second stipulation also served to strengthen the idea of state control over highways by requiring that it assume responsibility for maintaining roads financed under the act's provisions. Third, states could not act unilaterally with federal funding for roads, since the Secretary of Agriculture (who oversaw the Bureau of Public Roads, the agency charged with supervising highway projects crossing federal lands) had to approve project statements submitted by the highway departments.

The small amount of money appropriated by Congress in 1916 boosted the prospects of actually constructing roads in more remote areas only a little. It did, however, lead to reconnaissance surveys and other types of location work that functioned as precursors to the three stages of construction: grading, surfacing and maybe even paving approach roads outside park boundaries. More federal money for highways became available once Congress passed additional legislation in 1925. One result was that roads within national parks improved to handle an exploding number of visitors who could now reach these destinations by automobile. Roads ultimately became the defining visitor experience in many national parks, with the best tour routes landscaped to blend with their park settings. It was analogous to what some carriage roads achieved on private estates, though the scenic highways provided circulation at a far greater scale and accommodated much faster vehicle speeds.

New and better roads made battlefields and other historic sites more accessible, so they too could become part of a 'national park system'; so did parkways, landscaped corridors with controlled access points that sometimes extended over several hundred miles. Some, like the Blue Ridge Parkway, aimed for the ideal of showing only idyllic rural simplicity to motorists, who in turn had to strictly adhere to the posted speed limits in return for maintaining an Arcadian illusion. What these other types of units show is that the institution of national parks can be generally described, but it continues to resist precise definition. It is probably best understood as a convergence of forces, some ancient and some more modern, that eventually aligned to produce the most recognized and globally pervasive way where national governments support heritage. Although arguably first, the institutionalizing of national parks in the United States has much in common with similar actions taken in other countries – not only because of shared cultural perceptions, but also the way in which improvements to transportation infrastructure produced an ever-increasing number of constituents for the parks.

7 John Muir and William Gladstone Steel

Activists and the establishment of Yosemite and Crater Lake National Parks

Stephen R. Mark and C. Michael Hall

Activists have done much to stimulate legislative action for national parks and equivalent reserves since they were first established in the nineteenth century. Whether in the United States or in countries such as Australia, Canada or the United Kingdom, individuals who publicly come out in support of national park conservation have been essential in the development of the park concept. In the present day, such activism remains important, but it is often subsumed within the overall actions of conservation groups such as the National Parks and Conservation Association (NPCA), the Sierra Club or the Wilderness Society. Nevertheless, the actions of such individuals have important repercussions for how we see national parks in the present day and the role of tourism within them.

This chapter discusses the role of activists in the creation of American national parks. Their efforts have been a key factor in the National Park System's continued expansion, particularly with respect to natural areas located in the western US. This growth has continued to occur in spite of some government officials expressing the view that this category was 'rounded out' in 1940 by the establishment of Kings Canyon National Park in California. Interest groups such as the NPCA (a group that has lobbied Congress since 1919 to defend, promote and improve the National Park System) continue to press for new additions to the natural area branch of the System (Miles 1995). This process is ongoing. Recent examples include the 1980 creation of 15 national parks in Alaska, the 1987 establishment of the Great Basin National Park, Nevada after 25 years of advocacy by Ralph Starr Waite and the 1996 creation of the Tallgrass Prairie National Preserve after 50 years of advocacy by NPCA.

Extended periods of time before a proposal to set aside an area of land as a national park or equivalent is accepted, is not a new phenomena. Many of the oldest park units were a result of the efforts of a single person, with examples including Sequoia (1890) and Rocky Mountain (1915) (Buchholtz 1983). However, the campaigns for Sequoia and Rocky Mountain were comparatively short (7 years each), in contrast to the 17 years it took to include Yosemite Valley in Yosemite National Park (established 1890) and Crater Lake (1902). This chapter examines the creation of these latter two parks with a focus on the role of two activitists – John Muir at Yosemite and William Gladstone Steel at Crater Lake.

In comparing Muir's campaign to place Yosemite Valley under national park administration with Steel's effort to establish Crater Lake National Park, there are some broad similarities. Each man won deceptively easy battles soon after becoming an activist, but found their larger goals more elusive. Besides framing their proposals similarly, they shared some methods still employed by modern park proponents. And, like some of their modern counterparts, Muir and Steel were also able to adapt to changing political circumstances to finally achieve their aims.

John Muir and Yosemite

John Muir (1838–1914) has become an international figure for his writings on conservation. He was born in Dunbar (near Edinburgh), Scotland, where the Muir house is itself now a tourist attraction and Muir is presented as an iconic father figure of the Loch Lomond National Park, although he had no part to play in its conservation. After coming to California in 1868, he worked at various seasonal jobs in the Sierra Nevada before making a name for himself in the early 1870s as a writer. Although Muir mentioned national parks and preservation of forests in his early writings, he did not become an activist until 1889; there is some suggestion that Muir toyed with the idea of a national park at Yosemite as early as 1872 (Hadley 1956) and was willing to take steps publicly to further the cause of forest conservation (Hooker 1886).

The turning point came when Muir and an editor named Robert Underwood Johnson embarked upon a camping trip to Yosemite in 1889. On the second night of the trip, they sat in front of a campfire planning a campaign that would alter Muir's life and the face of Yosemite. As Johnson later recalled:

> It was at our campfire at the Tuolumne fall at the head of the canon that Muir let himself go in whimsical denunciation of the commissioners [appointed by the State of California to manage the state park in Yosemite Valley] who were doing so much to make ducks and drakes of the less rugged beauty of the Yosemite by ill-judged cutting and trimming of trees, arbitrary slashing of vistas, tolerating of pig-sties, and making room for hay-fields by cutting down laurels and under brush – the units by which the eye is enabled, in going from lower to higher and still higher trees, ultimately to get adequate grandeur of cliffs nearly three thousand feet high. It is an old scandal, and I only refer to it now because it was at this campfire that a practical beginning was made of a campaign which, after fifteen years, by the recent act of recession of the Valley to

the United States, we may confidently hope has ended an era of ignorant mismanagement.

(Johnson 1905: 303–4)

Two extremely well-timed magazine articles written by Muir for Johnson's *Century Magazine* on 'The Treasures of Yosemite' (Muir 1890a) and 'Features of the Proposed Yosemite National Park' (Muir 1890b) greatly aided passage of a bill creating a 2 million-acre forest reservation in the Yosemite region on 1 October 1890 (Johnson 1905: 304). Several open letters that Muir sent to Johnson may have also been a factor in the passage of this legislation (Sierra Club 1944). However, Yosemite Valley, originally reserved as a state park in 1864, remained under state control, while the 'forest reservation' was under federal management and became known as Yosemite National Park. Not until 11 June 1906 did Muir and Johnson realize the goal of getting the valley and surrounding national park under the unified administration of the federal government (Runte 1990).

To frame his proposal, Muir had to summarize how he would address the problem of park management in Yosemite Valley. He did this by centring on three main points, the first being that the valley was explicitly a national – not state – concern. Muir believed the federal government had the ability to provide more permanent improvements and policies than did the state through its appointed commissioners. Federal control would lead to increased appropriations for roads, trails and utilities, which would facilitate greater tourist travel. The federal authorities would also be in an economically disinterested position, thereby increasing the chances that appropriate development would be coordinated by a landscape architect (Sierra Club 1896). The focus on tourism was quite apt given Muir's own history of working as a summer tour guide in Yosemite in the 1870s and Muir's extensive writing for vicarious and actual tourists.

As with many conservationists, Muir regarded tourism as a less evil form of economic development than grazing or commercial clear cutting of forests. In the 1870s, his writing suggests 'that a growing tourist business might drive the more exploitative users' (Cohen 1984: 206) and especially, sheep interests, out of the Sierra and Yosemite in particular. Accordingly, 'Despite his suspicion that the path of moderation was not the best way to a true vision of Nature, he attempted to write moderate articles which would bring urban tourists' (Cohen 1984: 206). Indeed, Cohen goes on to argue that 'in a sense, all of Muir's writings were for the tourist, since they involved the question of how to see. Most tourists did not want to hear philosophy, but wanted to know exactly where to stop and look' (1984: 207). In some of Muir's writing this is extremely plain to see. For example, in *The Yosemite*, originally published in 1914, Chapter 12 is entitled, 'How Best to Spend One's Yosemite Time', providing instructions for two one-day excursions, two two-day excursions, a three-day excursion and longer routes, with the Upper Tuolumne excursion being 'the grandest of all the Yosemite excursions, one that requires at least two to three weeks' (Muir 1914: 155). These excursions, along with other advice on visiting the park, remain used by travellers to the park to the present day.

The second part of Muir's proposal was that resident authorities must have sufficient power to protect the entire park area. Galen Clark had been appointed guardian to the valley, but he had no assistants, little money for administration and was under the orders of commissioners who were often motivated by political considerations (Runte 1990). Muir preferred the use of the US Army to guard Yosemite Valley and the backcountry from trespass by sheep, damage caused by careless campers and the effects of forest fires. The latter was to prove especially troublesome while the park was under two jurisdictions because their representatives could not agree over who should pay for fire protection.

Muir's third point was that recession of the valley – ceding from state ownership back to the federal government – was tied to protecting surrounding forests whose primary importance was conservation of water supplies. He used the water supply argument to lobby against the Caminetti bill of 1895, which would have reduced Yosemite National Park by half and severely damaged the recession campaign (*San Francisco Examiner* 1895, Kimes and Kimes 1986). Muir's opposition to the bill also stemmed from the belief that the newly created federal forest reserves (which were later to become national forests) should not be compromised by inholdings. During this period thousands of acres of formerly public domain forestland slipped into private hands, often by fraudulent means. Once the timber was cut, there were aesthetic problems and difficulties in maintaining enough water for irrigation and municipal supplies. Without federal control, he saw the infamous 'stump forest' in Yosemite Valley being duplicated on a larger scale throughout the Sierra (Muir 1890a, 1901a, b), a point upon which Muir agreed with the utilitarians in the forestry movement (e.g. Pinchot 1900: 87) and subsequent US Forest Service research (Zon 1927).

William Gladstone Steel and Crater Lake

The components of Muir's campaign matched those of Steel's, though the beginning of the Crater Lake effort predated attempts at Yosemite recession. William Gladstone Steel (1854–1934), like Muir, enjoyed something of an early victory by seeing 10 townships around Crater Lake reserved from settlement in 1886. This was done as a necessary first step in the creation of a national park, but soon encountered the reluctance of many congressmen who viewed such reservations as a drain on the Treasury.

Born in Ohio, Steel finished high school in Portland, Oregon. He became a postal carrier after short stints as a newspaperman, railroad promoter and publisher. His first visit to Crater Lake came on a short vacation from the Portland post office in 1885. Steel and a friend went to southern Oregon to meet up with geologist Joseph LeConte, who was studying the volcanic features of the Pacific Coast. After seeing the lake for the first time, Steel wrote:

> Not a foot of the land about the lake had been touched or claimed. An overmastering conviction came to me that this wonderful spot must be saved, wild and beautiful, just as it was, for all future generations, and that it was up to me to do something. I then and there had the impression that in some way, I didn't know how, the lake ought to become a National Park. I was so burdened with the idea that I was distressed. [For] Many hours in Captain Dutton's tent [Dutton was head of a small military party assigned to accompany LeConte], we talked of plans to save the lake from private exploitation. We discussed its wonders, mystery and inspiring beauty, its forests and strange lava structure. The captain agreed with the idea that something ought to be done – and done at once if the lake was to be saved, and that it should be made a National Park.
> (Quoted in Unrau 1988: 27–8)

Steel's story with respect to the idea of a Crater Lake National Park has considerable similarity with the mythology of the creation of Yellowstone (Runte 1979), although Steel also provides other accounts of his first visit to Crater Lake (Steel 1886, 1925). Nevertheless, regarding the exact circumstances surrounding Steel's first trip, upon returning to Portland, Steel began circulating a petition that eventually found its way to the state legislature. It was favourably received and a resolution recommending a public park around Crater Lake was forwarded to the Secretary of the Interior Lucius Q.C. Lamar. As a result, 10 townships were withdrawn from entry by executive order of President Grover Cleveland on 1 February 1886 (Steel 1907).

Like Muir, Steel contended that his proposal was of national concern. He did not support either Dutton's or LeConte's views that efforts to establish a state park at Crater Lake might be more fruitful (Dutton 1886, LeConte 1886). In addition, Steel was convinced that Oregon could not afford proper maintenance and protection of Crater Lake, so he opposed the state park bills introduced to Congress in 1889, 1891 and 1893 (Unrau 1988).

Provision was made in Steel's proposal for enforcement of the regulations by resident authorities. Like Muir, uppermost in his mind was the damage caused by sheep. Their trampling had so destroyed the area's vegetation in the years before the park was established that the result could still be seen in the 1930s. Fires, whether started by lightning or sheepmen, were another nemesis, which Steel wanted controlled. Interestingly, Muir also mentioned fire's effect on the Crater Lake area in his journal entry of 31 August 1896 (see Wolfe 1938).

The Crater Lake park proposal was also tied to the larger goal of protecting forests in Oregon's Cascade Range. As with the Sierra, the primary justification for their retention in public ownership was the utilitarian argument of water supply. Steel fought for the establishment of a 300-mile-long forest reserve stretching from the Columbia River to the California border. This was proclaimed by President Cleveland in 1893 and included the Crater Lake reservation. The Cascade Forest Reserve was the largest in the nation and consisted of 4,883,588 acres when it was proclaimed on 28 September 1893 (next in size was the Sierra Forest Reserve, which was established on 14 February 1893 and had 4,096,000 acres). However, the Cascade Reserve was subsequently attacked by sheepmen and timber speculators (Unrau 1988).

Steel and a state Supreme Court justice named John Waldo (both of whom envisioned a reserve managed much like a national park) worked to defend it throughout the 1890s. Steel also saw the Cascade Reserve as giving Crater Lake another layer of protection. Without its creation, he feared the possibility that the Crater Lake townships reserved in 1886 would be restored to entry. This happened at Sequoia, where an order by the Secretary of Interior was revoked for a brief time in 1890, before park proponents succeeded in getting national park designation for the Giant Forest and other groves (Fry and White 1930).

Tactics used by Muir and Steel

Once the components of the Yosemite and Crater Lake proposals had been formulated, Muir and Steel used some remarkably similar methods to achieve their aims. Although the two men were only acquaintances, they did have common interests and were in intermittent contact from 1888 to 1912. Their first meeting was in 1888 when Muir climbed Mount Rainier. Both of them also attended the National Park Conference of 1912, held at Yosemite. This may perhaps explain some of the similarities, particularly with respect to the development and use of constituencies, including tourism and utilitarian conservationist interests, to back their proposals.

Both Muir and Steel obtained early local support, something that sustained them throughout their campaigns. The major cities of their respective states furnished each man's base of support – Muir in San Francisco and Steel in Portland. Having already emerged as a literary figure, Muir had many powerful friends in California who could provide him with introductions to useful contacts. Likewise, Steel was well situated within Oregon's Republican Party and had two brothers who were Portland financiers. Each man received the support of their states' major newspapers early in their campaigns. This move proved useful when sheep and timber interests tried to dismantle Yosemite National Park and the Cascade Forest Reserve. They also gave public lectures as a way to enhance their proposals' credibility, as well as creating interest in visiting the areas they spoke about. The fact that each man was a renowned climber and participant in the scientific study of mountain areas helped attendance. For example, Muir began giving public lectures in 1876 and throughout the next decade went to west coast cities to speak about glaciers, botany and his travels. By the time he became an activist, he was a popular speaker whose income from other sources allowed him to be very selective (Cohen 1984). Steel's career as a speaker began when he returned from Crater Lake in 1885 and broadened over time to include several lecturing trips across the country.

Both men started their campaigns by writing articles in literary magazines. Muir had a national audience while Steel's fame remained largely regional. Although Muir began his literary career by mostly writing for newspapers, he found the national literary magazines not only paid better, but were a more effective way of promoting his proposals (Fox 1981). Steel's writings, by contrast, were generally newspaper articles whose distribution was limited to the Pacific Northwest. Nevertheless, Steel was the first to write a book that he could use to promote his proposal. *The Mountains of Oregon* was published in 1890 as a loosely organized anthology of articles on mountaineering and proposed parks. Steel highlighted the longest piece, one about Crater Lake, when he mailed copies of the book to congressmen and other federal officials. The book's title is interesting in light of an acknowledgment that Muir wrote to Steel after receiving a copy:

> I thank you for a copy of your little book The Mountains of Oregon + congratulate you on the success with which you have brought together in handsome shape so much interesting + novel mountain material. With pleasant memories of my meeting with you the year I was on Mt. Rainier.
> (Muir 1892)

Muir's *The Mountains of California* was published in 1894. Far more cohesive than Steel's book (which was a hasty arrangement of material originally intended to be published in separate pamphlets), it enhanced Muir's reputation among scientists and brought him critical acclaim from the public. With the Caminetti bill looming over Yosemite in 1895 and the forest reserves threatened by hostile interests, Muir began to intensify his literary efforts. Ten of his essays were published in the *Atlantic Monthly*, starting in 1897, and later appeared as a book entitled *Our National Parks* in 1901. Six of the ten pieces were devoted to Yosemite, while three others focused upon the fate of the forest reserves.

Significantly, in terms of the mobilization of tourism and recreational interests, both men found that groups organized to enjoy the outdoors could form a useful constituency. Steel again predated Muir in this regard by organizing the Oregon Alpine Club on 14 September 1887. It was largely a social fraternity whose purpose was:

> to attract attention to the scenery of our [Pacific Northwest] mountain ranges. By late 1892, the expense of a mountaineering museum had bankrupted the club and personally cost Steel $1,000. Membership had dwindled to less than a hundred and most observers thought the club was dead.
> (*Portland Oregonian* 1892)

Steel eventually realized that an active mountaineering club might have a longer life. On 19 July 1894, amid great local publicity, 193 climbers ascended Mount Hood and became the first to be known as Mazamas. According to Steel, one of the group's aims was to make the Oregon Cascades famous and to sponsor regular outings (*Medford Mail* 1895). Article II of the Mazamas' constitution is precise:

> The objects of this organization shall be the exploration of snow-peaks and other mountains, especially those of the Pacific Northwest; the collection of scientific knowledge and other data concerning the same; the encouragement of annual expeditions with the above objects in view; the preservation of the forests and other features of mountain scenery as far as possible in their natural beauty and the dissemination of knowledge concerning the beauty and grandeur of the mountain scenery of the Pacific Northwest.

After being elected its first president, Steel organized an outing to Crater Lake in August 1896. The group gave it wide publicity and supplied the event with an interesting touch by christening the mountain that contains the lake 'Mazama', a name which is used to the present day (Scott 1969). In 1895, Muir and LeConte were among the first three honorary members to be elected by the Mazamas.

The Sierra Club was organized 25 May 1892, and evolved from a proposal that Robert Underwood Johnson made to Muir in 1889 regarding an 'association for preserving California's monuments and natural wonders' (quoted in Fox 1981: 106). The public meetings in San Francisco were heavily attended at first and the club began publishing a regular bulletin. As president, Muir's attendance at meetings was erratic so the organizing fell to other board members. Almost nonexistent by 1898, the club was revived when its new secretary William Colby sold the idea of sponsoring regular outings. The first was held from a base camp in Tuolumne Meadows in 1901 and was an immediate success. Aimed at attracting new members, the outings included organized hikes as well as natural history lectures by Muir and other club leaders (Greene 1987). Colby argued that such outings would be a good tool for the Club: 'An excursion of this sort, if properly conducted will do an infinite amount of good toward awakening the proper kind of interest in the forests and other natural features of our mountains, and will also tend to create a spirit of good fellowship among our members' (quoted in Cohen 1984: 311). Indeed, the Sierra Club's articles included the phrase 'render accessible':

> Buried in the phrase were the relationships among roads, developed accommodations in parks and reserves, and 'styles' of recreation. After all, ease of access, the comfort of accommodations, and the kind of recreational trails and facilities would determine the kind of ecological

consciousness produced by the parks and reserves. From the beginning the Sierra Club involved itself in decisions about access and development, advocating roads and trails in Yosemite and elsewhere, later encouraging private means of access by railroad and lobbying for improved and more extensive public roads. And the Club itself would become a means of access when it published information and organized outings.

(Cohen 1984: 306)

The differences between the Yosemite and Crater Lake proposals also shaped the way each group responded as a constituency. Muir aimed to provide better management for an area where there was substantial human impact, so the Sierra Club aimed at becoming a Yosemite Valley resident. As early as 1894, the Sierra Club's board of directors wanted to establish a patrol system in the valley to help enforce state park regulations. This would be 'the first step in the direction of preserving the Valley from the wanton destruction of visitors' (McAllister 1894).

What evolved was an information bureau housed in a refurbished wood frame cottage in Yosemite Valley from 1898 to 1902. In 1903, the bureau was moved to the newly completed LeConte Memorial at the base of Glacier Point. The structure's completion coincided with the chaos arising from a disastrous fire, which burned from the Wawona Road to Glacier Point. This happened largely because the state commissioners and US Army authorities could not agree who should fight the fire. The case for recession was further strengthened that summer when the state commissioners notified the transport companies not to allow more visitors to enter the valley until overcrowded conditions were relieved (Muir et al. 1905).

The Mazamas' response to its founder's proposal was different because Steel wanted national park status for a feature little known to science. As a result, the group fostered scientific investigation at Crater Lake on one occasion and used the findings to promote the proposal. Although their involvement was largely peripheral, the Mazamas' facilitation was important in allowing scientists to build upon what an earlier expedition had done at Crater Lake.

During the summer of 1886, the US Geological Survey sounded the lake and mapped the area's topography (USGS 1887). Much of its success was due to Steel, who, in his role as special assistant to the expedition, was responsible for transporting the boats and equipment. His role in the undertaking gave him credibility and allowed the Oregon Alpine Club to co-sponsor the O'Neal Expedition of the Olympic Mountains in 1890. Another success followed so Steel felt confident in organizing an even larger undertaking – the Mazamas outing of 1896. By arranging the trip so that the Mazamas were climbing nearby Mount McLoughlin while scientists from various government bureaus made their investigations, he hoped to give the proposal both scientific merit and wide publicity. After their climb and an excursion to Wizard Island, the Mazamas assembled on a site overlooking Crater Lake so that the findings could be presented. The outing also allowed the scientists to meet with members of the National Forestry Commission, a body whose purpose was to make recommendations about the disposition of the forest reserves. For this to happen, Steel cut his participation in the Mazamas' trip short so he could bring the commission to the lake less than a week later. Steel later recalled that he had to walk from Crater Lake to Medford (some 85 miles in two days) so he could escort the commission back to the lake. Although the Commission recommended Mount Rainier and Grand Canyon for national park status, they failed to reach a consensus about whether to include Crater Lake. Its members were: Charles S. Sargent (Harvard University), William H. Brewer (Yale University), Arnold Hague (USGS), Henry S. Abbott (US Engineer Corps), Alexander Agassiz (Coast and Geodetic Survey), Gifford Pinchot, and John Muir (Unrau 1988).

Neither Muir nor Steel were strangers to state and national politics by the time they finished their park campaigns. Both found ways to secure influence with businessmen, legislators and government officials through various lobbying techniques. In addition, each man chose an unexpected intermediary when his proposal reached a crucial stage.

After years of petitions, testimonials and localized legislative support, the proposals began to move toward realization when Theodore Roosevelt assumed the Presidency in 1901. It was Roosevelt's influence that allowed the Crater Lake bill to come up for debate in the House of Representatives in April of 1902 (Unrau 1988). Muir's most publicized lobbying for recession came when he and Roosevelt camped alone in Yosemite for three days in May 1903, with Muir postponing his long-planned trip to study big trees in Australia and New Zealand to help further justify the conservation of sequoia within federal reservations (Ryan 1985, Hall 1987, 1988a, 1993). According to Muir, 'he stuffed him [Roosevelt] pretty well regarding the timber thieves and the destructive work of the lumbermen, and other spoilers of the forests' (quoted in Wolfe 1938: 291). Despite a substantial divergence in the two men's outlook on forest use, the meeting had a great effect on Roosevelt's attitude towards conservation and encouraged the President's resolve to support the progressive conservation movement (Nash 1967). Muir's deep solicitude over the destruction of America's forests and mountain scenery had made a strong impression on the President's mind. 'Roosevelt had shown himself to be a great friend of the forests before this camping trip with Muir, but he came away with a greatly quickened conviction that vigorous action must be taken speedily, or it should be too late' (Badè 1924: vol. 2, 411). The result was the President's intervention when Senate cooperation was needed to add the valley to Yosemite National Park in 1906 (Runte 1990).

Although Roosevelt was a key figure in the adoption of both proposals, Muir and Steel had to use unusual intermediaries before the President could sign either bill. In Muir's case this proved to be E.H. Harriman, president of the Southern Pacific Railroad. Harriman made use of the railroad's influence on the California state legislature after Muir and William Colby did some hard lobbying for recession. When the measure came up for a vote in February of 1905, nine crucial votes turned the tide and it passed. About a year later Harriman came to the rescue again when a joint resolution accepting the valley stalled in the House (Fox 1981, Orsi 1985).

Although Harriman's actions can be explained largely by his friendship with Muir, the Southern Pacific also wanted control of transportation to Yosemite (Ise 1961). In spite of the railroad's ulterior motive, Muir accepted Harriman's assistance, particularly because of the company's role in bringing tourists to the park. Muir reasoned that federal control of the entire park area would lessen the destruction caused by the numerous concessioners (27 at the time of recession) and other entrenched interests. Furthermore, the Sierra Club's board declared that Yosemite's poorly-maintained toll roads and the valley's substandard accommodations were hurting California's economy (Muir *et al.* 1905).

Steel's intermediary was Gifford Pinchot. At first this seems strange, especially given the view that Pinchot's name never appeared in connection with the promotion of national parks (Ise 1961). But he did seem to have been more enthusiastic about Crater Lake than Muir, whose writing about his visit in 1896 indicated that the most impressive feature of southern Oregon was its variety of tree species (Muir 1897, Wolfe 1938). Pinchot camped with Muir at the lake and later wrote, 'we drove to Crater Lake, through the wonderful forests of the Cascade Range, while John Muir and Professor [William H.] Brewer made the journey short with talk worth crossing the continent to hear. Crater Lake seemed to me like a wonder of the world' (Pinchot 1947: 101).

A somewhat similar situation developed in February 1902 when Steel was eliciting testimonials for the bill that would establish Crater Lake National Park. Muir begged off in his response:

> I don't know the Crater Lake region well enough to answer the question 'Why should a national park be established to include Crater Lake.'
>
> You know this region much better than I do. I should try to show forth its beauty + usefulness explaining its features in detail + pointing out those which are novel + which require Government care in their preservation etc. . . .
>
> (Muir 1902: 46)

By contrast, Pinchot's reply was ecstatic:

> You ask me why a national park should be established around Crater Lake. There are many reasons. In the first place, Crater Lake is one of the great natural wonders of this continent. Secondly, it is a famous resort for the people of Oregon and of other States, which can best be protected and managed in the form of a national park. Thirdly, since its chief value

is for recreation and scenery and not for the production of timber, its use is distinctly that of a national park and not a forest reserve. Finally, in the present situation of affairs it could be more carefully guarded and protected as a park than as a reserve.

(Pinchot 1902a)

The bill was passed unanimously by the committee, but was opposed by the Speaker of the House, who refused to let it be debated. He relented only after Pinchot had spoken to Roosevelt about the bill (Tongue 1902). After it passed the Senate, Pinchot wrote Steel again: 'You give me more thanks than my small share in getting the Crater Lake bill passed deserves, but I am sincerely glad it has got along so far. There is no doubt, in my judgment, that the President will sign it' (Pinchot 1902b).

Steel's triumph came a week later on 22 May 1902 when Crater Lake became a national park. His ability to get along with Pinchot allowed the proposal to get over the final hurdle. This is in contrast to Muir, who had severed all ties with the forester in 1897 over the issue of sheep in the forest reserves (Runte 1979, Cohen 1984).

The best explanation for why Pinchot was willing to work with Steel might be common interest. Passage of the Crater Lake bill occurred three years before Pinchot created the US Forest Service and stimulated transfer of the reserves from control by the Interior Department's General Land Office to the Department of Agriculture. Steel started the first forestry organization in Oregon, the Oregon Forestry Association, in 1896 as another way to defend the Cascade Reserve, and had surveyed the Stehekin section of the Washington Reserve when Pinchot was 'special forest agent' for the Department of the Interior in 1897. Pinchot made it a point to visit Stehekin that summer after Steel failed to receive a patronage appointment as forest superintendent in Oregon. Steel, however, was more inclined towards forest recreation than was Pinchot (Steel 1898). They shared a vehement dislike for the General Land Office's administration of the reserves, and Steel had at one point begun to waver from his previous position on sheep. Wolfe (1938: 379-80), gives Muir's journal entry for 29 May 1899: 'Met Judge George. Had a long talk on forest protection, found him lukewarm. Mr. Steel uncertain on the same subject. Told him forest protection was the right side and he had better get on record on that side as soon as possible. He promised to do what he could against sheep pasture in the Rainier Park and also in the Cascade Reservation'. It was only when Pinchot attempted to bring the national parks under Forest Service administration in 1904 that this coalition began to wither.

Ramifications of the park campaigns

Although Muir and Steel at last saw their proposals favourably received by Congress, neither park retained all of what Steel obtained in 1886 and Muir won in 1890. Crater Lake National Park was established without the adjoining Diamond Lake area, which had been in the original reservation. The opposition generated by Pinchot's Forest Service has been successful in stopping Diamond Lake's incorporation into the park and all but two minor extensions. Indeed, no national parks have been established in Oregon since the Forest Service was created.

Yosemite National Park was reduced by boundary changes in 1905, which allowed some notable giant sugar pines to pass into private ownership. The trees were restored to the park in 1939 over the objection of the Forest Service, but they seemed small compensation for the part Pinchot played in damming the Hetch Hetchy Valley (Ise 1961, Runte 1990).

Perhaps the long campaigns waged by Muir and Steel also have a lesson. Park management continues to deal with problems that both men thought were going to be solved by enactment of their proposals. It may have saddened Muir to find the National Park Service having had substantial difficulty implementing its plan to reduce congestion in Yosemite Valley. The difficulty of managing present-day expectations of tourist access to Yosemite, and to many national parks in general is noted by Greene (1987):

> It will be interesting to note as the Yosemite story continues that despite their seemingly disparate interests at times today, conservationists and park managers succeeded in working alongside each other with little friction in the early years. Indeed parks flourished under their combined patronage. Both groups, in their zeal to find support for a national park system, recognized the advantages of developing tourism and commercial recreation. As long as crowds remained small and impact on the resources minimal, conservationist aims remained compatible with park use. Not until the twentieth century would preservationists become dismayed by the seeming impossibility of promoting park use without an adverse impact on natural resources and consequent physical deterioration of park facilities and the environment. Even worse, growing commercialism would threaten to overshadow the basic purpose of parks as areas of relaxation and contemplation.

A similar irony exists at Crater Lake where geothermal energy interest is growing again in the area surrounding the park (Oregon Department of Geology and Mineral Industries 2003). Nonetheless, as Steel expressed it in 1930, there is an intangible satisfaction:

> Plundering through this wilderness of sin and corruption, tasting of its wickedness, forgetting my duty to God and man, striving to catch bubbles of pleasure and the praise of men, guilty of many transgressions, I now look back on this my 76th birthday, and my heart bounds with joy and gladness, for I realize that I have been the cause of opening up this wonderful lake for the pleasure of mankind, millions of whom will come

and enjoy and unborn generations will profit by its glories. Money knows no charm like this and I am the favored one. Why should I not be happy?

(Steel 1930)

Conclusion

The literature on tourism and national parks is surprisingly brief when it comes to discussing the role of activists and the use of tourism as a justification for the creation of national parks, although its role is generally recognized, particularly in its modern day incarnation of 'ecotourism' (e.g. Hall and Page 2006). This chapter has shed further light on the role of activists in the park establishment process, who were willing, as few people have been, to carry a considerable burden for little material gain. In most cases (Muir is a notable exception) the reward of activists has been obscurity. Yet the high, almost iconic, profile of both parks in the United States park system and beyond, along with the ongoing role of the organizations established to lobby for national park status, has had substantial influence on national parks around the world.

As well as indicating the approach of two activists who helped lead to the establishments of two significant early national parks, this chapter has highlighted the way in which tourism, recreation and visitor access was from the very beginning integral to the case of creating parks. However, as Runte (1990: 1) pointed out, 'Among all of the debates affecting American national parks, the most enduring – and most intense – is where to draw the line between preservation and use'. This chapter has illustrated that this tension, particularly with respect to tourism, originates from well before the creation of the first national park systems and is instead very much a product of the political and social processes that led to the creation of the parks in the first place.

8 Tourism and the Canadian national park system
Protection, use and balance

Stephen W. Boyd and Richard W. Butler

Introduction

Canada was one of the first countries to embrace the notion of setting land aside to both protect and make use of by the public for their enjoyment. The movement to establish national parks in Canada was part of a larger preservation/conservation ethic that permeated North American society in the mid-nineteenth century (see Chapter 3). From Banff, Canada's first national park and still the most recognized and visited park in the country today (Parks Canada 2007c), a system of parks have emerged that embrace the diversity of eco-zones that constitute the Canadian landmass. This chapter charts the development of the Canadian system, examining the role that preservation, tourism and economic development played in how the system evolved. In particular, the chapter addresses the significance of the 1930 Act – Canada's first specific national park act – and the challenging terms 'unimpairment', 'enjoyment', and 'benefit', which created for the development a system of national parks. The tension between 'use' and 'protection' is further traced through the development of national park policies of 1964, 1979 and 1994 and how the changing emphasis within policy witnessed the pendulum swing over time away from development and towards full protection. This evolution of policy would eventually culminate in the passing of the *Canada National Parks Act of 2001* where priority was shifted beyond the notion of 'ensuring ecological integrity' to maintenance and restoration of ecological integrity through the protection of natural resources and natural processes. The emphasis of 'protection' over 'use', however, raises some challenging questions for many of Canada's national parks and the authors use Banff National Park as a case study to illustrate the complexity of this relationship and how it has shaped the development and use of the park over time. Banff, Canada's most visited park, albeit atypical in its development, embraced the tourist from its inception in 1885 and emerged as the 'jewel in the crown' for the Canadian system. The chapter traces the development of Banff National Park, how it balanced the dual mandate of 'use' and 'protection', and how with the recent emphasis of ensuring ecological integrity it promotes responsible tourism development. In particular, the authors examine the development of a heritage tourism strategy within the park and how this promotes sustainable tourism development, encouraging partnership arrangements with key stakeholders involved. The chapter offers discussion of the partnership arrangements in place with Aboriginal peoples that have facilitated the development of the system over time, the issues surrounding National Park Reserves and the implications for these when land claims are settled and if the land reverts to indigenous ownership. The chapter concludes by addressing the 'other' elements that Parks Canada is responsible for, namely Heritage Rivers, National Historic Sites and Canals, and raises questions over the sustainability of the wider Canadian national parks system.

The establishment of Banff as Canada's first national park

The establishment of Canada's first national park – Banff, in 1885 – mirrored the characteristics of the early national parks within New World countries; namely, it was established on worthless lands, often as a remnant of frontier regions, where preservation of wilderness spaces was the driving factor in safeguarding areas from commercial development and change (Boyd and Butler 2000, Boyd 2004; see, also, Chapters 1 and 3). The year 1885, according to Lothian (1977), was a momentous one for Canadians. First, it witnessed the completion of the transcontinental railway, linking eastern Canada with the Pacific Coast. Second, the frontier was pushed further westward with early settlement of the prairies. Third, it heralded the reservation for public use of mineral hot springs in the Rocky Mountains near the railway station of Banff. This reservation was the genesis of what would be later named Banff National Park by legislation in 1887.

Canada's first park was the result of several accidents: the route carved out for the transcontinental railway, the discovery by accident of the hot springs by rail workers who failed to press for title to their discovery, and the decisive action of officers of the Department of the Interior at Ottawa who saw the need to preserve the hot springs from private development. Lothian (1977: 5) cites that the Honourable David MacPherson (Minister of the Interior), while visiting the site in August of 1885, advised his Deputy Minister as follows:

> I have just returned from a visit to the Hot Springs at Banff and have made up my mind that it is important to reserve by Order in Council, the sections on which the springs are and those about them . . . it is important this should be done at once. What we may do with them afterwards can be considered when I get back.

The earlier attempts by the engineers working for the Canadian Pacific Railroad to commercialize the hot springs were clearly a cause of government concern, and led to the establishment of a federal reserve of 10 square miles around the hot springs in 1885. The reason given for the establishment of the reserve around the hot springs was public sanitation, as Nichols (1970: 1–2) notes when quoting the original dedication:

> Whereas, near the station of Banff . . . there have been discovered several hot mineral springs which promise to be of great sanitary advantage to the public, and in order that the proper control of the lands surrounding these springs may remain vested in the Crown, the said lands . . . are hereby reserved from sale or settlement or squatting.

While Banff was set aside to be free from sale, settlement or squatting, this would establish a precedent that other Canadian national parks would follow. Nevertheless, Banff was to develop in an atypical manner compared to other parks that would be declared later, as it became Canada's most 'developed' national park. Prior to legislation being passed in 1887 to establish the region as a national park, the Prime Minister, Sir John A. MacDonald, accompanied by his wife, had visited Banff the year prior and had taken the waters. The debate within Parliament in April 1887 extolled the beauties of the region. *Hansard* records his contribution in part as follows:

> I do not suppose in any portion of the world there can be found a spot, taken all together, which combines many attractions and which promises in as great a degree not only large pecuniary advantage to the Dominion, but much prestige to the whole country by attracting the population, not only on this continent, but of Europe to this place. It has all the qualifications necessary to make it a great place of resort . . . There is the beautiful scenery, there are the curative properties of the water, there is the genial climate . . . there is mountain sport; and I have no doubt that that will become a great watering-place.

The *Rocky Mountains Park Act of 1887* saw the expansion of the reserve extended from 10 square miles to 260 square miles, with the passing of the legislation creating the Rocky Mountains National Park. The preamble to the Act noted that the area was reserved 'as a public park and pleasure ground for the benefit, advantage and enjoyment of the people of Canada'. What the *Rocky Mountains Park Act* (1887) ensured was the continued emphasis on recreation and enjoyment by stating that there would 'not be any leases, licences or permits' that would 'impair the usefulness of the park for the purposes of public enjoyment and recreation'. There can be fewer clearer statements of the importance of the place of early elite tourism in a new park. The focus on early tourism saw the Canadian Pacific Railroad in 1887 start construction of the first of several buildings to be known as the Banff Springs Hotel, modelled after early nineteenth-century Scottish baronial-style castles. Located on a bench above the junction of the Bow and Spray Rivers, on a site selected by W.C. Van Horne, the Vice President of the Company, the hotel was opened in 1888. Standing five storeys tall, along with adjoining bathhouse with plunge baths and tubs fed with water from the Upper Hot Springs, the elite tourists enjoyed the finest accommodation and recreational pleasure the park could offer (Marty 1984, Dearden and Rollins 1993).

An early 'system' of parks established

The precedent was now set for the establishment of other parks. First, the Rocky Mountains Park Act established the political feasibility of setting aside tracts of land to be protected and enjoyed as part of national heritage. Second, it also helped demonstrate that private exploitation of natural resources was not necessarily the best public policy. Third, it set down the principle that responsibility for national parks was to be a federal and not provincial government's responsibility. The last was important as Ontario was toying with ideas of its own national parks; it particularly wanted a national park at Niagara Falls (Dearden and Rollins 1993, Eagles and McCool 2002). A federal system of national parks closed the door on the evolution of a decentralized system like Australia.

By 1930, Canada had a system of 14 national parks. The majority of these were located in the west, including Glacier and Yoho in the Province of British Columbia, both being established one year after Banff. These would be followed by Waterton Lakes (1895), Elk Island (1906) and Jasper (1907) in Alberta, and by Mount Revelstoke (1914) and Kootenay (1920) in British Columbia. Only three of the early Canadian parks were established in the east, all within the Province of Ontario: St Lawrence Islands, established in 1904 from crown (federal) land that had been initially under Indian ownership; Point Pelee converted from a naval reserve in 1918; and Georgian Bay Islands, established in 1929 (Nelson 1973, 1982).

Examination of the early parks reveals a number of characteristics (Marty 1984, Dearden and Rollins 1993). First, a similar location bias existed in park establishment, displaying an early spatial disparity between supply (parks) and demand (population), with the exception of the St Lawrence Islands, which was close to the population centres of Kingston and Ottawa. Early reports from the western parks revealed that visitors came either from the built-up areas along the eastern American seaboard, the metropolitan centres in eastern Canada or as far afield as Europe (Nelson 1973). Second, as a result of this, visitor numbers remained small, partly because of the distances travelled, but also because of the limited tourism infrastructure that had been built in the early parks. Marsh (1982) notes these factors in his discussion of the early recreation development of Glacier National Park, commenting on the primary role the Canadian Pacific Railroad played in both developing a hotel within the park (similar to what took place for Banff), and providing access for the early tourists. By 1910 onwards, the rise in ownership of the automobile was influential in enabling visitors to travel to the parks within the Rocky Mountains from growing settlements such as Calgary; important also as a staging post from which international travellers set out to travel to the western parks. Third, attractions were focused on nature and the spectacles it offered; the elite traveller enjoyed the alpine environments of the western Canadian parks. The exception was Banff; with its hot springs it developed as a spa in the European tradition of offering this pleasure against a spectacular natural backdrop. The emphasis in promoting Banff National Park was on the hot springs, the grand hotel and the therapeutic 'powers' and attraction the waters were believed to hold. While the Canadian government developed many of the early parks on what was termed 'worthless lands', in the sense that they had no economic value to be developed for other human uses, early visitors did not hold to such thinking. Apart from the opportunity to enjoy the majesty of nature, visits to the early parks had deeper and higher meanings attached to them. In the writings of early visitors to Glacier National Park, the mountains were said to have taken on a spiritual dimension, where the 'peaks were inspiring, dangerous, close to God and a reminder of His power and achievements' (Nelson 1982: 44).

The early system benefited from legislation to assist in national park creation and early park management. First, the *Dominion Lands Act of 1883* helped place the management of public lands under the auspices of the Minister of the Interior. This allowed the federal government to reserve forest lands and forest parks in a state of preservation. It was under this legislation that Waterton Lakes, Elk Island and Jasper National Parks in Alberta were established. Second, the *Dominion Forest Reserve and Parks Act of 1911* established the Dominion Parks Branch, which became the first Canadian national park service. Under the Act, all parks were placed under the jurisdiction of the park service, which in essence was the start of a national parks system (Eidsvik 1983). Under the legislation, which had strong preservationist tones, large areas were protected from mining and timbering activities. As a consequence, Mount Revelstoke, Kootenay in British Columbia, Wood Buffalo bordering Alberta and the Northwest Territories, Prince Albert in Saskatchewan and Riding Mountain in Manitoba were developed (Nelson 1973). Two other developments took place in 1930 that had long-lasting implications for future park developments. First, after 1930, the federal government no longer had control over the natural resources within the crown (federal) lands within the Canadian prairie provinces, and so by handing over control to the individual provinces, expansion of the park 'system' would be halted in the west until the 1970s. Second, Canada's first specific National Parks Act was passed that year. This defined the function of Canadian national parks:

> The Parks are hereby dedicated to the people of Canada for their benefit, education and enjoyment, subject to the provisions of the Act and Regulations, and such parks shall be maintained and made use of so as to leave them unimpaired for the enjoyment of future generations.
>
> (in Lothian, 1977: 8)

Although the early policymakers in Canada felt they had explicitly stated the function and purpose of national parks, problems in the wording chosen and the ambiguity within the Act allowed for varying interpretations of 'unimpaired', 'enjoyment' and 'benefit' to develop. These saw the emergence of a dual mandate of protection and the use to develop, a problem that would only be addressed by amendments to the Act in 1989, and the passing of a new *Canadian National Parks Act* in 2001 that would swing the pendulum towards protection as opposed to development. By the time the Act of 1930 had passed, the early park 'system' had enjoyed firm leadership under its first commissioner, J.B. Harkin, who held office from 1911 to 1936. His insight would be responsible for the development of recreation and tourism within the parks. Although a strong supporter of preservation and maintaining the parks in a true state of wilderness, he also had the foresight to recognize the potential that lay in encouraging tourism developments within parks as a viable means to provide the interest and funding needed if the park system was to develop into a truly 'Canadian' system (Marty 1984, Dearden and Rollins 1993).

Developing the park 'system'

A 'stop-go' policy existed after 1930 in how parks were added to the existing system (Boyd 2006). With federal legislation curtailing development of parks in the prairie provinces for four decades, the expansion of the 'system' took place in Eastern Canada and in the north (Table 8.1).

A number of influences were important in understanding the nature of the expansion of the park system. The postwar decades heralded in a time that saw increased economic prosperity, a buoyant economy, and an expanding middle class that had more leisure time and disposable income available to it.

Table 8.1 Expansion of the Canadian national park system 1936–1972

Park	Province/Territory	Year established
Cape Breton	Nova Scotia	1936
Prince Edward Island	PEI	1937
Fundy	New Brunswick	1948
Terra Nova	Newfoundland	1957
Kejimkujik	Nova Scotia	1969
Kouchibouguac	Quebec	1969
Pacific Rim	British Columbia	1970
Gros Morne	Newfoundland	1970
Forillon	Quebec	1970
La Maurice	Quebec	1970
Pukaskwa	Ontario	1971
Kluane	Yukon	1972
Nahanni	Northwest Territories	1972

Source: Modified from Boyd and Butler (2000).

This resulted in most of the parks enjoying increased numbers of visitors, thus shifting the focus away from the elite traveller and early tourist to appeal to a wider market who viewed the parks as destinations for summer vacations, camping and day-trip activities (Boyd and Butler 2000). A second key influence was the actions of Jean Chretian (then Minister of Indian Affairs and Northern Development), who almost doubled the existing system by establishing 10 new parks between 1969 and 1972. The primary purpose of establishing parks in the Maritime Provinces was for them to act as regional economic growth poles, as well as becoming popular recreation spaces for its citizens. In contrast, parks developed in the north were partly the result of being a political gesture of sovereignty, but also the emergence of including areas where native peoples lived and as an outcome started the discussion on how national reserves should be co-managed in line with national park policy. In this later period of park expansion, the numbers of visitors increased threefold from 5 million to 15 million, albeit with existing western parks attracting the majority of this increase (Boyd 1995).

The Act of 1930 had created the problem of meeting a dual mandate of protection along with park use, as it enshrined the thinking that a park should be left 'unimpaired', but at the same time be 'enjoyed' by visitors and 'benefit' the region it is located in; the latter came to be read as support for wider forms of development. The first national parks policy, tabled in 1964, while having strong development-use tones, helped to establish the preservation of significant natural features in national parks as its most fundamental and important obligation (Eidsvik 1983). In terms of philosophy, the 1964 policy introduced the concept of systems planning into the overall park system. It was only now that a true system of parks was put in place, for prior to the issuing of the policy each national park had been managed as an individual unit. With a systems approach, each park became part of the national whole. The first version of a systems plan was introduced in 1966, and was based on dividing Canada into a number of physiographic regions. This would be followed in 1972 by a more elaborate systems plan whereby Canada was divided into 39 terrestrial regions, based on biological, physical and geographic features (Parks Canada 1971). Into this system were placed those parks that had already been established. Areas that did not have any park representation were examined for potential future sites to ensure that all terrestrial regions would have national park representation. Only when all regions included at least one park would the system be viewed as complete. By inserting existing parks into these natural regions it was found that many areas had overrepresentation, particularly in the Rocky Mountains, while there was limited representation in natural regions in the north of the country.

Alongside the introduction of park policy and systems planning, a number of key institutional arrangements emerged to assist with how parks were managed. As with many developments, Banff led the way with the introduction of an early form of zoning, particularly for winter activities. This would be later followed up in the early 1970s with a more comprehensive zoning scheme developed within Canadian parks that allocated land-use priorities to different areas of individual parks (Murphy 1985). What it did offer park managers was a broad framework for land management that attempted to balance the twin mandates of protection and visitor enjoyment by setting aside small areas within parks for recreational and intensive development of tourist-related infrastructure, while ensuring that large sections of the backcountry within parks were set aside for primarily preservation purposes. Public hearings also emerged as the vehicle to promote public participation in national park planning and management; an early form of this was evident in the public pressure against plans in the 1960s to develop Lake Louise (in Banff) and a rejection of a proposed village service centre concept. By the late 1960s, specific management plans were in place for individual parks and public participation (in the form of public hearings) was used as the principal mechanism to control developments within parks (Nelson 1973). It could be argued that with the emergence of the first parks policy, a systems approach to park development, along with institutional arrangements and actions to control the type and level of development within individual parks, Canada had in place a 'developed' parks system that, while by no means complete, had mechanisms in place to promote its further growth (Dearden and Rollins 1993).

Development through to intended systems completion

The expansion of the system from 1973 to the present day has witnessed the development of parks, particularly to the north in Yukon, and the Northwest and Nunavut Territories (Ivvavik, Ellesmere Island, Vuntut, Quttinirpaaq, Sirmilik, Tuktut Nogait, Ukkusalik, Vuntut), with limited establishment of parks on the west coast (South Moresby, Gulf Islands Reserve, Gwaii Haanas) or in the east in Ontario (Bruce Peninsula) and Newfoundland and Labrador (Torngat Mountains Reserve). As of 2008, there were 42 parks within the system and it is currently 60 per cent complete, given the vast programme of park expansion in the north. This 'developed' system was accompanied by more extensive policy changes and processes. By 1979, a revised parks policy was released, which stressed ecological integrity over use, required that Environmental Impact Assessments be undertaken prior to the establishment of new parks, and ensured that the public would be involved in the early stages of park planning through a master planning process (Parks Canada 1985). The *National Parks Act* was amended in 1988 to introduce ecological integrity as the guiding principle in parks as a prerequisite to use. The pendulum continued to swing away from development and use when the introduction of the State of the Parks reports in the early 1990s concluded that many of the parks in the system were under increasing pressure from a range of internal and external threats (Boyd 2006). It is not surprising then with the release of the new parks policy in 1994 that the focus was shifting away from tourism (1964 policy) to protection of park ecosystems and the introduction of ecosystems-based management.

Despite a shift to ecosystems-based management, a chronology of Canadian national parks since 1998 reflects ongoing concern over use levels and pressures placed on parks by users (Boyd 2006). A report by the panel on the ecological integrity of Canada's national parks (Parks Canada 2000) stated that the ecological integrity of the parks was under threat from a varied number of sources, and this precipitated the passing of the *Canada Parks Act of 2001*, which went beyond ensuring that ecological integrity of the park environments took priority to requiring the maintenance and, where possible, the restoration of ecological integrity through the protection of natural resources and natural processes.

The programme of park establishment in the North saw its genesis after the passing of the Act, when the government in 2002 declared its intentions for 'the most ambitious expansion of national parks in over a hundred years' by aiming to establish 10 new parks by 2007, increasing the size of the national parks system by 50 per cent and seeing a system with over 40 parks and 60 per cent completion (Parks Canada 2002). Since 2002, an integrated strategy has been in place to restore the ecological health of Canada's national parks, namely understanding what is changing in parks, why these changes are taking place and what actions are needed to address changes. While the focus is primarily on scientific understanding of the parks, involving active monitoring to understand how human use affects park ecosystems, an element of importance is directed at building partnerships with local peoples within and near to parks to form a new type of park management that goes beyond the traditional approach of top-down management from Parks Canada, the federal body responsible for the management of the overall parks system. A pilot scheme run in Banff where the local residents are actively engaged alongside the traditional park managers may have wider implications across the system (Timothy and Boyd 2003). In addition, new approaches to management are needed for those parks established in the north as reserves where people of first nations have land claims and issues that need to be addressed. It is to these two areas that the chapter now turns.

Role of Banff in the system

Banff has been an atypical park within the overall system. It has enjoyed an elevated status as the first park to be established, but it has also had attention placed on it as it is one of the few parks where residents live inside the boundary of the park, as well as being the most visited park within the system with almost 3.3 million visitors in 2006–7 (Parks Canada 2007c). This level of visitation has brought new challenges to how the park should be managed.

As early as the mid-1990s, there was recognition that the use levels in Banff township were having a negative effect on the ecological integrity of the

Montane valley floor, and that wildlife and people were not able to easily coexist without having a detrimental impact on this ecosystem. A study was commissioned in 1994 to identify appropriate tourism use within the Banff–Bow valley. The study published its findings in 1996, and called for a heritage tourism strategy to be put in place. The strategy had four broad aims:

1 To make all visitors aware they are in a national park;
2 To encourage and develop opportunities, products, services that were consistent with heritage values;
3 To encourage environmental stewardship initiatives upon which sustainable heritage tourism depends; and
4 To strengthen employee orientation, training and accreditation programming as it relates to sharing heritage understanding with visitors (Banff Heritage Tourism Corporation 2004).

The foreword to the Banff National Park Management Plan (Parks Canada 1997) clearly states that, first and foremost, Banff is a place for nature in which ecological integrity is the cornerstone of the park as well as the key to its future, but that it is also a place for people, for heritage tourism, for community, for open management and for environmental stewardship. The plan was strategic as it offered a clear core vision with an integrated approach to decision-making. Under the orientation programme within the heritage tourism strategy, residents of Banff are informed about the history of the park, what makes it special and how they can share this when interacting with visitors. This form of agreed partnership alongside the traditional managers of the park has been promoted as an alternative management arrangement that encourages local residents to be active in the management of the park. The partnership is driven by the local community, and exists to assist in interpretation and shifting management towards appropriate activities and an ecosystems-based management. Although parks policy overall has shifted towards ensuring and restoring ecological integrity, in the end, it is tourism that will continue to drive the economy of Banff, and also ensure that the system overall is adequately financed.

Co-management

The relations between the various agencies in Canada with responsibilities for national parks and the first nations have varied over the past century and a quarter, but for the most part have avoided extreme negative attitudes (see, also, Chapter 19). In contrast, protests and opposition to the establishment of national parks in New Brunswick and Quebec (Kouchibougac and Forillon in particular) in the 1970s generated considerable opposition, and the management plans for the four mountain national parks, especially Banff, also aroused considerable local opposition (Nelson 1982). In the context of native peoples, the establishment of new parks, particularly in the then Northwest Territories, provoked much less opposition. Parks such as Baffin and Ellesmere Island did not face the same problem of land and resource ownership as those in settled provincial locations, and successful efforts were made to employ local residents in the national parks in the far north where possible. These parks did not represent a loss of land ownership (a concept not practised by the Inuit), and their establishment did not prevent the continuation of local harvesting of wildlife. As there were no local settlements within the parks, there were no relocation problems as experienced in Forillon or Kouchibougac (Nelson 1982).

One fundamental issue does exist, however, and it poses considerable problems for the future. A number of the newly established parks were established as 'National Park Reserves', where there were issues with land claims that had not been resolved at the time of creation of the parks (Parks Canada 1994). The intent was to resolve the land claim issue and then formally create a national park from the existing park reserve. In reality, however, few of the relevant outstanding land claims have been settled, and the park reserve status was never intended to be of long-term duration. Increased visitor use of reserves can be seen by some first nations as violation of their (claimed) territory on an increasing basis, and may harden their resolve and demands when it comes to settling land claims. Some first nation groups may decide not to approve the conversion of park reserves to national parks if such establishment removes the rights they have so far maintained over traditional land uses within the reserves. Thus it would be naïve to assume all park reserves would automatically and easily become national parks in their entirety or even in any smaller form. The assumptions of park officials and politicians in the 1980s and 1990s about the completion of the park system by converting the reserves in due course into parks may have been misplaced and overly optimistic, and the longer the land claims take to settle, the more difficult it may be to achieve such completion, based on the concept of reserve conversion. With hindsight the problem might have been anticipated, although undoubtedly at the time of the reserve establishment, such a process seemed a reasonable and promising method by which to achieve the goals of park expansion and system completion. This element of the national parks story is still to be played out, and the result is far from certain.

Conclusion: a wider system beyond parks

Since the establishment of Banff over 120 ago, a wider system of parks has emerged beyond areas just designated as national parks. Similar to many other national protected area systems, national parks are just one, albeit important, element which Parks Canada has responsibility for. A more complex system that includes National Historic Sites (over 800 sites of which Parks Canada is responsible for 145 sites), National Marine Conservation Areas, and Canadian Heritage Rivers (40), to name a few, has emerged where the management challenges are distinct to each sector and thus call into question the viability of the sustainability of this wider system of parks and protected spaces.

Despite the size of this wider system, it has been Canada's national parks that have played a pivotal role in helping to shape and manage Canada's protected landscapes, as well as offering special places for visitors and Canadians alike to enjoy, appreciate and use from the later decades of the nineteenth century to the present day. What has changed over the years has been the balance between protection and use as played out within its national parks.

9 The Great Barrier Reef Marine Park

Natural wonder and World Heritage area

Leanne White and Brian King

Reef profile and history

Australia's Great Barrier Reef (the Reef) is one of the world's most striking natural features from the perspectives of biodiversity, scale and visual appeal (Hundloe, Neumann and Halliburton 1988). Estimates of its age range from as recent as 50,000–100,000 years (Bowen 1994) to 30 million years (Coleman 1990). Spanish explorers were the first Europeans to encounter the Reef in the early seventeenth century (Lawrence, Kenchington and Woodley 2002). It is the world's most extensive reef system and incorporates three major elements, namely the fringe reef (along the shore), ribbon reef (long thin reef systems found north of Cairns and popular with divers and snorkellers), and platform reef (tall systems off the continental shelf and the most common). It may be divided into three regions – the inner reef, the outer reef and the Island reefs, which are a mix of inner and outer sites in relation to the coast. It covers 348,700 km^2 (134,633 square miles) – about 1.4 times the size of the United Kingdom – with most of the area administered by the Great Barrier Reef Marine Park Authority (GBRMPA). The Reef is actually a system of reefs, islands and cays extending for more than 2,000 kilometres (1,250 miles) along the Queensland continental shelf from the Tropic of Capricorn to the tip of Cape York (Mylne 2005: 1). Built from coral polyps and forming part of the Indo-Pacific reef system, it is the world's largest and most complex living structure (Colfelt 2004, Zell 2006). At its southern end the Reef lies up to 300 kilometres offshore, but it is closer to the coastline along its northern section.

In 1770, Captain James Cook sailed along the Reef until running aground near Cape Tribulation (Gillett 1980). The name 'Great Barrier Reef' is attributed to explorer Matthew Flinders (in 1803), and the term 'Barrier' was first used by Charles Darwin (Bennett 1971: 14). The term, however, is inaccurate, since the wider system is made up of thousands of reefs, and not a single barrier. Early European encounters helped to shape subsequent perceptions and attitudes towards the Reef and coincided with the formulation and subsequent dissemination of Darwin's Theory of Evolution. The latter challenged many canons of Western thought and drew substantially on fieldwork undertaken in the Pacific. The 'exotic' flora, fauna and ethnography of the South Pacific, including the Reef, challenged many previously unquestioned views about creation and emerged as a prominent feature in the Western imagination. Indicative of the resonance of the images transmitted by early explorers, art historian Bernard Smith (1992) has noted that the voyages of Cook and other Europeans from 1768 to 1850 led to the transmission of images of the Pacific, which were indicative of the preoccupations and dichotomies of the period, including human interactions with the land and sea, and Enlightenment attitudes to scientific investigation. Such romantic images laid the foundations for many of the present-day tourism images of the region.

A long history of progressively increasing awareness led ultimately to the achievement of protected status for the Reef. The first Reef photographs were exhibited in London in 1891, and raised awareness and appreciation (Lawrence *et al.* 2002). Accounts of the 'wonders' of the Reef proliferated over the course of the twentieth century, stretching from popular nature writing to romantic adventures (see, for example, Banfield 1913, Barrett 1930, 1943). The Reef is home to a profusion of species of fish, coral and other forms of marine life (GBRMPA 1994: 1). In *Australia: The Rough Guide* the Reef is described as being 'to Australia what rolling savannahs and game parks are to Africa' (Daly *et al.* 1999: 404). Water temperatures range between 23 and 26 degrees Celsius, which makes it ideal for a variety of marine-based recreational activities. The subtle dependence between the ecosystem and water temperatures does however make the formation highly vulnerable to a prospective rise in global sea temperatures (van Tiggelen 2007: 19).

An iconic World Heritage-listed attraction

In 1981, the Reef was the first Australian location to be granted World Heritage status along with Kakadu National Park (in the Northern Territory) and the less well-known Willandra Lakes Region (in New South Wales). The Reef is now one of 17 Australian locations inscribed on the World Heritage List, indicative of its 'outstanding universal value'. It is one of a handful of United Nations Educational and Scientific Cultural Organization (UNESCO) sites meeting all four natural heritage criteria, as defined by the World Heritage Convention. An outstanding example of major stages in the earth's evolution, it has superlative natural phenomena and plays an important part in the history of Aboriginal groups. It is widely acknowledged as the world's most significant marine park (Bowen and Bowen 2002). However, its vast scale and enormous appeal led to vulnerability. In 1991, the United Nations International Maritime Organisation declared the Reef to be the world's first 'particularly sensitive area' (Bowen 1994: 234). This warning was to be very prescient in view of the increasing awareness of the threat posed to low-lying natural formations by global warming.

The worldwide significance of the Reef has been a marketing opportunity for Australia and state and national tourism authorities to highlight the Reef as a 'must-see' destination for overseas visitors. Increasing accessibility accelerated the emergence of the Reef as a mass tourism destination when it received its first high-speed catamaran arrivals. In 1981, the year of its World Heritage listing, this innovative technology transformed the scale and scope of tourism in and around the Reef by enhancing the accessibility of the outer reef. Day-trip vessels bring passengers from a distance of up to 100 kilometres, providing these tourists with experiences that had previously been confined to those willing to undertake a real expedition. Predictably, this increased accessibility has compounded human pressures on the fragile ecosystem. As the volume of arrivals progressively increased, the introduction of a charge (currently $10 per person) was an attempt to strike a balance between access and conservation. Commercial operators are responsible for collecting the fee (or 'reef tax'), which is then passed on to tourists. The 'user pays' approach emphasizes the development of a partnership model between Park management and commercial operators. While this model depends ultimately on responsible visitor behaviour, it relies heavily on the tourism operator for monitoring and enforcement.

Reef-related tourism activity has been long-established. Of the various islands within the confines of the Reef, the most popular destination is Green Island, east of Cairns (Cronin 2000). The first huts intended for tourist use were constructed on the island in 1889, the earliest evidence of tourism

Figure 9.1 The Port Douglas Marina, base for visits to the Great Barrier Reef.

activity on the Reef. Pleasure cruising commenced in the following year (Hopley 1989). During the twentieth century, advocacy on behalf of conservation expanded in parallel with tourism, notably a campaign by underwater photographer Noel Monkman to preserve the Reef from 'souveniring' tourists who collect coral and shells. The world's first glass-bottom boat commenced operations in 1937, though the overall scale of operations remained modest. A daily ferry operation commenced in 1947 marking the start of 'regular reef-based tourist services' to Green Island, although the fee of 10 pounds per person was excessive for many potential tourists (Hundloe *et al.* 1988: 29). In 1954, an underwater observatory was opened making coral viewing widely accessible. Fitzroy Island is a lower-key though popular day-trip destination from Cairns, offering walking tracks and a more secluded environment. It combines impressive fringing reef with rainforest and woodlands, allowing visitors a wide range of land or water-based activities, including camping.

The Reef is affected by tourism activity along the adjacent mainland. Cairns is a major gateway city and attracts many tour groups and backpackers, while Port Douglas – located to the north – attracts a larger proportion of higher spending tourists. The Port Douglas area claims to be the world's only region where marine and terrestrial World Heritage sites meet – the Reef and the Daintree Rainforest (Mylne 2005: 93). The Wet Tropics region represents an important coalescence of the terrestrial national park and the marine park concepts (Zell 2006). Day trips to the outer reef are readily accessible. Indicative of its success as a tourism destination, Port Douglas claims Australia's 'highest number of restaurants per capita' (Zell 2006: 109). The Port Douglas example is a reminder that much of the Reef-dependent development is mainland-based.

The Low Isles, located off Port Douglas, are a popular day trip and are where the 'Crocodile Hunter' and sometime Australian tourism 'ambassador' Steve Irwin died after being speared by a stingray barb in 2006. It could be argued that the associated media profile surrounding Irwin's death as well as films set in the Reef (e.g. *Finding Nemo*) continue to reinforce the appeal of the Reef as a tourist destination.

The image and profile of the Reef was boosted during the Opening of the Sydney 2000 Olympics. The main Olympic arena was designed to conjure up Australia's beaches with a flock of seagulls commencing the soundtrack. During the 'Deep Sea Dreaming' of the opening ceremony, the central character lies on a huge beach towel and dreams about Australia's colourful marine world. In front of a massive global audience, Stadium Australia was transformed into a giant fishbowl featuring sea creatures including oversized jellyfish, seahorses, angelfish, sea cucumbers and moray eels. The event served to highlight the significance of the Reef as a part of Australia's international appeal.

Figure 9.2 The Low Isles, Great Barrier Reef.

The features and challenges of marine parks

Although Yellowstone was designated the world's first national park in 1872, the first marine parks were not declared until the 1960s – almost 100 years later. The much shorter marine park history has meant that the issues of managing conservation and tourist access are of quite recent origin (Eagles and McCool 2002). The belated introduction of a conservation regime in settings such as the Great Barrier Reef has waged a constant struggle to keep pace with the growth of commercial activity, including tourism. In recognition of the emerging conflict, the Australian Conservation Foundation (ACF) staged a symposium on the future of the Reef in 1969. The resulting report concluded that:

> The Reef has areas which are isolated, lonely and peaceful and others where there is happy communal recreation. Yet it is fragile and it can be damaged ... It will not be many years before we have lost certain areas of the Reef irrevocably, through over-population and consequent overuse, unless a rigorous control and overall plan of management is made.
> (ACF 1969: 69)

In the years prior to the ultimate establishment of the Marine Park, the threats were real. In 1967, the Queensland government permitted the leasing of over

20 million hectares of the Reef for oil exploration (Hutton and Connors 1999). In 1968, drilling leases were granted to two oil companies (Bowen 1994). Though regulations were subsequently introduced (in 1999) to prohibit operations for the recovery of minerals in any part of the Reef region not formally incorporated within the Park, it is indicative of the mounting commercial pressures on the Reef prior to the subsequent proliferation of tourism activity. Prior to the passage of the World Heritage-related legislation in 1983, the Queensland and commonwealth authorities were involved in a series of disagreements about management of the Park. The conflict over the exercise of authority arose because commonwealth jurisdiction was confined to the high water mark. This limited authority provided the state government with a justification for exercising its own authority. This occurred most frequently though not exclusively when there was a conservative Queensland Government and Labour Commonwealth Government in power. The passage of the 1983 legislation was important for introducing a more stable regulatory environment for the Park.

Though some individual islands had previously been declared as national parks, the Marine Park was finally proclaimed in 1975 with the passage of the Great Barrier Reef Marine Park Act to 'protect, preserve and manage' the Reef. The Park incorporates everything 'below the low tide limit, from the tip of Cape York right down to Bundaberg' (Colfelt 2004: 109). This was an important step forward, since state governments had previously laid claim to determining the appropriate uses of the resource. However, the date of establishment only marked the start of effective management of the various park zones and the mechanism to give these effects took a further 15 years. It was ironic that the Reef received park status so late, given that national parks in Queensland's rainforest areas had been established so early (Frost 2004). Hundloe and coauthors regard the formation of the GBRMPA as a watershed for the management of the Reef since the formation had previously been treated as open access common property with management regimes poorly defined and few restrictions imposed (1988: 9). Hundloe *et al.* argue that the Queensland Government (and the relevant government agencies responsible for reef administration prior to the establishment of the GBRMPA) 'have never really accepted the new management regime' (1988: 10). Though the proclamation was belated and could not ignore a range of long-established uses, which could potentially undermine the conservation values of the Park, it did have the advantage of being comprehensive rather than fragmented. The final agreement about the management arrangements for the marine park was signed between the Commonwealth and State (Queensland) governments in 1988 (Hutton and Connors 1999).

The Great Barrier Reef Marine Park Authority

The GBRMPA is responsible for managing the reef system and seeks to achieve balance between the various pressures, including commercial and recreational fishing and tourism activities such as snorkelling, scuba diving, scenic cruises and flights, sailing, and water skiing. Bowen (1994: 252) has argued that the Reef is 'the most managed environmental region in Australia' and that the GBRMPA's charter: 'To provide for the protection, wise use, understanding and enjoyment of the GBR in perpetuity through the development and care of the GBRMP', must be 'one of the most difficult on Earth'. One might argue that the accommodation of such diverse interests is inconsistent with the conservation principles and philosophies of marine and national parks. A typical defence of this approach is that commercial activities had been established a long time prior to the formation of the GBRMPA and that the pursuit of balance is an inevitable consequence of the Park being established at such a late date.

The various zones range from unrestricted access through to areas excluded from public access. In keeping with overall conservation principles, the zoning allows for all 'reasonable uses' of the Reef. The zones types are: general use; marine national park; habitat protection; estuarine conservation; scientific research; conservation park; and preservation (GBRMPA 1999). In the early 1990s, Hall pointed out that while the *Great Barrier Reef Marine Park Act 1975* allowed for the Commonwealth government to 'zone' an area as 'wilderness' if it so desired, 'no wilderness zones have been declared under the act' (Hall 1992: 42). For the Reef, the closest category that exists to a wilderness zone is the GBRMPA's 'preservation' zone where entry is prohibited – except in an emergency or where scientific research cannot be conducted elsewhere. In this sense, marine parks are unlike many 'national parks', which have a strong emphasis on wilderness. In 2004, the *Great Barrier Reef Marine Park Zoning Plan 2003* consolidated the zoning of the Park and significantly increased the area and level of protection (DEH 2006: 9).

The various zones within the Reef align to categories determined by the International Union for the Conservation of Nature (IUCN). As discussed, the zones include preservation and scientific research zones (IUCN category I), Marine National Park zones (category II) and General Use zones (categories IV and VI). The use of zoning to regulate access has been reflective of the reality that human activity predated the establishment of the GBRMPA. Resort development occurred with the Whitsunday Islands from an early date, within the confines of what was to become the Park. The progressively easier accessibility of these resorts to the Reef has added to the pressures of human impact. Some of the larger cruise operations responsible for transporting visitors to the Reef have established pontoon-type constructions adjacent to the coral. These structures have attempted to strike a balance between providing visitors with easy access to the formations and minimizing physical interactions with the fragile system of coral reefs. The GBRMPA exercises strict controls over the activities occurring in and around the pontoons. The application of a licensing system for operators is a key element of this control. The licensing system applies to tourism operators, watercraft, aircraft and structures such as pontoons (ANAO 1998).

The use of zoning has its critics. Unlike the prevailing practice in various Australian terrestrial national parks, the Reef allows for voluntary regulation rather than legally enforceable rules, and permits commercial activities such as fishing and tourism. Craik (1987: 155) has argued that the emphasis on voluntary codes 'furthered the trend towards large scale development of tourism through resorts, package deals and the like' on the basis that monitoring is easier when activity is occurring as part of an organized operation. In light of the breadth of its charter, the Authority places considerable emphasis on public education (Bowen 1994). The 'iconic' status of the Reef is critical because it provides an opportunity for the development of an emotional relationship between the Reef and its various stakeholders, including tourists. Reflective of the complexities associated with the longer standing and diverse uses, it took almost 10 years to introduce a comprehensive system of zoning for the Reef (Colfelt 2004: 33).

Engaging with tourism

National and marine parks are a 'vital element of Australia's international tourism promotion' (Hall 2000: 29). When commercial tourism images of Australia come to mind, the Reef tops the list of iconic attractions. Many international tourists undertake what is often referred to as 'the golden triangle' tour – Sydney, the Reef, Uluru and finally back to Sydney. The dependence of Australia's tourism image-making on 'pristine' natural attractions such as the Reef creates a strong interdependence between tourism and conservation. Tourist perceptions will be highly susceptible to any variation in the quality of the marine environment. While tourism has been central to the management of the Reef since the establishment of the Marine Park, the range of formal tourism mechanisms has been progressively expanded. In recognition of the need for the park management authorities to engage with tourism industry representatives as a key stakeholder group, the GBRMPA established a Tourism Advisory Group (TAG) in 1997 to provide advice on issues such as planning and the issuing of licences and permits. The TAG operates at a Park-wide strategic level and complements other consultive processes connecting the GBRMPA with local groups (Tourism Review Steering Committee 1997).

The Reef is a critical element of Queensland's appeal to domestic and international tourists (King 1997). Tourism is far and away the region's largest industry and is worth $5.8 billion annually and accounts for around 63,000 equivalent full-time jobs (Minchin 2007: 1). Tourism to the Reef increased by a multiple of 40 over the period 1946 to 1980 (Hopley 1989). Over the period 1995 to 2000, average annual visitation to the Park was approximately 1.2 million with a peak of 1.6 million in 1996. Tourism access is permitted in almost all areas (or 99.8 per cent) of the Reef (Lawrence *et al.* 2002). This is a notable characteristic, given the strong wilderness type ethos of many national parks. Most Reef-based tourism activity is concentrated in and around Cairns, and off Airlie Beach – north of Mackay (where many tourists embark for the Whitsunday Islands).

The various tourism resorts located within the park boundary are classed as Marine National Park 'A' Category. Each zone is subject to a management plan and 'A' Category aims to achieve 'protection of the resources of the park while allowing recreational activities and approved research' (Green and Lal 1991: 23). The 74 Whitsunday Islands, located within the Central Zone of the Park, highlight the diversity of resort provision within the Park. The group consists of some islands that are occupied almost exclusively by a single resort development (e.g. Daydream Island), larger scale islands housing a single resort (e.g. South Molle and Lindeman Islands), larger islands featuring multiple resorts (e.g. Long Island) and many with no resort development (the largest of the group, Whitsunday Island has no resort development). The style of tourism ranges from luxury (e.g. Hayman Island) to 'middle' market and 'budget' though the emphasis is on the middle to luxury segments. Progressively, the pressure of developments and shortage of opportunities has pushed the more budget-oriented development towards the mainland (e.g. Airlie Beach and the Cairns area). Lower income visitors tend to reach the Reef by day trip from the mainland, avoiding the higher accommodation prices prevalent in the island resorts, which are a result of the high costs of operation (Craik 1991).

The relatively high cost of access to the Reef has been a barrier to, and a protection from, mass tourism. The Reef is fairly remote from Australia's major population centres and though it is possible to access major stepping-off points by car, train, or coach, the long distances make this relatively inconvenient. In the case of air access the duopoly within Australia's air transport sector, which prevailed for much of the postwar years, kept airfares high. When the first jet airport within the GBRMPA area was established in 1983 on Hamilton Island in the Whitsundays, it was a monopoly for one of the carriers (Ansett) and the era of high airfares continued into the 1990s.

A pattern of cheaper access began during the 1990s, with improved international air access and the introduction of low-cost carriers on domestic routes. The first of these phenomena brought Japanese tourists into the Whitsundays and a range of European, Asian and American tourists into far north Queensland through Cairns. The latter trend has made airports such as Mackay, Hamilton Island, Rockhampton, Townsville and Cairns readily accessible through low-cost carriers such as Jetstar and Virgin Blue. A high proportion of the visitors arriving in these ports proceed to spend some of their time in and around the Reef. This is progressively transforming the phenomenon of Reef-based tourism from a minority pursuit to one that is available to millions, including a substantial proportion from overseas. In due course the conduct of tourist activities such as snorkelling on the Reef itself may become unsustainable as visitor numbers continue to rise and pressures grow.

Many commentators have viewed tourism as the principal threat to the sustainability of the Reef. The threats include activities in the Cairns/Port

Douglas region and in and around the Whitsunday Islands. Humpback whales, dugong, and nesting seabirds and turtles are some of the more vulnerable species said to be directly affected by increased tourism. Other researchers have, however, shown the value of tourist wildlife encounters as a means of raising awareness of endangered species (Valentine et al. 2004). Damage to some of the popular reef locations has also been noted in research undertaken by the Tourism Review Steering Committee (TRSC). Over the past decade or so, unsustainable levels of damage occurred because tourists accidentally walk (or hit their fins) on the coral. This has been noted at sites such as the heavily visited Michaelmas Cay off the coast of Cairns (TRSC 1997: 6).

The tourism appeal of the Reef has also prompted developers to exploit opportunities on the mainland. The Great Barrier Reef Wonderland in Townsville is described as 'a blend of illusion and reality that carries you to the outer reef or outer space within moments' (Daly et al. 1999: 433). This attraction allows visitors to appreciate the reef without the need to travel to the formation itself. If the deterioration of the Reef accelerates, such recreations may play an increasing role in achieving a balance between visitor access and conservation.

Not all tourism operations along the Reef have met with success. The world's first custom-built floating hotel, launched and moored at John Brewer Reef, 70 kilometres from Townsville in 1988, was short-lived. The nearly man-made 'Fantasy Island', a million-dollar circular platform incorporating a bar and theatrette, became submerged in a storm and collapsed in 1988 after two weeks of being anchored into position. The Bright Point condominium development and marina on Magnetic Island was a third tourism failure. A commonwealth government inquiry was held after the developers went into liquidation in 1990. Unfortunately, the inquiry 'raised more questions than it really answered' (Bowen 1994: 254). While conservationists are keen to ensure that inappropriate development does not take place in the region, developers are just as keen to see that it does. The GBRMPA will need to keep a watchful eye on balancing the needs and wants of eager entrepreneurs and tourism operators with those of future generations of reef visitors.

Indigenous issues

Natural heritage has dominated the ethos of the Reef, rather than human heritage. The Aboriginal occupation of the Reef area has been small-scale, despite the long history of interactions. Interactions between Aborigines and Torres Strait Islanders and the Reef do however play an important role in its significance. Fitzgerald has estimated that Queensland's Aboriginal population dropped from 100,000 in 1840 to 15,000 in 1900 (1982). Significant numbers of Aboriginal people (i.e. more than 1,000) are located in the major cities of Townsville, Cairns and Rockhampton, while concentrations of Torres Strait Islanders can be found in Cairns and Townsville (Benzaken, Smith and Williams 1997). It has been estimated that approximately 12,000 Aboriginal and Torres Strait Islanders live in the major cities and towns adjacent to the Reef, while another 11,000 live in the communities around the Cape York Peninsula (ANAO 1998: 74). There are more than 70 traditional owner groups along the coast from Bundaberg to the east Torres Strait islands (DEH 2006: 8). From the perspective of Reef tourism, the indigenous population can play a part in management, tour guiding, entertainment and interpretation. Many indigenous Australians, however, view the concept of wilderness as 'a negation of prior occupation and property rights' and a form of dispossession – a notion that some argue should be reconsidered (Benzaken *et al.* 1997: 476).

The major 'mass-market' interaction between tourism and Aboriginal culture in the Reef area occurs at Tjapukai Aboriginal Centre, located at the base of the Kuranda Skyrail, near Cairns. Here, tourists are exposed to boomerang throwing, didgeridoo playing, dancing and tales of the Aboriginal Dreamtime. Aboriginal and Torres Strait Islanders differ from Europeans in their relationship with the land and sea. The relationship involves a sense of kinship with the environment as opposed to land ownership. Smyth has argued that kinship forms the basis of Aboriginal maritime culture and involves 'an inherited relationship to places on land and sea, other people, animals, plants, sacred and cultural sites and Dreaming tracks' (Smyth 1997: 496). Hunting (especially for turtle and dugong) and fishing play an important part in the traditional culture and lifestyle of many indigenous Australians (Benzaken *et al.* 1997). Torres Strait Islanders have also had a long involvement in pearl fishing, which has brought them into close contact with the Reef.

In 1992, the High Court of Australia handed down the historic Mabo decision, which recognized that the traditional owners of Murray Island in the Torres Strait exercised Native Title. Hollinsworth has noted that Mabo had 'enormous political and symbolic impact' (1998: 193). According to Kelly, the Mabo decision was not just about land but 'about confronting the moral vacuum at the heart of Australian nationhood' (2001: 185). The passage of the *Native Title Act* in 1993 through the Federal Parliament marked the end of the notion of 'terra nullius' – that no one occupied the land prior to European settlement.

The recognition of indigenous issues progressively gained momentum with prospective impacts for the management of the Reef. The International Year of the World's Indigenous People (1993) was followed by the High Court's Wik decision (1996). This decision ruled that native title outweighed the granting of pastoral leases. The Mabo and Wik rulings prompted native title claims along coastal Queensland. This wider context has reflected that the indigenous population and the tourists are increasingly shaping the policy and management agenda around the Reef. The issue of indigenous disadvantage, which characterizes communities across Northern Australia, is evident around the Reef. Having been prohibited for many years, Torres Strait Islanders returned to live on Thursday Island after the Second World War (Finlay, Armstrong and Wheeler 1998: 241). Just 3 square kilometres in size, the island has had a history of social problems and welfare dependency (Daly *et al.* 1999). The realities of life in the island are in sharp contrast to the upmarket resort islands along the Reef. The persistence of entrenched disadvantage in such adjoining communities is relevant to the future of the Marine Park as management considers the relationship with adjoining communities.

Pressure has grown to increase indigenous participation in management of the Reef. A report prepared by Bergin (1993) for the GBRMPA proposed an increase in Aboriginal participation in the activities of the Authority. The report observed that Aboriginal involvement was modest, relative to Kakadu and Uluru National Parks where the indigenous population is involved in management, operational and interpretative roles. It noted the significant Aboriginal populations in areas adjoining the Reef at Thursday Island, Park Island and Yarrabah and the prospects for building stronger partnerships with these communities. The GBRMPA's 25-year objective for Aboriginal and Torres Strait Islanders is to allow them to:

> pursue their own lifestyle and culture, and exercise responsibility for issues, areas of land and sea, and resources relevant to their heritage within the bounds of ecologically sustainable use and consistent with our obligations under the World Heritage Convention and other Commonwealth and State laws.
>
> (GBRMPA 1994: 35)

Reflective of growing indigenous awareness, the Sea Forum group was formed in 1997 to represent the interests of the traditional owners in GBRMPA decisions. Sea Forum has contributed to indigenous land-use policies that impact upon the Reef (Lawrence *et al.* 2002). One Government-nominated member of the Authority Board of Directors has explicit responsibility to represent the interests of indigenous communities in and around the Reef.

Despite the strong history of indigenous engagement with the Reef, involvement in the major tourist development has been minimal, with a few notable exceptions. Historically, Torres Strait Islander musicians and other performers played an important role in the in-resort entertainment in the Whitsunday region (Hayward 2004). These performances have subsequently been discontinued, largely reflective of changing visitor preferences, but are a reminder of an important indigenous connection during the earlier days of tourism along the Reef. Awareness of the importance of authenticity in indigenous entertainment has increased and the material presented by these early performers fell out of favour. However, as indigenous communities seek to negotiate sustainable partnerships within the tourism sector, the prospects of a revival of such activity cannot be entirely ruled out. Indigenous populations in Pacific island countries remain actively involved in resort-based performances and the absence of such indigenous activity within Australian resorts may be viewed as an anomaly.

The future

Because of its vast dimensions, the Reef is vulnerable to shipping accidents. As the size of the largest ocean-going vessels has continued to grow, the potential for large-scale disaster has increased, albeit in a context where contemporary vessels are generally more stable and less accident-prone than their predecessors. While some argue that the presence of overseas vessels has also been responsible for the Crown-of-Thorns Starfish infestation, others believe that the outbreaks may be a natural part of the Reef's lifecycle (Finlay *et al.* 1998: 16, Bowen and Bowen 2002: 378). Regardless of the origins of the pest, the starfish has had a devastating impact on affected parts of the Reef. It progressively eliminates the other elements of the biosystem, leaving the Reef colourless and dead. Commercial fishing also negatively impacts on the ecology of the Reef. During the 1970s and 1980s, large numbers of giant clams were killed by foreign fishing crews, due to its high commercial value from sale in up-market restaurants. The clam plays a vital role in protecting the ecology of the reef because of its filtering activity. However, despite such practices, a report commissioned in 1990 found that despite the prevalence of the contrary view, the Reef has generally not been seriously 'overfished' (Craik 1990).

Though the Reef retains its status as the richest marine habitat on earth, coral bleaching has become an increasing challenge in the face of global awareness about the risks of rising sea temperatures. The more alarmist projections anticipate the virtual elimination of the Reef as a living organism over the forthcoming 15 years. Activities undertaken outside the Park boundaries constitute another threat to the sustainability of the Reef. The Reef's sensitivity and vast scale compound the effect and make it harder to control either cause or effect. The extensive use of chemicals on the 42 million hectares adjacent to the mainland is a major concern, notably in the case of nitrogen, phosphorus and potassium, which are applied for sugar, fruit and vegetable growing. Historical data reveals that in 1990 alone, 83,000 tonnes of nitrogen was added to the land adjacent to the Reef – an amount equivalent to what had been applied cumulatively from the beginnings of farming until 1945 (Pulsford 1993). Since 1990, the situation has continued to worsen as evidenced by the findings of a 1997 report on chemical pollutants caused by agriculture (Bowen and Bowen 2002: 407).

The Marine Park Authority has a strong reputation for research and monitoring activities. The Department of the Environment and Heritage undertook a review of the 1975 Act and noted a number of areas for improvement (DEH 2006). First, it noted that the emphasis on compliance to the 1975 Act needs to be strengthened to bring it more closely into line with the *Environment Protection and Biodiversity Conservation Act 1999*. Secondly, stronger coordination is required between the Commonwealth and State governments. While the Reef has a reputation as among the world's best managed marine parks, vulnerability to global forces limits the capacity of the GBRMPA and the Australian government to ensure a sustainable future.

It is ironic that this interdependency is similar to the challenge facing many low-lying sovereign states in the Indian and Pacific Oceans. In such cases, governments are looking to the international community, including Australia, to help them confront the challenges. The extent of such interdependency is a reminder that effective management within Australia may be insufficient to ensure the future of the Reef.

10 'Welcome to Aboriginal Land'
The Uluru-Kata Tjuta National Park

Tamara Young

> This is Aboriginal land and you are welcome. Look around and learn, in order to understand Aboriginal people and also understand that Aboriginal culture is strong and alive.
> (Nellie Patterson, traditional owner, in Parks Australia 2005: 4)

> National parks in Australia, despite cherished Euro-Australian beliefs which conceive of them as pristine wilderness or natural landscapes which must be preserved for all human activity, are in fact cultural landscapes brought about by thousands of years of Aboriginal management.
> (Birckhead and Smith 1991: 4)

Uluru-Kata Tjuta National Park is one of Australia's most high-profile tourist destinations. Each year, approximately 400,000 tourists visit the national park; around half are international visitors. Situated in the western desert of Central Australia in the Northern Territory, the arid landscape of the 1,325 square kilometres national park is visually dominated by the single red monolith of Uluru and the nearby mountainous domes of Kata Tjuta. From a Western perspective, Uluru is well known as Australia's most famous natural attraction; 348 metres in height and 3.6 kilometres in length, it is popularly regarded as the world's largest rock. From an indigenous perspective, Uluru-Kata Tjuta and the surrounding areas are of enormous cultural and spiritual significance to Anangu, the Aboriginal traditional owners of the land. The Western and indigenous perspectives are internationally recognized in the listing of the Uluru-Kata Tjuta National Park as a World Heritage Area for both its natural value and cultural significance.

This chapter examines the history of Uluru-Kata Tjuta National Park, focusing on the contestations involved in the construction of this destination as an attraction for tourists. This chapter argues that this symbolic tourist site is a contested space where multiple layers of meanings are embedded, particularly in relation to the often conflicting discourses of Western tourism and indigenous culture. While Uluru-Kata Tjuta is owned by Anangu, these areas are leased back to the Australian Government as a national park and, thus, visitor access to this highly significant and iconic site is ensured. The intention of the Aboriginal traditional landowners to facilitate a greater awareness of indigenous culture (as exemplified in the lead quote above) is accounted for by the joint management arrangement of the site. Accordingly, the value of the Uluru-Kata Tjuta National Park as having living cultural heritage is central to the contemporary management agenda. However, this particular interpretation of value can sit uneasily alongside its quite considerable economic and environmental values. The outcome is that conflicts of interest arise in relation to meaning, access and usage of the site. The chapter highlights that the contradictory discourses of tourism and Aboriginal culture, and the tensions between them, are exemplified in the popular tourist activity of climbing Uluru. Further, the contested usage of the Uluru-Kata Tjuta National Park as a tourist destination reflects the tensions that are implicit in the joint management structure.

Narratives of the Uluru-Kata Tjuta National Park

Until the mid-twentieth century, the area was relatively unknown except to the local Aboriginal people. To the Central Desert Aboriginal people, the Pitjantjatjara and Yankunytjatjara people (collectively known as Anangu) Uluru-Kata Tjuta and the surrounding lands are culturally and spiritually significant. For Anangu, who have lived in the area for thousands of generations,

Figure 10.1 Uluru, the iconic image.

Uluru is at the crossroads of several ancestral groups of dreaming tracks. During the creation time, known by Anangu as the Tjukurpa, the landscape was created by powerful ancestral beings, or Tjukuritija. As Burnam (1988: 260) explains, Uluru is a 'living record of many hundreds of Dreamtime events, permanently retained within its huge body. Every crack, mark, stain and indentation has an explanation in the Dreaming'. The following explanation of the Tjukurpa as the basis of Anangu knowledge, extracted from the *Uluru Kata–Tjuta National Park Visitor Guide* (Parks Australia 2005), provides a glimpse into the history and complexity of indigenous perspectives relating to landscapes such as Uluru and Kata Tjuta:

> Tjukurpa is the foundation of our culture. Just as a house needs to stand on strong foundations, so our way of life stands on Tjukurpa. It is our traditional law guiding us today ... Tjukurpa is our religious heritage, explaining our existence and guiding our daily life. Like religions anywhere in the world, the Tjukurpa provides answers to important questions, the rules for behaviour and for living together. It is the law for caring for one another and for the land that supports our existence. Tjukurpa tells of the relationships between the people, plants, animals and the physical features of the land. Knowledge of how these relationships came to be, what they mean and how they must be carried out is explained in the Tjukurpa. Tjukurpa refers to the past, the present and the future at the same time. It refers to the time when Tjukuritja (ancestral beings), created the world as we know it. Tjukurpa also refers to Anangu religion, law, relationships and moral systems. Anangu life today revolves around Tjukurpa.
>
> (Parks Australia 2005: 20)

Considered within non-Aboriginal discourse, Anangu are the traditional owners of Uluru-Kata Tjuta. However, Anangu do not actually consider their relationship with these landscapes as one of Western-style ownership. Rather, they see themselves as being part of the landscape and it a part of themselves (Hill 1994). This view is significantly different to that of non-indigenous people who tend to appropriate natural landscapes as terrain to be conquered as well as the location of potential resources. Natural landscapes are also frequently appropriated as icons of national identity.

As the quintessential image of the outback, Uluru is incorporated into the national identity of settler Australia. Uluru is seen as sacred to settler Australians (Hamilton 1984, Marcus 1997). Described as the 'symbolic centre of the Australian nation' (McGrath 1991: 115), the power of this symbolism is illustrated by the metaphoric terms used to describe Uluru, such as the 'Red Centre' or 'Heart' of Australia (Fiske, Hodge and Turner 1987: 123). McGrath (1991: 122), in writing of the outback myth, argues that it provides Australians with a 'distant past' that is related to the uniqueness of the desert landscape and its authenticity as an ancient Aboriginal land. It is also this uniqueness that contributes to the significance of Uluru as an international tourist attraction.

It has been argued that the appropriation of Uluru into national consciousness as a spiritual centre for settler Australians results in a transformation of meaning for Aboriginal people who are left with no unique cultural focus (Haynes 1998), including the loss of 'identification of themselves with their country' (Marcus 1997: 29). Central to the construction of Uluru as a national icon is its identification as a natural, rather than a cultural, landscape. For instance, James (2007: 402) states that the discourse on Uluru as a central site in settler mythology 'focuses on Uluru as an *a priori* natural site, constructing it as a national icon rather than as an Aboriginal cultural landscape. This construction seeks to legitimise control over Uluru by "national" interests such as tourism'. Indeed, the history of tourism development at Uluru has, until recently, demonstrated little regard for the significance of this area as an Aboriginal landscape of living cultural heritage.

The colonization of Uluru-Kata Tjuta by tourism: the national park as a contested space

The European discovery and appropriation of Uluru and Kata Tjuta took place in the late nineteenth century, at almost the same time as the establishment of Yellowstone National Park (see Chapter 1). In 1872, the explorer Ernest Giles was reportedly the first European to see Kata Tjuta, which he subsequently named Mt Olga. The following year, William Gosse was the first European to sight Uluru, which he named Ayers Rock after Sir Henry Ayers, the chief secretary (or premier) of South Australia. Gosse's report, published in the colonial parliamentary papers, included the first visual image of Uluru (reproduced in Horne 2005: 10). However, due to the isolation and aridity of the area, no European development resulted at the time.

In 1920, a parcel of land was set aside as the Petermann Reserve, also known as the Great Central Australian Aboriginal Reserve. Due to their growing popularity with tourists, the landmarks of Ayers Rock and Mt Olga were taken out of the reserve in 1958 and were set aside within a Tourist and Wildlife Reserve. This area was subsequently declared the Ayers Rock–Mount Olga National Park on 24 May 1977 under the Commonwealth *National Parks and Wildlife Conservation Act 1975*. Despite having lived in the area for at least 22,000 years (Brown 1999), the local Aboriginal people were not consulted about the establishment of the national park, nor were they allowed to live in the area with the Northern Territory government evicting them from areas in the vicinity of Uluru (Berzins 1998, Digance 2003).

On 26 October 1985, following an historic and controversial Native Land Claim, Ayers Rock was renamed Uluru and the title deeds to the land were 'handed back' to the Aboriginal people of Central Australia (the Uluru-Kata Tjuta Aboriginal Land Trust) (Breedon 1994, Hill 1994). The national park was then leased back to (what is currently known as) Parks Australia for a period of 99 years. Today, Anangu and Parks Australia jointly manage the national park. In 1993, the name of the park was officially changed from Uluru (Ayers Rock–Mt Olga) National Park to the Uluru-Kata Tjuta National Park.

In relation to the transferring of title, Galarrwuy Yunupingu (1997: 11), Chairman of the Northern Land Council, explains that:

> Under the framework of the Land Rights Act, Uluru . . . [was] handed back to Aboriginal people who then leased these areas back to the Australian people through its Federal Government for all the world to enjoy, forever. It is important to remember these places are big tourist drawcards, bringing dollars into the Australian economy.

Uluru has long been marketed by the tourism industry and travel guidebooks as part of the unique Australian experience. As the quintessential symbol of the Australian outback, Uluru is one of Australia's best-known icons and, as an internationally renowned tourism attraction, is constructed as worthy of a once-in-a lifetime-visit. Images of Uluru pervade Australian travel and tourism media as well as other popular cultural forms such as film, literature, music and poetry. Haynes (1998: 261) notes that 'The Red Centre has been embraced so warmly by popular culture that it has become the most exported Australian landscape . . . Uluru has not only achieved international iconic status but become a site of modern pilgrimage'.

The first recorded tourist visit to Uluru was in 1935 (Davidson and Spearitt 2000). The 1940s saw the area become accessible as a tourist destination, which was made possible by the construction of a track through to Uluru in 1948. However, popular tourism to the area did not really develop until the 1950s when travel to Central Australia became easier and more appealing. Prior to this time, the wildness and harshness of the desert, and its isolation from civilization, served as a deterrent for tourists (Berzins 1998, Haynes 1998). The beginning of organized tourism was marked by the appointment in 1957 of Bill Harney as the first ranger for the then reserve (Harney 1963, Hill 1994). At around the same time, an airstrip was developed there, and in 1959 the first motel leases were granted. The establishment and improvement of accommodation and transport infrastructure resulted in the area quickly becoming the 'best-known tourist destination' in the Northern Territory (Berzins 1998: 74). Visitation to Uluru continued to steadily increase throughout the 1960s, with a rapid growth in visitor numbers since the 1970s (Davidson and Spearritt 2000). Haynes (1998: 262) suggests a reason for the recasting of perceptions regarding travel to the desert was the 'technological breakthrough' of four-wheel drive vehicles that became readily available in Australia in the 1980s and allowed travellers to 'negotiate the terrors and monotony of the desert', either independently or on organized tours.

Early tourism developments at Uluru, however, were poorly managed and resulted in numerous negative impacts. For instance, the crude construction of motels and camping grounds at the base of Uluṟu resulted in the desecration of the natural environment. Other negative impacts that caused particular concern included tourists trespassing onto Aboriginal sacred sites and the vandalism and defiling of Aboriginal cave paintings. In response, a government task force recommended the development of a new town located 20 kilometres away from Uluṟu. In 1984, the town of Yulara was opened, which offers a range of accommodation styles including five-star resorts, backpacker dormitories and camping grounds. Yulara, owned by the Northern Territory Government and private sector interests and branded as Ayers Rock Resort, is now the closest site where tourists visiting the national park can stay.

A major issue in the development of tourism at the Uluṟu-Kata Tjuṯa National Park is the Uluṟu climb, which has long been promoted by the tourism industry as the reason to visit the national park (Wells 1996, McKercher and du Cros 1998, Davidson and Spearritt 2000). Conquering the steep monolith was made possible by the hammering of a metal fence into the face of Uluṟu, an act of 'colonizing mentality' that was authorized by Bill Harney during his time as ranger between 1957 and 1962 (Waitt, Figueroa and McGee 2007: 254). The visitor compulsion to climb Uluṟu is considered by Aṉangu as a sign of disrespect to their traditional culture and heritage. For Aṉangu, climbing is only sanctioned on special occasions by initiated men (Breeden 1994). However, generally disregarding this indigenous value, research in the 1990s suggested that most tourists went to Uluṟu for the specific purpose of climbing (McKercher and du Cros 1998, Brown 1999). According to McKercher and du Cros (1998), the average length of stay at the nearby town of Yulara is less than two nights, with many people visiting the national park for less than a day on a package tour. As a result, the visitor experience at Uluṟu is usually reduced to activities that can be undertaken in a short period of time. Consequently, the viewing of sunrise and sunset over Uluṟu at strategically situated viewing areas and climbing Uluṟu are the most popular activities pursued by visitors to the area.

What is apparent from this discussion of tourism developments at the Uluṟu-Kata Tjuṯa National Park is that tensions exist between multiple layers of meaning and value. This is evident in Western interpretations of value that are grounded in the unfettered right to experience the natural landscape, and indigenous interpretations of significance based in a spiritual and cultural connection to land. These tensions, manifested in the above examples, demonstrate a lineage of management decisions, which typically prioritize touristic imperatives above indigenous values. The Uluṟu-Kata Tjuṯa National Park can thus be described as a contested site – it is a 'highly contested space where localized, deep-rooted aspirations of the indigenous population come into sharp play with the demands of global postmodern tourists' (Tribe 2008: 248). Conflicts of interest between the Aboriginal owners and commercial tourism arise at the Uluṟu-Kata Tjuṯa National Park, particularly in relation to access and usage of the site. Digance (2003: 144) defines a 'contested site' as a:

134 *Tourism and National Parks*

Sacred location where there is contest over access and usage by any number of groups or individuals who have an interest in being able to freely enter and move around the site. There may also be elements of conflict between those who own and those who manage it on their behalf, or perhaps those who depend on it for their livelihood.

The primary issue of conflict is the tourist activity of climbing Uluru. The contradictory discourses of tourism and Aboriginal culture, and the tensions between them, are exemplified in this popular activity that flagrantly contravenes the cultural and spiritual beliefs of Anangu. The Aboriginal traditional landowners and managers of the national park ask that visitors not climb Uluru, claiming it is 'painful' to see tourists climbing on the Rock (Whittaker 1994: 317). More recent strategies have been to emphasize the danger. As of the time of writing, 35 tourists have died from falls and heart attacks while climbing Uluru (for a comparison with tourist deaths in Yellowstone, see Whittlesey 1995). Other strategies are to close access on days of high temperature or winds and promote other activities such as walking around the base of Uluru and the Valley of the Winds walk at Kata Tjuta. Such strategies have been successful in reducing the numbers of climbers.

Research conducted on visitor decision-making in relation to climbing Uluru has suggested that the representation of the climb in travel and tourism media influences visitor attitudes, perceptions and ultimately, decisions, in

Figure 10.2 Visitors embarking on the Uluru climb.

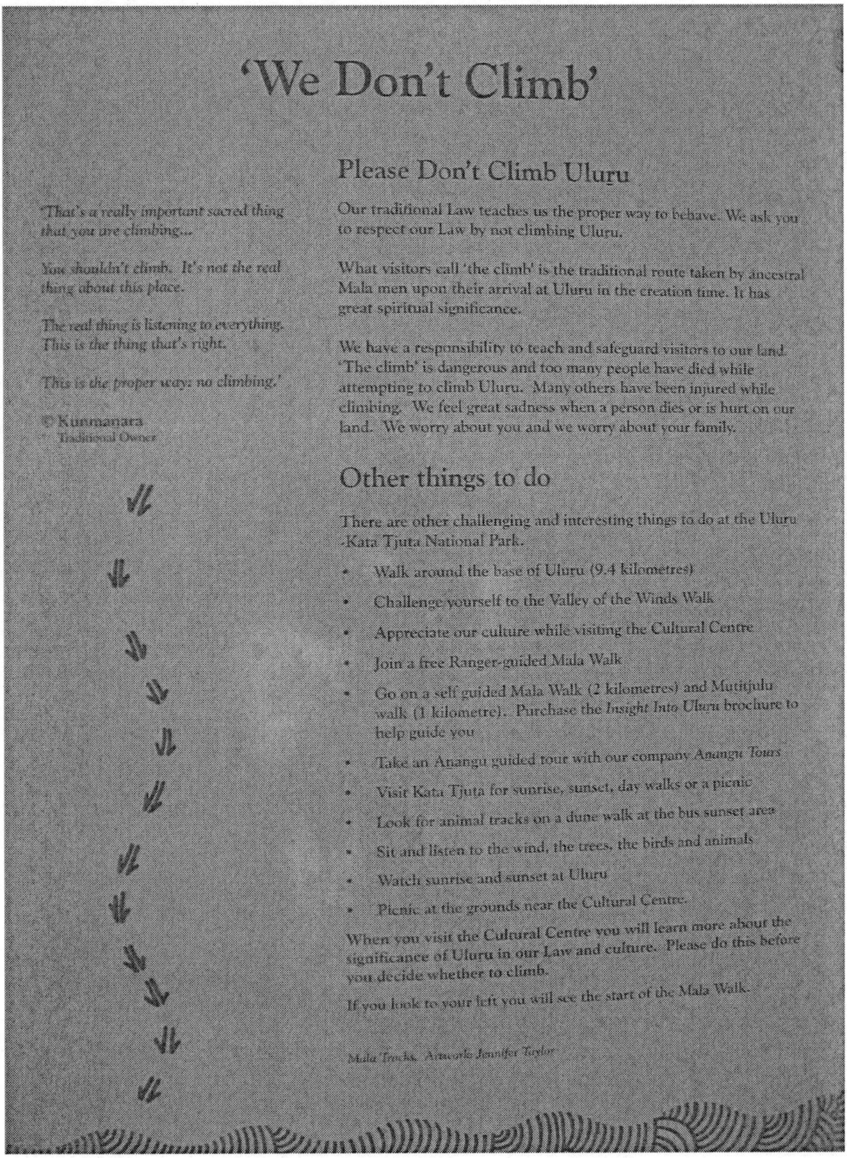

'We Don't Climb'

Please Don't Climb Uluṟu

'That's a really important sacred thing that you are climbing...

You shouldn't climb. It's not the real thing about this place.

The real thing is listening to everything. This is the thing that's right.

This is the proper way: no climbing.'

© Kunmanara
  Traditional Owner

Our traditional Law teaches us the proper way to behave. We ask you to respect our Law by not climbing Uluṟu.

What visitors call 'the climb' is the traditional route taken by ancestral Mala men upon their arrival at Uluṟu in the creation time. It has great spiritual significance.

We have a responsibility to teach and safeguard visitors to our land. 'The climb' is dangerous and too many people have died while attempting to climb Uluṟu. Many others have been injured while climbing. We feel great sadness when a person dies or is hurt on our land. We worry about you and we worry about your family.

Other things to do

There are other challenging and interesting things to do at the Uluṟu -Kata Tjuṯa National Park.

- Walk around the base of Uluṟu (9.4 kilometres)
- Challenge yourself to the Valley of the Winds Walk
- Appreciate our culture while visiting the Cultural Centre
- Join a free Ranger-guided Mala Walk
- Go on a self guided Mala Walk (2 kilometres) and Mutitjulu walk (1 kilometre). Purchase the *Insight Into Uluṟu* brochure to help guide you
- Take an Anangu guided tour with our company Anangu Tours
- Visit Kata Tjuṯa for sunrise, sunset, day walks or a picnic
- Look for animal tracks on a dune walk at the bus sunset area
- Sit and listen to the wind, the trees, the birds and animals
- Watch sunrise and sunset at Uluṟu
- Picnic at the grounds near the Cultural Centre.

When you visit the Cultural Centre you will learn more about the significance of Uluṟu in our Law and culture. Please do this before you decide whether to climb.

If you look to your left you will see the start of the Mala Walk.

Mala Tracks Artwork: Jennifer Taylor

Figure 10.3 Sign at the base of the Uluṟu climb. The left hand side of the sign warns visitors about issues of safety when climbing Uluṟu.

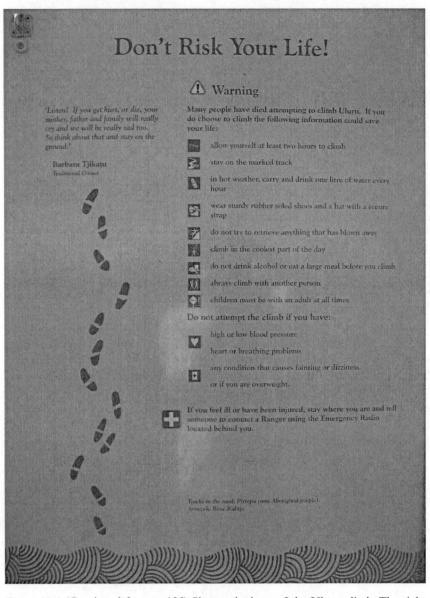

Figure 10.4 (Continued from p. 135) Sign at the base of the Uluṟu climb. The right hand side of the sign welcomes visitors to Aboriginal land and explains the Aṉangu request not to climb Uluṟu.

relation to undertaking the climb (e.g. Young 2005a, Young 2005b, James 2007). The conflict between the discourses of tourism and Aboriginal culture was evident in media coverage of the decision by Anangu to close the climb for a period of 20 days in May 2001. This was the first time that a ban had been enforced on climbing and it was put in place by Anangu as a mark of respect for the death of a traditional elder. The response to the closure was heated. The Northern Territory Chief Minister, Denis Burke, was reported as saying that the closure of the climb would send 'enormous shockwaves through the whole of the tourism industry' (Toohey 2001: 3). A concern with the effect that closing the climb would have on the tourism industry and visitor numbers to the national park has continued to overshadow the cultural and spiritual beliefs of Anangu.

Thus, the dominant discourses of tourism and Aboriginal culture that frame and shape the visitor experience at the Uluru-Kata Tjuta National Park are contradictory. According to Edensor (1998: 7), symbolic tourist sites are spaces where national, political, cultural and spiritual identities can be expressed and imagined. The various layers of meanings, therefore, can result in contradictory and contested notions of what these places mean (Edensor 1998). The sacredness of Uluru to Anangu is contrasted with the commercial interests of the tourism industry. Uluru, on one hand, is an Aboriginal sacred site of cultural and spiritual significance but, on the other hand, it is a natural site that seemingly belongs to everybody. Based on these opposing meanings, the visitor experience at Uluru, and their attitudes towards climbing it are marked by ideals of 'respect' versus 'conquest' (Young 2005a). Arguably, the way that visitors perceive the site – either as a natural landscape or as a cultural landscape – influences their decision on whether or not to climb (Young 2005b, James 2007). The ways by which the Uluru-Kata Tjuta National Park is constructed and managed can, however, play an important role in influencing perceptions of visitors, particularly through visitor education and Anangu interpretation. Joint management of the national park provides a way for the living culture and heritage of Uluru-Kata Tjuta to be central to the visitor experience.

Managing cultural-natural landscapes: joint management at the Uluru-Kata Tjuta National Park

Australia has a long history of recognizing the value of natural areas through the creation of national parks; however, it is only relatively recently that national parks have been recognized as also facilitating the protection of cultural heritage (see Chapter 5). The development of national parks in Australia as 'public recreation' reserves commenced in 1879 with the creation of the Royal National Park (see Chapter 1). National parks provided spaces for public health, recreation and enjoyment (Pigram and Jenkins 2006). However, the creation of some Australian national parks was about more than just the protection of the environment and providing a place for recreation – it was about positioning Australia as a world-class tourist destination with an iconic natural environment, and recognition of the role that the tourism industry can play in providing economic benefits. Drawing on the Uluru-Kata Tjuta National Park as an example, a report by the Commonwealth of Australia (2007) states:

> The protection and marketing of iconic locations – sometimes referred to as monumentalism – has always been an important motivation behind the creation of parks ... The preservation of such iconic natural assets pointed to a growing recognition by government and society of the value of preserving such places not only for their natural heritage value but also for their international recognition value, effectively putting Australia more firmly on the world map.
> (Commonwealth of Australia 2007: 10–11)

The Uluru-Kata Tjuta National Park is internationally recognized as a World Heritage Area of outstanding universal value. The Uluru-Kata Tjuta National Park was first inscribed as a World Heritage Area in 1987, gaining international recognition for its natural value. Significantly, the World Heritage listing was updated in 1994 to include recognition of the cultural heritage of the area. At that time, the park was only the second in the world to be recognized as a cultural landscape with living cultural heritage, thus making the rare acknowledgement of the relationship between Aboriginal people and the natural environment (Brown 1999).

In terms of management, the Uluru-Kata Tjuta National Park is distinctive for two reasons. First, it is one of only six Commonwealth National Parks declared under the *Environmental Protection and Biodiversity Conservation Act 1999* (the majority of Australia's 600 national parks are managed by state agencies; see Chapter 3 for more on this anomaly). Second, the national park is owned by the Aboriginal traditional landowners and leased back to the Commonwealth agency Parks Australia. A consequence of the lease-back arrangement is that the national park is jointly managed by the Aboriginal landowners and non-indigenous agencies (similar joint management arrangements are in place in two other Commonwealth National Parks, Booderee National Park in New South Wales and Kakadu National Park in the Northern Territory).

Joint management is a process in which indigenous landowners and non-indigenous agencies have shared responsibility in national park management. In Australia, joint management regimes have played a significant role in national park policy and decision-making since the 1970s (Wearing and Huyskens 2001). Since that time, joint management has become 'an increasingly common practice to help resolve tensions arising from colonization' (Waitt, Figueroa and McGee 2007). Joint management evolved out of lease-back arrangements arising from land claims under the *Aboriginal Land Rights (Northern Territory) Act 1976*. Such 'compulsory' lease-back arrangements mean that the 'government recognises aboriginal ownership of the land, yet requires the immediate lease-back of the land and the formation of joint management arrangements' (Wearing and Huyskens 2001: 191).

The joint management of Uluru-Kata Tjuta National Park commenced in 1985 following the 'handback' to the traditional landowners. As noted above, the granting of legal title to the traditional landowners subsequently resulted in a 99-year lease-back agreement. The national park is managed by Anangu (represented on the Uluru-Kata Tjuta Board of Management), and Parks Australia. Anangu have a majority representation on the 12-person Board of Management (Commonwealth of Australia 2008), thus ensuring that Aboriginal culture is a central consideration in the management of the natural and cultural heritage of the national park, as well as in the interpretation of the site for visitors (Brown 1999). The Board of Management and the Director of National Parks are responsible for making management and policy decisions, and for the preparation of the Uluru-Kata Tjuta National Park Plan of Management (see Commonwealth of Australia 2000).

Importantly, joint management provides for the involvement of Aboriginal people in the management of national parks. According to Craig (1993: 137), such involvement represents a 'cross cultural approach to land use and management in protected areas', which provides Aboriginal people with the autonomy to implement 'political and cultural power over decisions affecting their lands and their lives'. As noted in the Uluru-Kata Tjuta Plan of Management, the traditional law of the Tjukurpa is central to the contemporary joint management of the national park as a cultural landscape (Commonwealth of Australia 2000), as reflected in the subtitle of the plan *Tjukurpa Katutja Ngarantja: Tjukurpa Above All Else*. The management policy that is guided by Tjukurpa principles maintains Anangu culture and heritage, conserves and protects the ecology, and provides opportunities for visitor enjoyment and learning within the national park (Commonwealth of Australia 2008).

There are a number of ways that visitor education of Tjukurpa is provided by Anangu. Aboriginal tours and interpretations of the landscape include some ranger-guided tours of the base of Uluru and organized tours with the company Anangu Tours which is owned and operated by the local Aboriginal community. Indeed, Anangu promote their tours as 'the only way to understand the real Uluru' (Anangu Tours, no date: 1). Anangu interpretations of the landscape are also presented in the Uluru-Kata Tjuta Cultural Centre. This Aboriginal-owned and -operated Cultural Centre, which opened in 1995, is located about one kilometre away from Uluru. Anangu and Parks Australia recommend that visitors to the national park visit the Cultural Centre before going to Uluru and Kata Tjuta to learn about Anangu living culture and the Tjukurpa.

The success of the Cultural Centre and Aboriginal owned and guided tours indicates that clearly the intertwining of multiple layers of value is not always productive of negative management decisions, as was evidenced in the past. Increasingly, Aṉangu culture is being made available as a central component of the tourist experience at Uluṟu-Kata Tjuṯa National Park. In turn, this is producing an enhanced level of awareness and respect for Aṉangu perspectives. Further, the joint management structure itself is facilitating an increased empathy and understanding between Western and indigenous perspectives and interests at the level of governance. According to Waitt, Figueroa and McGee (2007: 248), joint management within this framework 'requires indigenous and non-indigenous people working together, and assists in the process of reconciliation by enhancing mutual respect between Aṉangu and visitors'. Yet, the climb remains and, thus, sentiments such as this reflect the ideal. Joint management implies joint responsibility and, yet, for Aṉangu an apparent incapacity to implement behavioural guidelines representing their own value systems perhaps reflects a structural impediment within the joint management arrangement.

Joint management regimes in Australia have been criticized as being a Eurocentric approach, based on Western cultural models of management (Wearing and Huyskens 2001). Further, Brown (1999: 678) points out that 'culture conflict' can result when 'dominantly Eurocentric or Western tourism utilises Indigenous heritage sites which are still occupied and controlled by the local host community'. When examined in this context, it is clear that joint management can be 'plagued by contradictions' (Waitt, Figueroa and McGee 2007: 254). Although the Aṉangu request for visitors not to climb Uluṟu has become more widespread through its representation in various tourism media (Young 2005b), the climb still remains, and the visitor car park remains located at the base of the climb path, which continues to be marked out by a metal fence up the face of Uluṟu. Thus, the contested usage of the site reflects the tensions that are inherent in the joint management structure.

Mild criticisms of this nature may have their place, but it is equally important to remind ourselves that joint management involving indigenous input is a relatively new and bold experiment in the running of national parks. Indeed, the joint management structure at the Uluṟu-Kata Tjuṯa National Park has been the recipient of numerous accolades and global attention for its successes in integrating indigenous living cultural heritage with more Westernised interests in the preservation of the environment (Commonwealth of Australia 2000). With perennial tourism fascination as the driving force, perhaps joint management at the Uluṟu-Kata Tjuṯa National Park will provide a pathway for the development of power sharing arrangements that are capable of integrating the primarily oppositional indigenous and Western narratives of natural and cultural value and significance.

Part III
Old World perspectives

11 The national park concept in Spain

Patriotism, education, romanticism and tourism

Jose Somoza Medina

In present-day Spain, the concept of a national park is much more closely linked to the field of tourism than to that of conservation or environmental education. Although the Spanish Ministry of the Environment and the similar institutions in each Autonomous Community are the bodies charged with managing them, the figure seen as most important by politicians and the general public as a whole is the number of visitors to a given park during the course of the most recent year. From the point of view of many Spaniards, parks are just another tourist attraction, with potential to bring in investments, so when proposals are made to list a new natural space as protected, the debate always revolves around possible limitations on the development of tourist activities or raising new buildings.

However, tourism was not the only reason for the setting up of the first Spanish parks at the beginning of the twentieth century, or even the chief reason. At that time a number of social, cultural, educational and even personal circumstances came together to lead to some of the earliest conservationist legislation in Europe, based specifically on the creation of national parks.

The social, economic and cultural context of the birth of the first parks

At the end of the nineteenth century Spain was a country that was taking its first timid steps towards industrial development. The aristocracy still dominated the corridors of power, and arable and pastoral farming was the livelihood of the vast majority of Spanish families. In the *Cortes* [Parliament] in Madrid, Liberals and Conservatives alternated in government, although the elected representatives of both these parties were alike in coming largely from rich families, being in a position to dedicate themselves exclusively to politics thanks to accumulated wealth. The merchant and industrial middle classes had some role in the society and local politics of urban centres. These, with the exception of a few cases, had scarcely experienced any major change over the previous two centuries. For their part, the common people were characterized by relative apathy and ignorance about the country's political, economic or cultural situation. Their main, almost their only, preoccupation was to manage to scrape a living, within a system whose roots went back into the distant past and which suffered from great inertia, preventing the regeneration for which intellectuals were pushing. Hence, there was an official Spain, made up of the aristocracy and the wealthiest part of the middle class, clinging to dreams of grandeur by looking to a glorious past, and a real Spain, marked by poverty and industrial backwardness. Among the factors that put a brake on any economic progress was the poor state of communication infrastructures, which prevented any improved fluidity of commercial interchanges.

The bigger cities were beginning to grow, as isolated enclaves, but there were still only a few such cases: Madrid, Barcelona, Bilbao, Valencia. In the meantime, the countryside was showing symptoms of crisis as a consequence of increased population pressures and low productivity levels, leading to the emigration of thousands of people to Latin-American countries such as Argentina, Cuba, Mexico or Venezuela. The rural exodus to urban areas was not to occur until the development of the medium-sized Spanish cities in the middle decades of the twentieth century.

In the culture of this period two different features may be noted that have a direct relationship to the appearance of the first national parks. On the one hand there was the so-called 'Generation of '98' and on the other the establishment and expansion of the *Institución Libre de Enseñanza* (ILE) or Free Institution of Education established in 1876. In the first case it was a question of a literary movement that tried to overcome the end-of-century crisis into which the final loss of its last overseas colonies after the war with the US plunged Spain, by turning its gaze back towards the purest elements in the national character. Spain's glorious past, warlike exploits, landscapes that inspired patriotism, a heavy stress on folklore and tradition dominated the literary scene. The second case was that of a prestigious educational organization that promoted direct contact with nature, following the theories of the German philosopher Karl Krause, to improve educational systems and inspire the population with respect for natural landscapes. To that end it ran excursions to mountain areas and organized holiday camps for schoolchildren on the coast (at San Vicente de la Barquera in Santander Province). Both the ILE and the Generation of '98 took the nation's scenery as a new orientation, after the loss of the last colonies of the Spanish Empire (Cuba, Puerto Rico and the Philippines).

At the time the Head of the Spanish State was Alfonso XIII, ninth monarch in the Bourbon dynastic line and the grandfather of Spain's current King Juan Carlos I. Alfonso XIII was the quintessence of the official Spain described above. He allied himself with the most conservative sectors of society; for instance during the Dictatorship of Miguel Primo de Rivera, he took an active hand in national politics, turning his back on the role of merely being an impartial arbiter that he should have maintained in a constitutional monarchy. Despite the national economy's strong growth rate during the

First World War, thanks to Spain's neutrality, allowing her to sell freely to the belligerent powers, the political instability arising from the exhaustion of an atavistic system divorced from reality prevented this period of economic growth from serving to create solid foundations on which to base the socio-economic development of the country.

The philosophic and conceptual bases leading to the emergence of national parks

Among the causes explaining the early appearance of this sort of conservation in Spain three philosophic or cultural trends may be cited: Romanticism, Krausism and Regenerationism. Mendoza (1998) considers the impact of Romantic ideas about nature and landscape on the origins of Spain's national parks and argued that the choices for the first two parks responded to the preferences of Romantic travellers and intellectuals, who associated wild mountains with the greatest symbolic load, as a sublime exaltation of nature. The symbolism of mountain scenery overrode other ecological factors such as biodiversity or the fragility of ecosystems when protected areas were being delimited in this first period. According to Mendoza (1998), this would be the principal reason for there having been for many years just two national parks in Spain, set in the Alpine system running across the north of the country, with none elsewhere. Thus, while the most characteristic Spanish landscape is Mediterranean in type, one of the areas of greatest ecological richness in Europe (Doñana) was not declared a national park until 1969.

Mendoza cites works by Humboldt, Simmel and Goethe, highlighting that for them the concept of landscape was seen as an aesthetic representation of nature, revealed to those who contemplated their surroundings with feeling. She adds that this feeling was never stronger for them than on top of a mountain. In the case both of the Picos de Europa and of Ordesa, influential figures had praised their characteristics before ever there was any institutional declaration. Victor Hugo, Jules Verne, J. M. Keynes, Lucien Briet, Baron Ramond de Carbonnières, Count Saint-Saud and Gustav Schulze spoke highly of the aesthetic quality of their scenery, this international recognition reinforcing the Romantic halo that surrounded both places.

Krausism was the ideological foundation of the academically prestigious ILE. In their publications a number of its members analysed the concept of landscape and the advantages of studying in contact with nature. The area that the ILE particularly studied was the Sierra de Guadarrama, near Madrid, which for the Institution's founder, Giner de los Ríos, was a life experience comparable to gazing on the cave paintings of Altamira. Following the example of the ILE, other organizations and societies promoted excursions at the beginning of the twentieth century, with the intention of uniting education and nature.

The ILE was born as a consequence of the expulsion of a group of professors and lecturers from the Central University of Madrid because they defended academic freedom against moral or religious dogmas. One of their basic principles was non-religious education, something that ran against the convictions of the population in a country as deeply Catholic as Spain. Thus the ideas of Krause were adopted and developed by the Institution's staff, especially pantheism or harmonious rationalism, the autonomy of human spheres of action, like science or education, and the defence of nature. In accordance with the paedagogical application of these postulates, students should base their learning on direct experience, fleeing dogmatism. So, in the ILE excursions and experimental work were strongly encouraged. Nonetheless, despite the persistence of Romantic ideas lauding the beauty of mountains and despite the activities of the ILE, precursors of environmental education, it remains true that unless public authorities had been seeking to regenerate Spain, the earliest national parks would not have been set up at that time.

The setting up of national parks often implies a nationalist affirmation, aimed at signalling to the outside world sovereignty over territory, or directed towards one's own population as a factor for cohesion in overcoming a period of crisis in the values defining the homeland (see Chapter 5). In Spain the second was the case, within a movement termed Regenerationism that had repercussions in many fields: in the literary world, politics, music, art, society and also in nature conservation. Regenerationism attempted to find an explanation for the loss of international power by Spain in the last days of its Empire. The causes were many: late industrialization excessively concentrated in a few places, a subsistence agricultural economy in the grip of virtually feudal patterns, an archaic and corrupt political system that held back necessary reforms, the limited importance of the commercial and industrial middle classes, among others. It was necessary to modernize Spain, copy the formulae for progress that were triumphing in other European countries and abandon forever the idea of the Spanish Empire and its great heroes. According to Joaquín Costa – one of the ideologues of this movement – the ills of the nation could be summed up in five points: lack of patriotism, disdain for its own, absence of shared interests, lack of a concept of independence and scorn for tradition. For this teacher at the Central University of Madrid, who resigned from his post, following the other academic staff who had been expelled, and who collaborated with the ILE, the formulae that would regenerate the country were 'school and larder' and 'double-locking the Cid's tomb, so he would not ride again'.

Regenerationism was a political, socioeconomic and cultural desire held by practically all the chief figures in Spain's public life in the early twentieth century, from King Alfonso XIII down to the humblest schoolmaster, including liberal and conservative politicians, writers, intellectuals and journalists. It had parallels in several regions of Spain with their own culture and language, like Catalonia (*Renaixença*) or Galicia (*Rexurdimento*). The ultimate aim was so laudable that any action taken to regenerate the country received immediate applause from society, for example the creation of the first national parks.

The leaders of conservation policy: Pedro Pidal and Eduardo Hernández Pacheco

In a socioeconomic context like that characteristic of Spain in the early twentieth century, the philosophical and conceptual bases just sketched would not in themselves have been enough to bring about the momentum for the first declarations of national parks without the decisive participation of individuals such as Pedro Pidal y Bernaldo de Quirós, the Marquis of Villaviciosa de Asturias. Similarly, they would never have survived down to the present day without the later contribution of rigorous scientific management, initially directed by Eduardo Hernández Pacheco.

If the origin of the Spanish parks must be attributed to one pioneer, there is no doubt this would have to be Pedro Pidal, called by some writers 'Spain's John Muir'. The story of Pedro Pidal is that of an unusual aristocrat. Born in Somió in Asturias in 1870, he was a life senator for the Conservative Party, a sportsman, traveller, businessman, journalist, hunter, minister, writer and personal friend of King Alfonso XIII, with whom he went hunting in Doñana, Cabañeros and the Picos de Europa (all now national parks). His restless personality took him to the Paris Universal Exposition in 1900. Once there, he participated in the Olympic Games in the trap shooting contest. In 1904, he was the first person to climb the Naranjo de Bulnes, one of the great feats of Spanish mountaineering, and now an emblematic site in the Picos de Europa National Park. Moreover, he wrote down all his adventures and thoughts about nature in short publications and articles in the newspapers of the period.

At the start of the twentieth century, Pedro Pidal visited the United States and the Yellowstone and Yosemite National Parks, gathering information on their origin and management. This was at the suggestion of Alfonso XIII, as Pidal claimed in an article. It is easy to imagine that on his return he would make comments to the King on his impressions of the parks he had visited. With the aim of preserving his beloved Covadonga Mountain, he began to encourage from within the Senate legislation to regulate this form of protection of nature. Finally, on 14 June 1916 he brought a *Bill for the Creation of National Parks* into the Senate, listing the precedents in other parts of the world and explaining in greater detail the United States model. In his speech (quoted in Fernández and Pradas 1996: 31), the following words are worth emphasizing: 'Are there not sanctuaries for art? Why should there not be sanctuaries for nature?'

The Law was approved by the King on 7 December 1916 and has just three articles. The first lays down that national parks are to be created in Spain. The second defines them as those exceptionally picturesque, wooded or rustic sites or areas in the national territory so designated by the State by means of a declaration of their status as such, with the sole purpose of favouring access to them by suitable routes and of respecting and ensuring there is respect for the natural beauty of their landscape, the richness of their fauna and the geological and hydrological curiosities they contain, hence effectively avoiding any act of destruction, damage or disfigurement by the hand of man. The third article outlines the procedure for declaration, saying that the Ministry for Development shall create national parks in agreement with the owners of the lands affected, regulate those that are created, and make available in its budgets the sums necessary for access routes and for maintenance.

In this law, national parks were seen as sanctuaries that should remain unchanged and which it should be possible to reach on good routes, so any visitor could contemplate their beauty. The third article includes one of the chief obstacles to the wider development of this early conservation legislation, as it lays down that in order to declare a national park there should first be an agreement among the owners of the lands affected. In Spain, parks were not State property, but rather hills, meadows and woods that were common land or privately owned, which by virtue of the declaration would have to be kept exactly as they were, with no possibility of changing anything. For example, of the Covadonga Mountain National Park, 95 per cent was communal land.

In February 1917 a Royal Decree requested district chief engineers to send in a list of noteworthy sites that should be protected. In the reports received, the Covadonga Mountain and the Ordesa Valley were both prominent. As was to be expected, the first national park was the Mountain that Pidal adored, this park being inaugurated by the King and Queen on 8 September 1918, exactly the day when the twelfth centenary of the Battle of Covadonga, start of the so-called Reconquest (the name given to the period between the eighth and the fifteenth centuries during which the Christian kingdoms of northern Spain slowly drove out the Moslem population), was being celebrated. Among the inaugural acts the King planted a tree and made an official speech whose Regenerationist tone is clear from his words referring to the Covadonga Basilica, inaugurated in 1901, to commemorate the popular tradition that the Virgin had aided the troops led by Pelagius in his battle against the Moslems:

> We are going to do something unique in the world: to join the art of Nature with Religion and History at the place where a nation was born. And this is something that is not just a question of money, especially when so many centuries have set their seal upon it. This is Covadonga: Spain.
>
> (quoted in Fernández and Pradas 1996: 83)

The Ordesa National Park was inaugurated by Pedro Pidal as the King's representative on 14 August 1920, without so much ceremony and after the resolution of a number of lawsuits brought by the proprietors of one particular privately owned hill area.

In later years, up to his death in 1941, Pedro Pidal's character led to clashes with Eduardo Hernández Pacheco, the man the Government set at the head of the body charged with managing national parks. His remains were carried by mountaineers to the Ordiales Belvedere, in the Picos de Europa National Park, where they lie beneath a simple tombstone. On the stone his epitaph is engraved:

> Enchanted by the Covadonga Mountain National Park, it is there that I would wish to live, to die and to rest for all eternity. However, this last I wish to do at Ordiales, in the magic kingdom of the chamois and the eagles, there where I became acquainted with the happiness of Heaven and Earth, there where I spent unforgettable hours of admiration, emotion, dreams and rapture, there where I adored God in his works as Supreme Craftsman, there where Nature truly appeared to me as a temple.

If Pedro Pidal was an idealistic politician, Eduardo Hernández Pacheco may be defined as a pragmatic scientist who proved able to convince various different political regimes with his ideas, since he remained in key posts relating to nature conservation both during the periods of military dictatorship and under the Republic. Since 2004, the Spanish Ministry of the Environment has been awarding two prizes for the best publications on Spanish national parks, named after Pedro Pidal and Eduardo Hernández Pacheco, although surprisingly, in view of their interests, the first is for scientific studies, the second for literary works.

Born in Madrid in 1872, he was a pupil and then a teacher at the ILE. Appointed Professor of Geology at the Central University of Madrid in 1907 (also Professor of Physical Geography from 1923 on), he began to appear on numerous commissions and committees relating to nature. As spokesman of the Spanish Central Board for National Parks, in 1927 he put forward alternative forms of protection, more limited in their size and more in tune, as he saw it, with the situation in Spain. These were termed Natural Sites of National Interest and Natural Monuments of Natural Interest. The first declaration in these categories took place in 1930, in the area most closely linked to the ILE – the Sierra de Guadarrama. However, instead of setting out an extensive protected zone, three distinctive sites were chosen: an area covered with large boulders, a pinewood and the glacial cirque at the summit. In a piece published in 1933 he claimed that in Spain, unlike in the United States, there were no large wild areas where a conservation policy could create natural sanctuaries. Rather, its territory was one of human settlement, governed by a multitude of varying and long-standing local rights relating to exploitation for forestry or grazing, rendering such an initiative impossible (Hernández 1933). Moreover, he criticized the choice of national parks declared at the behest of Pedro Pidal and stated that the country had a large number of sites, scattered all over its territory, with distinctive ecological and landscape characteristics, which deserved to be protected by the State (Casado 2000).

The outcome of the acceptance of these ideas by the various governments down to the start of the Franco dictatorship was the declaration of numerous Natural Sites and Monuments of National Interest throughout Spain, but also that the next national park was not opened until 1954.

The attitude of the population towards the first national park declarations

As noted above, at the start of the twentieth century there were two Spains – one official, one real – the latter more concerned with getting enough to eat than with cultural movements or advances in nature conservation. As Joaquín Costa put it, what the country needed was 'school and larder', that is, better education and no need to worry about where the next day's meals would come from, since only in this way could it attain the same level of development as other nations.

In the files on the earliest national parks there are abundant notes about protests and lawsuits brought by residents and settlements affected by the declaration (Fernández and Pradas 1996). (In fact, this attitude has been repeated in practically all the national parks since they were first created.) The population of these rural territories was accustomed to gleaning from the common-land meadows, woods and hills, numerous resources that eked out their limited income. The regulations put in place in the early twentieth century prohibited these uses and triggered a reaction of rejection towards the implications of a protected natural space reserved solely and exclusively for contemplation by visitors from distant parts. Furthermore, the provision of public funds envisaged by the 1916 Law was never sufficient.

In view of the impossibility of expropriating lands, the choice was made to permit certain activities in a controlled way, such as grazing livestock or collecting firewood, although lawsuits between the Parks Board and the affected population continued. For example, in 1995 when the first national park was extended and renamed Picos de Europa Park, growing from the previous 16,925 hectares to the current 64,660, the protests from the population directly affected (20 settlements in 10 different municipalities [civil parishes]) were reflected in all the mass media.

On this point the categories of Natural Site or Monument were much more useful. This was because they were limited to protecting only individual sites or specific elements of the landscape or geology, flora or fauna of special interest and gave responsibility for their care to local councils.

Later on, further laws were passed modifying the controversial third article of the 1916 Law, declaring all lands included within the limits of a national park were to be considered of public utility, this making their expropriation by the State easier. All the same, it is difficult to achieve this when entire villages are to be found within the boundaries of some of the parks. For example, one of the most striking cases is the Sierra Nevada National Park, the largest, with an area of 86,208 hectares. These lie in no fewer than

44 different municipalities [civil parishes], with several complete villages standing within or on the edge of the protected zone.

In fact, as stated by Francisco Ortuño (1980), founder of the National Institute for Nature Conservation, the creation of the first parks took place thanks to the concerns of a select group of people, advanced for their time, because the country was not ready either economically or culturally for a policy of this sort.

Later developments: Spanish national parks and tourism

As noted above, the establishment of the third Spanish national park had to wait almost 40 years after the declaration of the first. In 1954 two new parks were set up, this time covering volcanic landscapes in what was then the part of Spain most attractive to international tourism – the Canary Islands. The first was the Teide Park on the island of Tenerife in January of that year; then in October the Taburiente Crater Park on the island of La Palma. The local authorities saw clearly that one of the main aims in promoting such a declaration was the publicity and the attraction for Spanish and foreign tourists that the new national parks would generate. Indeed, on Tenerife, proposals were immediately made for the construction of a *parador nacional* (a luxury state-owned hotel) and a cable car, to make it easy for tourists to climb to the top of the volcano. This objective may be considered well and truly achieved, since the Teide National Park is by far the most frequently visited protected natural space in Spain, with 3.5 million visitors on average. Its maximum capacity has been established as more than 4.3 million.

The fifth national park was set up in October 1955. In this case it was at the specific wish of the country's dictator, Franco (Fernández and Pradas 1996), who had several times visited Aigüestortes and Lake Sant Maurici. The 1957 *Law on Wastelands* repealed the old legislation from 1916 and it was under this new disposition that in 1969, the Doñana National Park was declared, followed by the parks of Tablas de Daimiel in 1974 and Timanfaya in 1975. In the first case, the decree founding it stated that among its purposes it could be used as an educational resource. Since that time this magnificent biodiversity reserve has been the setting for a range of environmental interpretation and education programmes.

In 1975, the *Law on Protected Natural Spaces* was approved. However, transfer to the Autonomous Communities (regions) of powers in the field of nature conservation triggered a series of legal disputes initiated by these regional governments, which were intent on creating their own 'national' parks, in some cases trying once more to give a 'patriotic' sense to this concept. A resolution of the Constitutional Tribunal in 2004 confirmed the exclusive right of the central State to declare national parks, but granted powers for their normal day-to-day running to the Autonomous Communities. For example, in the Picos de Europa National Park, management has to be coordinated by the authorities of the three different Autonomous Communities within whose boundaries parts of the park lie.

To bring legislation into line with the ruling of the Constitutional Tribunal, in April 2007, a new law was passed concerning the Network of National Parks. In this, besides laying down that the initial proposal for a declaration must come from Autonomous Communities with the definitive declaration taking the form of a Law passed by the national *Cortes* [Parliament], national parks are defined as:

> natural spaces of considerable ecological and cultural value, little changed by human exploitation or activity, which, thanks to the beauty of their scenery, the representative status of their ecosystems or the singular nature of their flora, fauna, geology or land formations, are of outstanding ecological, aesthetic, cultural, educational or scientific merit such as to make their conservation worthy of preferential attention and of being declared of general to the State.
>
> (Law 5/2007, Article 3)

Since then six new parks have been established: Garajonay (1981), the Cabrera Archipelago (1991), Cabañeros (1995), Sierra Nevada (1999), the Atlantic Islands of Galicia (2002) and Monfragüe (2007).

In total, the current network of national parks in Spain is composed of 14 zones. Of these, two are combined land and sea (the Cabrera Archipelago and Atlantic Islands of Galicia), four Alpine mountain areas (Picos de Europa, Sierra Nevada, Ordesa and Aigüestortes), three of volcanic origin (Teide, Taburiente Crater and Timanfaya), two wetlands (Doñana and Tablas de Daimiel), two Mediterranean woodlands (Cabañeros and Monfragüe) and one subtropical forest of lauraceous species (Garajonay). Together they amount to 3,109.29 square kilometres, equating to 0.6 per cent of the country's territory. The biggest park is Sierra Nevada, at 86,208 hectares, followed by Picos de Europa at 64,660, while the third largest, Doñana, covers 54,252 hectares. At the opposite end of the scale comes the Land and Sea National Park of the Atlantic Islands of Galicia; while its total extent is 8,480 hectares, only 1,195 are dry land. It is followed by the Tablas de Daimiel Park, at 1,928 hectares and Garajonay, at 3,986.

In 2006, there were more than 11 million visits to the National Park Network. These involved considerable differences between the park with the biggest flow, Teide, averaging 9,000 visitors each day, and the least visited, Cabañeros, reaching only 74,000 visitors over the whole of the year.

A different statistic allows the number of visitors to be linked to the extent of the parks so as to gauge appropriately the effect of visits on a park's ecosystem. From this viewpoint, the four parks in the Canary Islands and the Atlantic Islands National Park are those with the greatest pressure of tourists. The highest density is in the Timanfaya National Park, with 350 visitors per hectare per year, compared to much lower figures for Doñana

(6.94 visitors per hectare per year), the Cabrera Archipelago (8.45), or Sierra Nevada (8.56).

The Spanish National Park Network has been governed since 1999 by a Master Plan. This lists a number of general objectives, including the following:

- To consolidate the National Park Network and to encourage its internal cohesion;
- To contribute to the Spanish system for the protection and conservation of nature by including the national parks in the set of conservation strategies;
- To establish the necessary guidelines for conservation, public use, research, training, education, social awareness raising and sustainable development;
- To favour the development of public awareness of the value of national parks and channel social participation into the decision-making process;
- To define and expand the framework for cooperation and collaboration with other administrative bodies;
- To improve the image and public profile of the National Park Network.

In this list there are two surprises. One is the lack of any mention of a link between tourism and the National Park Network, when it is obvious that one of the clearly stated objectives in the respective Master Plans for Use and Management is to attract visitors. The other is that the fourth point explicitly draws attention to the lack of public awareness (specifically at a local level) of the advantages offered by the National Park Network.

Overview

Over the course of this chapter it has been demonstrated how the concept of national park has varied in Spain from the first declaration in 1918 down to the most recent in 2007. Nonetheless, patriotic, educational and tourist values have always been present in the successive declarations, even if with different individual weightings in each period.

In the early twentieth century the action of a few gave rise to the first Spanish national parks, based on Romantic alpine archetypes and the United States management model. At that period the concept was seen as synonymous with sanctuary. Nevertheless, alongside the general objective of preserving their beauty from human tampering, the law gave priority to improving communication routes so as to favour access by anybody who wished to view at close quarters the exceptional landscapes of the Spanish homeland.

In later declarations the explicit objective was to attract visitors and at this point the national park concept became more like a resource for tourism. Now it was not just necessary to have new access routes, but also accommodation, car parks, cable cars and the like.

Successive laws and the management of national parks have given rise to another change of concept in recent years. Currently, priority is given to the representative nature of ecosystems and their educational potential. The Network of National Parks is now like a catalogue of natural spaces of great beauty, intended to define the rich variety of landscape in Spain.

However, the zoning established by the Use and Management Plans for each park allows the coexistence of varying uses and meanings. On the one hand, there are reserve zones (sanctuaries), access to which is permitted only for scientists; then there are zones of restricted use ('nature's classrooms'), where public access is limited because of their mixture of aesthetic, ecological and educational interest; and, finally, there are zones for moderate use and special use (a tourist resource), where the park buildings, visitor services and the network of walks are concentrated. Reality is always complex and admits of many meanings, not just those assigned by the dictionary, and always hampered by limitations.

12 The English Lake District – national park or playground?

Richard Sharpley

Introduction

In 1810, the poet William Wordsworth concluded his *Guide to the Lakes* with his now widely cited wish that the English Lake District should be deemed 'a sort of national property in which every man has a right and interest who has an eye to perceive and a heart to enjoy' (Wordsworth 1810: 92). In so doing, he was, arguably, one of the first to propose the concept of national parks (see Chapter 2), yet it was to be a century and a half (and some seventy years after the creation of the world's first national park) before his dream was realized. While numerous national parks were designated elsewhere in the world, it was only in the 1950s that, along with nine other areas in England and Wales, the Lake District achieved similar status, by which time it was firmly established both nationally and internationally as a tourist destination. Indeed, it is ironic that Wordsworth was, and continues to be, a major influence in the emergence and popularity of the Lake District as a destination. Although his principal concern was with the protection of the area (particularly from a tourist 'invasion' facilitated by the construction of the railways), his poetry contributed significantly to both the creation of an enduring place-myth and the related rise of the Lake District as a tourist 'playground' from the early nineteenth century onwards (O'Neill and Walton 2004: 22).

Given its long history as a tourist destination, it is perhaps not surprising that the designation of the Lake District as a national park was controversial (Shoard 1999). Indeed, despite the more recent designation of a further four national parks in England and Scotland, the creation and role of national parks in the UK has long been, and remains, the subject of intense debate (MacEwan and MacEwan 1982, 1987, Blunden and Curry 1990). Since their designation, not only have the planning and management structures of the parks been subjected to criticism, scrutiny and change, but also their two statutory objectives – namely, conservation and the promotion of tourism and recreation – have increasingly come into conflict (FNNPE 1993). Moreover, the socioeconomic characteristics of the parks as inhabited, multipurpose and principally privately owned landscapes (less than a quarter of the total land area of national parks in the UK is under public or quasi-public ownership, while less than 3 per cent is owned by National Park Authorities (Eade 1987, Sharpley 2007)) have led many to question the relevance of national park status in the UK context. It is frequently noted, for example, that no national park in the UK is formally recognized as such by the IUCN, while the Lake District in particular has consistently failed to achieve UNESCO World Heritage Site status.

Against this background, the purpose of this chapter is to explore the development of the English Lake District as a national park, highlighting in particular the challenges of planning and managing the area as a resource for tourism within the broader sustainability/conservation imperative explicit in national park status. In so doing, it not only considers the extent to which national park status remains an appropriate framework for addressing the contemporary challenges facing the Lake District, but also exemplifies the development and role of national parks in the UK as a whole. The first task, however, is to review briefly the characteristics of the Lake District and the historical development of tourism in the area before focusing specifically on the establishment and effectiveness of its national park status.

The Lake District

Situated in the northwest of England within the county of Cumbria, the Lake District, covering an area of some 2,280 sq km, is England's largest national park. It is renowned for its unique juxtaposition of lakes, tarns, valleys and mountains, the result of 500 million years of complex geomorphology and glacial activity, and within the boundaries of the national park can be found England's highest mountain (Scafell Pike, 977 metres), and her longest (Windermere, 20 kilometres long) and deepest bodies of water (Wast Water, 70 metres depth) (Wyatt 1987, LDNPA 2004a). Yet, although the Lake District is a place 'endowed by nature with a richness of opportunities for varied visual experiences' (Appleton 1984: 118), its landscape has also been nurtured and shaped by centuries of human activity (Rollinson 1967). Settlement patterns and built heritage reflect the agrarian socioeconomic tradition of the area and, currently, some 55 per cent of the national park's area is registered as agricultural land (NWDA 2005) – as noted shortly, however, it is the development of tourism that, since the early nineteenth century, has arguably had a greater influence on the environment and built heritage (Berry and Beard 1980). Many farms are owned by the National Trust (the UK's largest conservation organization) which owns 25 per cent of the land area of the Lake District; some 65 per cent of the land area is under private ownership while other quasi-public bodies, such as the Forestry Commission, own the remainder.

In common with all other national parks in the UK, therefore, the Lake District is a living, working landscape. According to latest (2001) census data, the total resident population within the national park is 41,650, representing

8.5 per cent of the population of Cumbria as a whole (LDNPA 2004b). Approximately 37 per cent of the park's population lives in the main urban centres, which include the principal tourism 'honeypots' of Windermere/ Bowness, Ambleside and Keswick (see Table 12.1). These towns provide a significant proportion of the Lake District's tourism facilities, amenities and attractions and, as Hind (2004: 55) notes, their 'population . . . increases significantly during the main holiday season'. Interestingly, over 29 per cent of the Lake District's permanent population is over sixty years of age, significantly higher than the county average, reflecting the relatively high proportion of retired people living within the park.

Conversely, the proportion of the population working full-time is below the county average – less than 40 per cent of men are in full-time employment – while women account for almost 60 per cent of those employed (Hind 2004: 60). These figures are indicative of the importance of tourism to the local economy; in 2004, there were an estimated 13,452 tourism-related jobs in the Lake District (LDNPA 2006), or some 43 per cent of all employment. Other important sources of employment include: manufacturing (9.3 per cent of all employment), agriculture and forestry (7.6 per cent) and education (8.09 per cent) (LDNPA 2004b). In 2004, total tourist spending in the national park amounted to £602 million; estimates of tourism's contribution to the national park's GDP are not available (Sharpley 2004) although, by way of comparison, agricultural holdings within the park generated a total income of £59 million in 2002, roughly one-tenth the value of tourism that year (NWDA 2005). It is also important to note in the context of this chapter that over one-fifth of the housing stock in the Lake District is accounted for by second or holiday homes.

Tourism in the Lake District: brief history

Records of early visitors to the Lake District can be traced back to the late seventeenth and early eighteenth centuries, of particular note being Daniel Defoe's journey through the region during his tour of Great Britain: he found the Lake District to be 'a country eminent only for being the wildest, most

Table 12.1 Cumbria and the Lake District populations

	Population
Lake District Total	41,650
– *Ambleside*	*3,017*
– *Keswick*	*4,966*
– *Windermere / Bowness*	*6,682*
– *Coniston*	*798*
Rest of Park	26,252
Cumbria Total	487,608

Source: Adapted from LDNPA (2004b).

barren and frightful of any that I have passed over in England, or even Wales itself' (Berry and Beard 1980: 1). In 1769, the poet Thomas Grey undertook a short 10-day excursion in the Lake District and is commonly regarded as the first genuine 'tourist' (Rollinson 1967: 133). His narrative was to inspire writers and artists alike to visit the region; visitors included Turner, Constable and Gainsborough, capturing the sublime beauty of the mountains and lakes in their paintings. However, it was the Romantic poets, particularly Wordsworth, who did most to create an enduring 'place-myth' of the Lake District, producing an imagined, literary landscape that remains an attraction for present-day tourists (O'Neill and Walton 2004). As Squire (1988: 237) observes, 'hordes of visitors, anxious to recreate the emotional experiences in places described by a literary idol, still descend on areas immortalized in poetry or prose'. Nevertheless, the Lakeland fells also attracted early walkers and climbers, the first recorded ascent of many peaks being made by Butterworth in the 1790s (Rollinson 1967), though it was not until a century later that fell walking and rock-climbing, now popular activities in the Lake District, became more established.

As the road network improved during the early nineteenth century, inns for tourists opened in the less accessible dales while the development of the three main tourist centres referred to above was accelerating; during the 1830s, their populations grew at between 'two and three times the rate for Cumbria as a whole' (O'Neill and Walton 2004: 28). Nevertheless, tourism was, at this time, relatively limited – it was the arrival of the railways in the mid-1800s that was to have the greatest impact on the Lake District, providing opportunities for tourists to visit in their thousands rather than hundreds, transforming the built environment and establishing the foundations of the contemporary tourism industry. In 1847, the line to Windermere was completed and, in its first year of operation, carried 120,000 passengers (about a quarter of whom were leisure tourists). This was followed by lines to Coniston (1859), Keswick (1865) and Lake Side (1869). As a consequence, the railhead towns grew rapidly; the population of Windermere/Bowness, for example, doubled between 1841 and 1861, establishing the town as the principal tourism hub in the Lake District (O'Neill and Walton 2004). Moreover, although the railways opened up the area to day-trippers from the northern industrial towns – on Whit Monday (Pentecost Monday) in 1883, over ten thousand day-trippers visited Windermere (Berry and Beard 1980: 2) – 'the expansion of the Lake District tourist market came to depend increasingly on the growing band of middle-class visitors who were holidaymakers first, sight-seers second and devotees of romantic mountain solitude hardly at all' (Marshall and Walton 1981: 186). One tangible outcome of this trend was that, from this period onwards, second-home ownership in the Lake District began to increase significantly as holiday homes were developed by the wealthier middle-class.

During the first half of the twentieth century, tourism in the Lake District grew steadily, the main influence being improved road accessibility and increasing car ownership, although the popularity of cycling in the 1920s, followed by the hiking craze of the 1930s (supported by the development of cheap accommodation, such as that provided by the Youth Hostel Association), were significant influences on the growth of tourism in the area. However, it was from the 1950s onwards that the Lake District experienced the most dramatic growth in tourism, the principal factors being greater mobility underpinned by a rapid increase in car ownership, along with increases in leisure time and disposable income, and the growing popularity of outdoor recreation. By the early 1960s, not only had mass tourism arrived in the Lake District but also, as one commentator at the time observed, 'tourism is becoming the dominant industry in the area . . . today the visitor is beginning to own the place and, if we are not careful, the whole area will be turned into one vast holiday camp' (Nicholson 1963: 181). Recent history demonstrates that this fear was misplaced although, as the following section reveals, tourism remains a significant activity in the Lake District.

Contemporary tourism

Some 20 million people live within a three-hour drive to the Lake District, an undoubted factor in its popularity as a tourist destination both in its own right and in comparison to other national parks in the UK. Tourism statistics tend to vary according to calculation method, but it is estimated that between 15 and 20 million visitor days are spent in the Lake District annually. The great majority of visitors are on day trips. In 2001, for example, over 80 per cent of all visitors were day-trippers; conversely, just 16 per cent of tourists were staying for at least one night, yet these overnight tourists accounted for 42 per cent of visitor days and 65 per cent of tourist revenue (Table 12.2).

For overnight visitors, the Lake District offers 61,300 bedspaces, 72 per cent of which are in non-serviced (self-catering) accommodation. There are also around 2,500 touring caravan and 4,500 static caravan pitches available, plus 3,000 approved camping pitches; almost a quarter of overnight visitors stay in tents or caravans (LDNPA 2006).

Over 80 per cent of all visitors arrive by car and continue to use their cars

Table 12.2 Day and overnight tourists in the Lake District

	No. of visits (mn)	% of all visits	Tourist days (mn)	% of total tourist days	Revenue (£mn)	% of total revenue
Day visitors	11.5	84	11.5	58	195.5	35
Overnight tourists	2.2	16	8.4	42	355.8	65
Total	13.7	100	19.9	100	551.3	100

Source: Adapted from Calway (2001).

to travel around the park. Indeed, a recent survey of tourist behaviour in the Lake District found that, although the natural beauty of the area is a major attraction, the majority of visitors participate in more passive activities, such as 'visiting towns, shopping, visiting restaurants and pubs and driving around by car' (Creative Research 2002: 10). Hence, tourist activity is centred primarily on the main honeypots referred to previously. Those who are more active go on short walks or visit an attraction, with literary 'shrines', such as Wordsworth's Dove Cottage or Beatrix Potter's home at Hill Top, being particularly popular; fewer than 10 per cent of visitors venture into the fells (Hind 2004). Thus, despite the natural attractions of the Lake District and the opportunities for outdoor recreational activities they offer, these are not the principal draw for the 'typical' Lake District visitor. As this chapter now considers, this appears to contradict the very purpose of the Lake District's designation as a national park.

The Lake District: national park designation

The Lake District was designated as a national park on 9 May 1951. It was one of 10 national parks that, following the enactment of the 1949 *National Parks and Access to the Countryside Act*, were established in England and Wales between 1951 and 1957. The 1949 Act itself was the outcome of a process that was, arguably, initiated in the Lake District itself where the UK's conservation movement first emerged (Berry and Beard 1980). Following the then Manchester City Corporation's successful application to convert Thirlmere into a reservoir, the Lake District Defence Society was formed in 1883 to oppose further industrial incursion into and exploitation of the area, including the building of more railways. One of its members, Canon Hardwicke Rawnsley, vicar of Low Wray Church in Hawkshead and, subsequently, Crosthwaite, near Keswick, went on to head the Kendal and District Footpath Preservation Society and subsequently became a founding member of the National Trust in 1895. It was from these beginnings that national conservation organizations, such as the Council for the Protection of Rural England (CPRE), were to emerge during the 1930s, as well as the Friends of the Lake District (FLD), an organization founded in 1934 with the express purpose of achieving national park status for the Lake District (see Sandbach 1978).

Also during the 1930s, walking and rambling were rapidly growing in popularity, particularly among the working populations of England's northern industrial cities, yet access to much open land, particularly privately owned grouse moors, was limited. Consequently, the access lobby, represented by organizations such as the Ramblers Association, sought to secure greater access to the countryside through so-called 'freedom to roam' campaigns (Shoard 1999). Thus, there was pressure on two fronts: the preservationist groups working for the protection and conservation of the countryside and the access lobby demanding greater freedom of access. Indeed, representatives from both sides joined forces in the 1930s to campaign for national parks in the UK.

A complete review of the subsequent processes leading up to the 1949 *National Parks and Access to the Countryside Act* and the eventual designation of the Lake District, and other areas in England and Wales, are beyond the scope of this chapter (see Sheail 1975, MacEwan and MacEwan 1982, Mair and Delafons 2001, Sharpley 2007). However, the twin themes of conservation and access were evident in the original statutory purposes of the national parks of England and Wales as defined in the 1949 Act; that is, national parks were to be designated to 'preserve and enhance the natural beauty of the areas specified, and for the purposes of promoting their enjoyment by the public'. Since designation, however, these two purposes have increasingly come into conflict and, as a consequence, subsequent governmental reviews and legislation have redefined the purposes of national parks, giving precedence to conservation over tourism and recreation. The implications of this for the Lake District in general, and the management of tourism in the park in particular, are discussed shortly but, first, it is important to consider two widespread criticisms of national parks in general in the UK: namely, that the areas designated are neither 'parks' nor 'national'; and that the areas designated do not, for the most part, boast a natural landscape.

Strictly speaking, the word 'park' is usually applied to an area that is enclosed and used for a specific purpose such as recreation or nature conservation; it also implies that access into the area can be regulated or restricted (see Chapters 1 and 2). This cannot, of course, be said for the Lake District and other national parks in the UK. They are all extensive areas of open countryside in which large numbers of people live and work and into which access cannot be regulated. Thus, boundaries to the park simply reflect the area within which the National Park Authority (the organization endowed with the responsibility for managing the national park) has jurisdiction. At the same time, the 'natural' beauty that national park status seeks to preserve is anything but natural; it is the artificial result of centuries of human habitation and influence on the landscape, particularly grazing and agricultural clearing. Indeed, most national parks throughout Europe, unlike those in North America and elsewhere, 'have not emerged from a wilderness tradition but reflect a concern for the protection of environments that are strongly marked by human activities' (Hoggart *et al.* 1995: 244). This is certainly the case with the Lake District (if not more so than other national parks in the UK), where the landscape has not only been materially altered by generations of farming and forestry but also carries the hallmarks of modern developments, including railways, roads, quarries, urban centres and, of course, a substantial tourism infrastructure.

Also important is the fact that the majority of the landscape within national parks is not 'national' in the sense that it is not publicly owned. Reference has already been made to patterns of land ownership in the Lake

District, where some two-thirds of the land is privately owned and a further quarter is owned by the National Trust, a voluntary-sector membership-based conservation body that seeks to preserve and maintain access to the country's natural and built heritage. However, this does not imply a lack of public access to the land within the park. There are over 3,000 kilometres of public rights of way (footpaths, bridleways and byways) along which the public have a right of access, although this has significant implications with respect to the erosion and repair of footpaths and other rights of way. Interestingly, this is not a recent problem; it has been observed that, 'in 1819, a Lakeland traveller arriving . . . in Langdale, via Snake Pass from Borrowdale, complained that the route he had just travelled was seriously eroded and in a worse condition than when travelled 10 years previously' (LDNPA 2000). However, since the 1970s, a variety of factors have contributed to erosion becoming a significant environmental problem, including the lengthening of the tourist season and newer activities, such as mountain-biking and legal use of all-terrain 4-wheel drive vehicles. Thus, a survey in 1999 suggested that some 41,690 person-days of work, costing £4,656,512, would be needed to repair the most seriously eroded paths in the Lake District (Sharpley 2004). Moreover, as a result of more recent legislation (*Countryside and Rights of Way Act 2000*), over 55 per cent of the national park is accessible as of right on foot, though subject to certain restrictions, potentially exacerbating the erosion problem. There is also controlled access to four of the lakes, although the conservation imperative limits the types of water sports that are permitted. Therefore, the Lake District is a 'national' park in a somewhat vague sense that it is an area of national importance, not owned *by* the nation but protected *for* the nation's enjoyment. The question now to be addressed is: how is this role as a national park and, in particular, the conflict between tourism and conservation, managed?

Managing the Lake District

The principal organization with the responsibility for the Lake District is the Lake District National Park Authority (LDNPA), a statutory local government body established for the purpose of managing the national park on behalf of the nation. Responsible to the Ministry of Environment, Food and Rural Affairs (DEFRA), its role is 'to secure the purposes of National Park designation' (Darrall *et al.* 2004: 107) and it is supported in this function with a grant of some £7 million of taxpayer funding. The LDNPA has 26 executive members, 19 of whom are elected members of local government bodies while the remaining seven are national members appointed by the Secretary of State. Of particular significance, the LDNPA has few if any powers available to it, despite the centrality of its role in the management of the national park. That is, although it acts as the planning authority within the national park, it must work within existing national planning laws and beyond this, 'it secures the purposes [of designation] by influencing the actions of others and by working in partnership. It must nurture relationships and gain respect' (Darrell *et al.* 2004: 108).

Thus, the LDNPA, as all other National Park Authorities in the UK, is not vested with the powers that might be imagined. This reflects, on the one hand, the rather complex political and administrative structures within which the national park is located. That is, in common with the governance of the wider UK countryside as a whole, a large number of agencies and organizations from the public, private and tertiary sectors are involved in the planning and management of the park in general and of tourism in particular (Sharpley 2003). Each of these organizations operates according to its own mission, ideology and objectives and, hence, may potentially conflict with others. The Lake District, for example, falls within the county of Cumbria, but also embraces four local authority districts (LADs) and numerous parish councils. The LADs play a role in tourism development and marketing, yet the Cumbria Tourist Board is responsible for countywide tourism development and promotion (including areas outwith the national park) and is itself responsible to the Northwest Regional Development Agency, a quasi-governmental body overseeing economic development throughout the northwest of England. A variety of other governmental and quasi-governmental bodies also have roles within the national park while a number of environmental and conservation organizations work with commercial tourism sector to balance the economic benefits of tourism with environmental protection. It is among all these organizations that the LDNPA must seek consensus and cooperation.

On the other hand, the potentially weak position of the LDNPA also reflects the somewhat hazy role assigned to it since the initial designation of the national park in 1951 and, perhaps, ongoing uncertainty regarding the role of national parks within the UK, particularly with respect to achieving a balance between tourism and conservation. Over the years, the structure and role of the LDNPA has been altered, primarily to reflect the need for a more proactive approach to planning and management; all National Park Authorities are now required by law to prepare a National Park Management Plan and to review it every five years. Inevitably, a section of this plan focuses on the development and management of tourism; the current plan seeks, among other things, to 'develop more ways in which the tourism industry can be sustainable, contributing directly to the conservation and enhancement of the environment on which it relies' (LDNPA 2004a: 55). The LDNPA has also published a separate sustainable tourism policy document (LDNPA 2005). This, in turn, reflects the evolving statutory role of national parks in the UK as a response to the ever-greater conflicts between the twin objectives of landscape conservation and the promotion of tourism. For example, the first National Parks Review Committee (the so-called *Sandford Report*) in 1974 concluded that, where conservation and recreation came into conflict, then priority should be given to preserving the natural beauty of the landscape, a recommendation that was later accepted by the government. This view was endorsed by a subsequent review in 1991 (Edwards 1991), which recommended that the purposes of the national parks should be redefined as being to (a) protect, maintain and enhance the scenic beauty, natural systems and land forms, and the wildlife and cultural heritage of the area; and (b) promote the quiet enjoyment and understanding of the area, insofar as it is not in conflict with the primary purpose of conservation. As a result of the report's proposals, the *Environment Act 1995* s 61 redefined national parks' purpose as: 'to conserve and enhance the natural beauty, wildlife and cultural heritage of such areas; and to promote opportunities for the understanding and enjoyment of the special qualities of those areas by the public'. Section 62 of the Act gives clear weight to the conservation imperative, while National Park Authorities also have a wider duty to foster the economic and social well-being of local communities.

Thus, the LDNPA's role is, essentially, to manage the sustainable development of the Lake District, a fundamental element of which is the management of sustainable tourism and recreation as the area's principal economic sector. As the planning authority, it controls all forms of development, including tourism, according to national policy guidance; it also has the statutory power to employ rangers and wardens to 'help visitors to understand and enjoy the special qualities of the National Park' (Darrall *et al.* 2004: 109), and to create byelaws for managing the recreational use of the lakes. Most controversial of these has been the recent imposition of a 10 mph speed limit for powered craft on Windermere, effectively banning water-skiing and other higher speed water sports. This is evidence of a specific – and arguably extreme – response to the Edwards (1991) recommendation for a focus on 'quiet enjoyment'; research has shown, however, that understanding of the term 'quiet enjoyment' and subsequent management responses varies significantly in different national parks (Caffyn and Prosser 1998). The LDNPA also enjoys limited delegated powers with respect to rights of way and traffic management; in common with other National Park Authorities, it also benefits from an annual £200,000 Sustainable Development Fund grant to support local sustainable initiatives, including tourism projects (LDNPA 2005). For the most part, however, it is dependent upon the cooperation of, and partnerships with, other organizations for the successful achievement of its statutory objectives. Consequently, and as noted elsewhere (Sharpley and Pearce 2007), the sustainability imperative as defined in UK national policy documents and elsewhere (for example, ODPM 2004) is not always manifested in practice in the national parks, although there are a number of examples of successful partnerships between the LDNPA and relevant organizations. Conversely, a plan to introduce a 'park and ride' scheme to reduce traffic levels in one popular valley were dropped following opposition from local hoteliers and traders who feared the measure would lead to a loss of business.

What conclusions, then, can be drawn from this chapter with respect to the Lake District's status as a national park?

Conclusion: national park or playground?

The overall purpose of this chapter has been to explore the development of the Lake District as a national park, thereby providing a basis not only for exemplifying, within the context of this book, the interpretation and implementation of the national park concept in the UK, but also for considering the extent to which national park status provides an appropriate framework for addressing the contemporary challenges facing the Lake District. From the necessarily brief review, a number of points have emerged. First, as often observed, the Lake District does not conform to the IUCN definition of a 'national park' in terms of landscape character, ownership or purpose. Thus, the term 'national park' is applied in the UK context to areas of living, working countryside that possess special or unique physical and, in the case of the Lake District, cultural characteristics, the latter relating to the literary landscape created by the Romantic Lakeland poets. In other words, national parks in the UK are in essence, simply areas of 'national' importance. Secondly, and related, the Lake District contains three urban centres plus associated paraphernalia of modern development: roads, railway lines, street lights, shopping centres, and so on; it is interesting to note that, in a recent consultancy report, it was recommended that 'there is no case for recommending the main towns [in a World Heritage Site bid] as they are not part of the cultural landscape and do not merit designation in their own right' (NWDA 2005: 8). Implicitly, these towns sit uneasily within the area's national park status. Thirdly, the main towns, plus much other built heritage, reflect the Lake District's long history as a popular tourist destination; the towns are, in effect, tourism resorts, developed to meet the needs of ever-increasing numbers of tourists; the cultural landscape of the Lake District is, therefore, very much a tourism landscape. Fourthly, a majority of tourists participate in urban, 'playful' (boat rides on Windermere, rides on steam trains), or attraction-based activities, primarily within the road corridor linking the three urban centres. Though walking is a popular activity, only a minority of visitors access the fells and open countryside (though still in sufficient numbers to inflict significant damage on the natural environment). Finally, the statutory definition of national parks in the UK (and consequential duties of National Park Authorities) have been reactively adapted to meet, on the one hand, the accepted need to manage the areas sustainably but, on the other hand, to impose what is arguably an increasingly anachronistic and atavistic vision of appropriate (that is, traditional and 'quiet') recreational uses of the landscape.

Therefore, it is logical to conclude that the Lake District is, in effect, a tourist playground, albeit framed within a landscape that is of particular quality and endowed with cultural significance. Certainly, there is a need to manage effectively that (physical) landscape, and its use by tourists, to maintain those special qualities; moreover, given the diversity of organizations with an interest in the area, there is undoubtedly a need for that management role to be undertaken by a coordinating body such as the LDNPA. However, managing the entire area as a national park within increasingly rigorous, culturally defined parameters not only contradicts the activities of the majority of visitors, but also leaves the Lake District in danger of becoming, over time, a living museum.

13 The Peak District National Park UK

Contemporary complexities and challenges

David Crouch, Duncan Marson, Geoff Shirt, Richard Tresidder and Peter Wiltshier

Introduction

In 1951, the first national park in the UK was designated. It is located in the Peak District, close to the middle of England. Its purpose was clear: conservation and popular recreation for peace and quiet. Its area included not only hill farming as well as some more lowland agricultural tracts, but small-scale industries and some nationally significant industries – limestone-quarrying – a peculiarly English national park. At that time these components were containable through the frameworks and powers of the park and the constituent, if several, local councils. Since that time the complexities of all these uses have increased significantly. This chapter includes a brief background of the park's response to popular access, and acknowledges the growing and changing demands for use. It focuses upon different dimensions of the current tourist/visitor debate and a consideration of the current complexity of interests involved, as stakeholders, in the park. This chapter exemplifies the considerable complexity and challenges that the park as a whole seeks to cope with. First, however, a review of the background debate of what makes the park so magnetic an attraction, in a section called Embedding the Peak.

The Peak District National Park (PDNP), established through the United Kingdom's National Parks and Access to the Countryside legislation in 1951, is the second most visited national park in the world, with over 24 million visitors each year. It is also a national park within less than two hours' journey of 50 per cent of Britain's population and immediately adjacent to the large industrial cities and residents of Lancashire and Yorkshire. As a national park it is a paradox of conservation and recreation and rejuvenation as it now represents an opportunity for politicians to build upon the twin agendas, dichotomous as they are, of landscape protection and of human personal development through leisure, recreation and the outdoors.

In addition to being governed in a paradoxical fashion for the natural and human factors enshrined within the structures ruling the PDNP, the destination has complications arising from multiple layers of governance and an appreciation of the difficulties and anomalies that arise from multiple stakeholders use and governance is expressed in this chapter. The park itself has fewer than 40,000 residents scattered through more than 100 small villages over 550 square miles. The surrounding hinterland has in excess of 180,000 people resident in a combination of small villages, market towns and dormitory towns for the metropolitan centres of middle and north England. From hill and dale walking to fishing and water sports, the range of activities engaging visitors to, and residents of, the environment is diverse, full of opportunity, but, as the chapter later focuses, also complexity.

The chapter opens with a conceptually driven consideration of the power of national parks generally in drawing popular interest. The subsequent section examines first the still-informing historical development of access for this park. Historically – at least for a century – health and the availability of space to enjoy and to engage in healthy recreation, or recreation-as-health was a focal impulse in pursuing better access to these spaces. The notion of national parks as one type of open space for exploring and making effective the notion of health, exercise, recreation and the energy of nature fits into the contemporary keenness for enhancing health and well-being, increasingly known as 'green gyms' (British Trust for Conservation Volunteers 2007). As this and its following section demonstrate, there are issues and challenges that confront delivering the potential of accessibility and inclusivity aspects for the green gym to be democratic.

The activity of bouldering exemplifies how this demand is drawn into a consideration of what the green gym means in terms of consumption and commodification: the green gym as iconic of the national parks' natural, original idea of peace and quiet, and/or style in leisure pursuits. This discussion necessarily echoes the earlier, early twenty-first century debates, and confrontations, concerning the accessibility, relative protection, and contemporary complexity with which national parks are used, given meaning and made possible as green gyms, not least in terms of the democracy of their accessibility and the character and cohabitive of its use.

Contemporary efforts to respond to this tangle of tourism, recreation more generally, and the range of stakeholders and popular interest are brought together in the final section of this chapter, that raises too the concerns and negotiations that surround them.

Embedding the Peak

This section provides a brief overview of the embedded meanings and definition of the Peak District. It is these embedded definitions that not only inform the marketing and promotion of the Peak, but also inform the visitors' motivation for visiting the Peak and how they interpret it once they arrive. As such, the debate offers an alternative definition of why tourists persist in visiting the park and what they consume when they arrive.

The individual tourist's interpretation and expectations of the Peak District are governed by a complex dichotomy of influences contrived by

Figure 13.1 The village of Edale, valley farms and high moors, Peak District National Park.

their own cultural origins (Garlick 2002, Robinson 2002). Theoretically however, the interpretation of the Peak may still be dependent on the 'interactive participant' (tourist) understanding and recognizing the discourses (Kress and Van Leeuwen 1996) and embedded meanings that historically have come to define the National Park. The individual development of a touristic knowledge (which enables the individual to formulate their interpretation of the destination) is created through the construction of individual expectations and hegemonic geographical stereotypes (Relph 1985). In the case of tourism it can be perceived that this process is reinforced by the semiotic appropriation and representations of culture, society, landscape and indigenous populations by the tourism industry (Mowforth and Munt 2003); within the designation of the 'represented participant', this process can be expressed in many forms of media materials, including the tourism brochure, television, literature (Crouch and Lubbren 2003, Crouch *et al.* 2005). The images of the National Park as presented in such television 'soap' series programmes as *Peak Practice* (1993–2002) or the BBC production of *Jane Eyre* (2006) can be seen to define, and socially and culturally embed representations of the Peak District and as a consequence, signpost the interpretation and consumption process of the individual tourist.

Critically, the consumption and interpretation of the Peak District is

Figure 13.2 Recreational walkers, Edale, Peak District National Park. Hardly dressed for going beyond made roads.

'signposted' (Jenkins 2003) through the social and cultural embedding of that destination within the media. Conceptually, this process is underpinned by the choice of a particular set of signs and images that may be socially and culturally recognized as a semiotic 'language of tourism' (Dann 1996a, Selby 1996). This language provides a Lingua Franca that encompasses the means of defining both place and tourist experience (or expected experience). Within this language is a set of signifiers that may be seen to present tourism as the consumption of sacred time and destinations such as the Peak Park as constituting sacred space.

It can be argued that the signs and images used to represent the Peak District often promise a refuge from modernity in which the individual can find the organic, the primitive, the original, the authentic and the expressive (Urry 1990). The globalization, commodification and homogenization of contemporary culture reduce instances of cultural differentiation in everyday

Figure 13.3 Grindslow Knoll, High Peak, Peak District National Park. High moors with rough walking tracks.

life and critically blur place and individual identity (Harvey 1989; Lash and Urry 1994). The return to the organic and the reliance on history and indigenous culture reconstruct the ego and provide roots in a rootless society. It is argued that the constructed representations of the National Park as the 'represented participant' within the media and literature provide a refuge for the tourist in which to escape, to belong (Selby 1996, Morley 2001), a place in which to conceptually rejuvenate and cleanse the soul.

It is through this differentiation that the significance of tourism and destinations such as the Peak District are heightened. The social and cultural embedding of the National Park offers a time and space to the tourist that is removed from everyday lived experience. The social and cultural construction of the park and what it has come to represent, reinforce the significance of tourism within contemporary society, while simultaneously defining the phenomenon itself, thus in a way, creating a 'circle of representation' (Jenkins 2003) in which escape and the search for meaning may be found. Theoretically, the construction of the landscapes of tourism and definition of the Peak District establishes a 'configuration' of time, space and power (Jokinen and McKie 1997), which defines the Peak as different from the everyday.

The media utilize a particular semiotic language (Culler 1981a, Dann 1996b, Selby 1996) that generates an embedded configuration of touristic time and space that differentiates the Peak from other more urban or developed landscapes, to create a destination that has become seen as 'sacred'. The reason this phraseology has been chosen is that, first, the sacred as a distinguishing element of time and space can be seen to represent the 'extraordinary' nature of tourism in contemporary society and secondly, that it enables the recognition of the distinctive time and space as represented within the media. For Rojek (1995:52), 'A tourist sight may be defined as a spatial location which is distinguished from everyday life by virtue of its natural, historical or cultural extraordinariness.' Conceptually, it is this configuration of time and space and the extraordinariness of touristic landscapes which reinforce the binary relationship between the ordinary/everyday and the extraordinary (Urry 1990). The media signposts, reinforces and defines this binary opposition (Tresidder 1999, 2001), by signifying timeless and spaceless images of the Peak District. These images utilized within marketing campaigns represent the Peak as overtly rural, often without human or technological representations, which when considered in juxtaposition with the city are seen to be 'extraordinary' in comparison to everyday lived experience.

For Durkheim (1995) the conception of the sacred and profane is socially generated and underlines the distinction between social and ordinary experience. Although Caillois (1988) recognized that the two worlds of the mutually exclusive domains of the sacred and profane do not mingle in unmediated ways, that is, in the absence of collectively recognized rites of passage and acknowledged risks of admixture. 'He took great care to outline how the profane needs the sacred, and the regulation, through rites, of the process of consecration in the passage into the sacred from the profane' (Genosko 2003: 75). The semiotic touristic landscapes of the National Park contained within the contemporary media are socially produced and consumed through various social and cultural influences. It can be suggested that the sacred or extraordinary status of destinations such as the Peak Park is reinforced, not only by its social and cultural production and consumption, but also by the status of its symbolic authority (Meyrowitz 1992, Couldry 2000) as signifying something 'extraordinary' or the 'sacred sphere of excess' (Caillois 1988: 282). Theoretically, it is within this discussion, that the sacred and profane is a useful analogy for the distinction between the semiotic world of tourism and the ordinary world of everyday lived experience (see Sheldrake 2001). Durkheim's (1995) account of the sacred and profane helps to identify the particular features of the tourism/ordinary distinction, yet it is not a social fact. However, it can be reasoned that the signs and images contained within the tourism brochure reshape social reality; they mystify the definition of touristic landscapes as being somehow apart from the ordinary.

Although conceptually, the notion of the sacred is a problematic category, by accepting the division between the sacred and the profane as a distinction, it allows the analysis of the differentiation of both time and space and the use of tourism as a means of distinction. What was sacred for Durkheim (1995) was society itself, and involved the sacralization of the social and the object of worship was society itself, as he states 'anything ... can be sacred' (1995: 35). What distinguishes the division between the sacred and profane is not content, nor its use to rank objects and people, but simply that the division is absolute (Couldry 2001: 160). On one hand, it could be perceived that religion and the sacred have a functional role in that they place the social on an unquestionable level, and on the other that religion and the sacred perform a cognitive role in providing the basic categories of thought within society.

Thus the definition of both tourism as a social activity and the Peak District as a destination become embedded in contemporary society by media representations and historical discourses that are forwarded within literature. This embedding provides an activity and location in which time and space become differentiated from that of everyday life, and as such, witnesses the elevation of the Peak District to that of the 'sacred'. It is this classification that provides the motivation of tourists to visit and informs their interpretation of the park during their visit. As such, the millstone markers located at the entrance to the National Park signify not only a geographical change, but also a psychological one.

From public access to the inclusive green gym

The relationship between national parks and recreation is nothing new, but in recent years has been reactivated in several ways. While always being a matter of popular, rather than specialist access, the recent turn of focus has centred around the efficacious use of open, green spaces out of town, in particular for exercise, linked explicitly with health and well-being. These terms are not distant from the nineteenth-century popular reaction against cramped industrial urbanization, but are emerging in different forms. On the one hand, national parks are among different kinds of spaces as opportunities and resources to enable popular exercise outdoors, with a particular emphasis on their being more inclusive for individuals with physical and mental disabilities. On the other, new demands and audiences for newer kinds of physical and mental well-being and health have accelerated, exemplified in terms of adventure activities, as popular versions of adventure sports.

The idea of the green gym is one of these interventions, particularly geared to actively encouraging exercise with the objective of extending opportunities for health and well-being. Various in audience, scope and character, and agency participation, these 'gyms' have the innovation of focusing possibilities and opportunities for connecting well-being with the idea of the value of greenspaces. Some prioritize responses to well-being needs for those socially excluded, tackling inequalities in health and social well-being and creating environments that will help people maintain good health and well-being. The UK Government recently has developed a major funding programme for a range of linked interventions, responding to interests of a range of particular groups in connecting available, accessible green space, anywhere, with the interest in well-being through outdoor exercise (Department for

Communities and Local Government 2007, for example, British Trust for Conservation Volunteers).

In the British context national parks offer one kind of space for delivering this in a way that can enhance accessibility. In general terms, this possibility is enhanced in these areas by the character of land as held in common; the diversity of surfaces, various levels of difficulty of slope and surface; the availability of tracks; lack, or lower incidence of pollution; the utility of legally backed measures and funds to facilitate access, information and interpretation. Other spaces exist, many with some preferential attributes: city parks (depending on pollution), urban walks (possibly affected by matters of safety); gardening on public land (for example, community gardens, allotments, but of increasingly pressured availability due to increased popularity combined with developer and institutional development pressure). Indoor facilities contribute significantly to opportunity for health, but lack the 'green'.

Many individuals visiting national parks do not use them as green gyms; they may be used for fairly sedentary hunting, driving, and short-walk visits to particular sites. However, in continuing a century-old identity as sites and spaces of popular access, British national parks increasingly are responding to possibilities of enhancing components of their accessibility to enlarge the opportunity of the green gym, for example in terms of disability and at the same time, adventure sports.

It has been argued that 'access is synonymous with conflict' (Curry 1994). Most national park stakeholders and researchers agree (Pye-Smith and Hall 1997, National Farmers Union, Association of National Park Authorities, Mercer 1998). The perspective risks polarizing and stereotyping stakeholders, and flies in the face of, for example, evidence that the financial interests of landowners are not put at risk by the recreational activities of walkers (Sidaway 1990). Of course, the central concern is a lack of mutual acknowledgement and trust. But the stakes, and the power, were, and partly remain, very unequal: a mutual distrust can exist between landowner and walker; exclusive owner and potential nuisance; and lack of civil liberties (either side). Very briefly, this is embedded in the distinctive story of national parks in England and other parts of the UK: a particular framework for making sense of the contemporary debates over access, and the prospects of achieving a 'green gym', that embraces the notion of a fair degree of freedom of access.

Over three centuries, larger landowners in the UK appropriated and fenced-off previously available, accessible common land (Curry 1994). Those excluded sought other lands for access in walking, having 'lost' their own land for working. The Enclosure Acts and subsequent Game Laws introduced during the 1870s effectively served to restrict access to 'open' countryside, thus depriving many families of their livelihoods. By the nineteenth century mill-owners encouraged recreation outdoors for their workers to improve their health and workability, compensating working conditions with planned holidays – 'Wakes Weeks' – and work organized coach trips, later to be realized more politically through the acknowledgement of enjoyment on individuals' own ability, seeking the right to roam, a movement that included mass trespassers where landowners and gamekeepers sought to resist popular access, as those driving the industrial revolution continued to remove land, used previously for recreational purposes, from the public domain at an alarming rate (Waterson 1994). Partly in response to this loss of access, philanthropists (e.g. Octavia Hill, Robert Hunter, and Canon Rawnsley) formed The National Trust at the end of the nineteenth century.

There were other stimuli affecting demand for recreation in the countryside before the end of the nineteenth century. A critical one came in the form of the Commons and Open Spaces Preservation Society (COSPS) – a group of individuals who became the first 'pressure group' to lobby for access to the countryside. They introduced a completely new dimension to an already complex situation, the status of class within the context of a recreational user. The formation coincided with Parliament designating the first Bank Holiday in 1871 (Curry 1994), an Act theoretically encouraging the use of leisure time by everyone. The COSPS saw a main purpose of its role to save the countryside for the learned scholar, thus largely excluding the workingman and woman. Its early members included such well-known members of the Arts and Crafts Movement as William Morris, social reformers like John Ruskin and the poet William Wordsworth who, among others, promoted and reinforced links with the Romantics. Poetry, literature, music and paintings had already begun to promote the countryside in a way that appealed strongly to the educated person.

So powerful was the 'high-brow' rural image they portrayed engraved upon the subconscious mind of today's visitor, the subsequent 'removal process' has presented postmodern countryside planners wishing to promote policies of 'a countryside for all' with perhaps their most difficult psychological barrier to date. Albeit, coming from different directions, by the end of the nineteenth and early part of the twentieth century, there were several 'pro-access' pressure groups lobbying the government. Their common aim was to highlight the direct relationship between continued industrialization, decreasing levels of rural accessibility and the effects of social exclusion upon the national workforce. In an attempt to restore the balance and represent their own interests, landowners formed their own groups – ones invariably associated with limiting access, exemplified respectively in rambling and caravanning clubs, and the Field Sports Society, mainly of landowners for their 'game'.

In response to conservation bodies rather than pressure from recreation groups, the government began to examine the possibility of adapting the US version of national parks to the UK setting before the Second World War. However, US parks were very much based around land as wilderness, with overall access, for recreation/leisure more generally associated with calm and a lack of wildlife disturbance (see Chapters 1–3 this volume). But in the UK, the eventual selected areas were populated lands owned in part for production, and also industrial, creating challenges for accessibility. Ironically, perhaps, the particular areas selected were sites of lower population, and generally less intensely cultivated land, that made the issue of accessibility less politically conflictual than it might otherwise have been, although current debates in the UK towards new national parks, especially in southern England, confront more politically contested ownerships. The eventual post-war parks provided a compromise. Full access, 'right to roam' for potentially everybody, led by the large-membership Ramblers Association, did not transpire. Instead, access was gained through identified and agreed pathways and other specially allocated areas.

Responding to a continually changing environment and increased public pressure, the 1968 *Countryside Act* included enhanced and extended access to wider areas of land, partly to shift the pressure upon relatively more accessible parkland. Not until the late 1990s did access supported by law rather than landowners' voluntary concessions generally occur in the UK countryside, and particularly in the kinds of extensively farmed land in the national parks. The 2000 *Countryside Rights of Way Act (CROW Act)*, significantly responded to the popular pressure for full access to all land with respect for areas of privacy, cultivation and stock, and provides for unimpeded access to all open land, so long as it is situated more than 600 metres above sea level, as is typical of most UK national parks.

Farmers and landowners indicated their concern that extending access may not increase usage as claimed; rather simply allow those already using rural space more freedom. However, a longitudinal study undertaken in the Yorkshire Dales National Park in 1998, 1999 and in 2001 investigated this fear and phenomenon. In the village of Burnsall, a significant access point to a network of established paths in the heart of the Yorkshire Dales and at several information centres scattered across the national park, it revealed a rising trend of 'new users'. Yet this was not the case from within Bradford, close to the southern boundary of the park. Walkers reaching the park by car became more numerous; those reliant upon public transport did not: bus services accessing the park area from the city had declined.

Of course, the notion of the 'green gym' applies to everyone. As the previous section has discussed, the idea of access has for well over a century been a quest across a wide population. Only in more recent years has the opportunity for the green gym been extended further. Indeed, the *CROW Act* (2000) contains reference to 'disability', or more politically correct 'accessibility'. Its provisions need to be considered alongside the more generally applicable *Disability Discrimination Act* (1995). Together, these Acts require national parks to make 'reasonable adjustments' for individuals using the parkland who have disabilities, now enhanced by moving the responsibility of evidencing shortfall in facilities from the users to the agencies themselves.

At and around the Starkholmes car park at the entrance to the Derwent Valley, close to Sheffield and known locally as 'Little Switzerland', innovations for better accessibility for those with disability included a park refreshment cabin with a low counter for serving wheelchair users, improved signage and prominent parking spaces for drivers with a physical disability; a range of cycles on offer in the hire station catered for an increased range of bikes modified for physically disabled cyclists, and the tandem, ideal for the blind user. Subsequent investment has further brought the facility in line with the latest technology and contemporary acknowledgement of experience.

So-called 'film tourism' provided another boost for parks access, but with interesting consequences for disability access. The BBC's *Pride and Prejudice* (1995) was filmed at Lyme Park, one of the four National Trust properties within the PDNP. Very wet ground inhibited disability access. The National Trust provided a much shorter circular walk around the much smaller lake car park, made with an entirely hard surface, wide enough for two people to walk side by side (essential for blind people), passing places for wheelchairs, seating, a footbridge with open slats that facilitated the sound and humidity associated with water running over stones.

In terms of disability, landowners' concerns have been largely allayed, such as the perceived requirement to remove features they feared might be considered obstructive for disability. However, rather than to grade the user with regards to what s/he can manage, the countryside becomes graded in terms of degrees of accessibility. A charitable body, The Fieldfare Trust, has recently produced a *Countryside for All Good Practice Guide* CD-ROM that provides the latest guide for managers as to how they can achieve least restrictive access, how to deliver accessible countryside path networks and how to survey and audit paths in relation to the needs of (physically) disabled people.

Current challenges/opportunities for adventure activities in national parks

While the generality of recreation and leisure has remained an enduring feature of the early intention of national park designation, they have become the focus of much contested debate in terms of one of their most significant more recent challenges – adventure activities. Most recently, British national parks have become attractive as one kind of 'green gym', in more popular pursuit, energetically and less so, of exercise linked with purposive health achievements. Their diversity of surface, intensity of slope and relatively unpolluted environs contribute to this attractiveness. This shift of interest has been accompanied by the process of commodification, combined with, as this section considers, significant new opportunities and challenges in the arena of national parks management.

Research into the role of recreational activities within national parks is diverse and incorporates a plethora of research areas. Issues such as the economic impact of specific activities within a park setting (Moran *et al.* 2006), the perception of recreational activities within national park boundaries (Miller *et al.* 2001), user interaction within different recreational activities

(Carothers *et al.* 2001) and more general management features relating to the paradox of conservation over socioeconomic development (McCarthy *et al.* 2002) show an ever-increasing body of discussion. In many ways, this has similar connotations to what Nichols (2000) adapts from Priest and Gass as the 'Central paradox for outdoor leaders' (1997: 122). They refer to a phenomenon where on the one hand, you have the importance of risk and on the other you have a necessity to focus on safety.

In this section, particular attention is given to the central paradox for national parks that is composed of three factors that allow increased challenge when considering and deciding on practice and policy. Individuals and groups seeking to use national parks have become increasingly diverse and fragmented from the early policy intention to provide recreation characterized by 'peace and quiet'. This 'triangle of recreation' involves the traditional focus of conservation and environmental protection, the planned socioeconomic incentives and the ability to allow each user group's freedom to participate in their chosen activity within the boundary of regulation a park provides. In this sense, it is little wonder that management of recreation within national parks has become more of a contested issue in recent years.

The fragmentation observed within specific adventure activities is an interesting addition to this discussion and raises new dimensions in relation to the future opportunities and challenges faced certainly within British national parks. An example of this can be seen in rock climbing and specific 'offshoots' that can be seen within all national parks in the UK (the authors here use their own observational data conducted for the Moors). For example, rock climbing has morphed into a variety of disciplines in national parks, and the Peak Park is no exception, perhaps even the more intense due to the proximity of significant urban–industrial populations. This contrasts significantly with the initial crag exploration of the 1940s and 1950s.

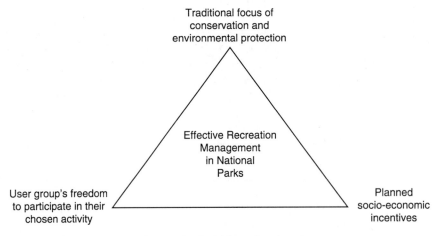

Figure 13.4 The 'triangle of recreation' within national parks.

Traditional rock climbing using the placement of gear (or not as with the early cases) made way for sport climbing and bolted routes of increased intensity and challenge in the 1970s and 1980s. This in turn allowed for the development of bouldering – an initial form of training that incorporates intensive moves at a low level without the use of ropes and needing a padded mat, rock boots and chalk bag (no specific rope skill required). Interestingly, this focus on intensive skill of movement and technique without ropes has also been one of the reasons for the increased growth in harder traditional forms of climbing. Mountain biking too has seen a growth within the PDNP, from cross country/'enduro' to downhill and freestyle movements. Even the distinction between the rambler and scrambler is becoming blurred in some cases as adventure is combined with softer exploration.

It is possible in this respect to identify that the fragmentation of certain outdoor/adventure activities is inevitable and necessary within the constant evolution of adventure recreation. The use of the term 'green gym' is surprisingly appropriate here, and like the constant innovation within fitness and exercise machines in modern gyms (fragmenting the workout of individuals), the modern British national park too has to contend with the innovation of adventure recreation. Why does this fragmentation exist in such a poignant state? Is it due to the significance of user-groups to assign and cement identity and how spatial practices help reinforce this? Is it the general evolution of consumer behaviour proposed by Foxall (1993) focusing on the adoption and reinforcement of recreational activities? As with classic motivational theory, the reasons are varied and as a result a pinch of each could be more effective. The proposition offered here is that understanding the significance of this evolution and the impact that it has on the 'Triangle' helps initially understand the opportunities and challenges national park authorities face with modern adventure.

Challenges of commodification

The examples provided above allow for further debate into the issue of where commodification fits into the management equation. This in essence is one of the core reasons why recreational activities increase in popularity and why the future of managing adventure in national parks needs to take into consideration why demand increases. Focus should not only be on the commodification of nature (through the construction of national parks), but on the commodification of adventure located within them (Shepherd 2002). Castree's (2003) discussion of 'capitalist commodifcation' refers appropriately to nature as an object and reinforces the complex discussion surrounding the management of recreation. With adventure recreation we have the potential socioeconomic benefits attached to a commodified adventure activity. The evolution of the activities above has allowed a business model-approach to be applied: incorporating issues such as competitive advantage, funding through sponsorship and promotional vigour. At the same time a balance has to exist between this and the process of how commodification impacts on group behaviour and furthermore, the interaction with other recreational users who use the same space. This could account for the increased variety of mountain biking and bouldering areas within the PDNP. Thus the value, meaning and identity-significance individuals feel, associated with national parks underlies substantial change in the character and intensity of opportunities sought from national parks.

Community participation in planning activity

In the year 2007, the PDNP welcomed in excess of 32 million day visitors or excursionists and 3 million longer stay visitors. The economic impact of these visitors on the host community has been estimated at £1.285 billion. There are currently in excess of 24,000 people directly involved with delivering the tourism product to these visitors within the Peak District itself (Destination Management Partnership Annual Report 2007). It should also be recognized that the National Park itself is characterized by an ageing population with a diminished set of skills and capabilities required by regeneration and diversification needs. The community itself is risk-averse, protective of its physical, social and cultural environment, and has difficulty in articulating the often confusing and contrary expectations of cross-border governments; the PDNP comprises territory of the regional development agencies of East and West Midlands as well as Yorkshire and Southwest Lancashire.

The Destination Management Partnership (DMP) was established in 2005 and was preceded by two committees operated through hybrid public–private sector organizations: the Peak District Sustainable Tourism Forum and the Derbyshire Sustainable Tourism Forum. The Peak District Sustainable Tourism Forum was largely comprised of private sector industry and academic representatives; the Derbyshire Forum was largely the domain of the local government and public sector.

The budgets for the operation of the DMP were largely reassigned from the district council tourism provision and from the Derbyshire and District Economic Partnership Rural Action Zone to the DMP. The DMP principally exists to promulgate strategic marketing and to ensure quality and uniformity of product delivery to visitors and by operators. Research to support these strategic initiatives is largely provided by consultants who are working on a national level to develop solutions to dataset blanks (for example LJ Forecaster of Edinburgh who provide data for the accommodation sector on occupancy).

The DMP do not currently charge a membership or subscription fee to their subscribing tourism industry members. Upwards of 40 per cent of current accommodation providers within the DMP's domain have obtained some quality assurance to assist listing with Visit Britain. There is little question that the current value of the DMP is provision of secure destination management distribution systems (through the local government funded tourism information centre network) and in promulgation of quality assurance for the accommodation sector.

The immediate issue is provision of research and strategic policy resources to support the stakeholders providing products for day visitors and excursionists. These visitors form 95 per cent of the total demand within the National Park. The former 'non-government organizations' supporting the Tourism Forums no longer exist and therefore funding of research and development for the non-accommodation sector has largely been abandoned to the LTAs with such ventures as Business Link and Pathfinder, two examples of schemes in English counties to assist and advise businesses to enhance their product to the market. However, the provision of services under these two products has been curtailed or substantially revamped, thereby limiting the support being offered to micro-business.

Local communities have relied for many years on public sector provision in training, quality assurance, marketing and planning. Aspects of this provision, with the exception of marketing and distribution, are now in doubt. Local communities may therefore need a new injection of human and financial resource to support the tourism industry stakeholders currently delivering to the excursionist. For example, a European Union funded activity from 2001 has included enterprise and community development through small business. The Derby and Derbyshire Economic Partnership (DDEP) was established to oversee projects and funding through a further body, the Rural Action Zone (RAZ). The strategy has been to provide mechanisms to support several projects, which include supporting new initiatives in sustainable agricultural pursuits, a new environmental economy and the Pathfinder project, which was designed to explore and experiment with new ways of delivering public-sector funded services to support business in rural communities. Specifically, the former project has funded a Peak District foods campaign and the latter has funded business advisory services and attempted to simplify procedures for obtaining grants to fund diversification. Both these examples broadly meet the goals of strategic regeneration activities and exemplify the local area's focus on building a store of available competence and capacity.

However, in the past two decades some enterprising new migrants decided that they could no longer wait for diversification and regeneration activities derived from the public purse and specifically from local territorial authorities. One of the entrepreneurs decided to establish a farm holidays bureau for the myriad of small bed-and-breakfast and emerging organic food operators in the National Park. The bureau has proven to have been a success and upwards of 50 operators now offer more than 1,000 beds during the year using an online booking system. Not only does the service provide an entry point for new entrepreneurs and micro-business owners, it also supports the organic foods of the Peak District, which has another, publicly funded marketing and promotion outlet. The RAZ has also linked these projects to quality assurance action plans through the DMP.

Ray (1998) wrote that endogenous development in rural locations is dependent on communities linking actors and agencies to recognize and celebrate milestones in a development cycle, thereby creating identity and fomenting unity in that community. The first phase of Ray's model is territorial control of products or commoditization, which has largely been achieved within the National Park. Mode II is promotion to outsiders, and this has largely been accomplished by the DMP. Mode III identifies that local community participation in developing and promoting products is widespread and that the local community recognizes that level of achievement, perhaps through Leader projects or the local Pathfinder funded capital building project. Again, this has largely been successful in the National Park, although the authors observe no room for complacency or abrupt termination of community-based initiatives. Ray's mode IV is the key. At this phase the community cooperates fully and identifies with the products and confirms the local identity. The community accepts and valorizes local knowledge. In terms of the Peak District National Park, such products as 'Bakewell pudding' (a locally produced sweet delicacy) or Hartington Stilton (a local blue cheese) become registered and iconic products for visitors in the same fashion as the French utilize the *appellation d'origine contrôlée*.

This community acknowledges territoriality, cultural artefacts and, in the United Kingdom context, the role that regionalism, postmodern complexity and European-ness (EU role) has taken and played. However, the PDNP has some way to go before the community stakeholders can identify and concur with Ray's Mode IV outcomes. The PDNP could examine structures and policies to afford greater support to micro-businesses in the core areas identified in: skills enhanced, social capital expanded, and business supported and quality assured that essentially support the needs of the majority of the 32 million excursionists. Research by Kneafsey *et al.* (2000) endorses the endogenous approach to building social capital and the ways in which natural and cultural resources are valorized through several options involving the concepts of embeddedness, complementarity and empowerment. In the PDNP it is perhaps salutary to note that cultural identity is still to be confirmed, recognized by the community and clearly articulated through integrated marketing. What will be important is that enthusiasm remains contagious and funding is continuously available to allow the community to clearly articulate the identity in future. This has previously been recognized in the underestimation of practical problems and occasional patchy support for the endogenous approach to destination development agendas (Foley and Martin 2000).

A community learning approach to this valorization and dissemination is called for and clear strategic plans recommended for facilitation of the learning that has occurred through both formal and informal process in the PDNP (Wiltshier 2007). In addition to the capacity building through 'learning destinations', there is a parallel need for economic impact assessment as the regeneration and diversification activities become embedded here. This second phase will not only validate the community learning process, but will independently validate the performance outcomes that the public sector funding organizations expect. This phased capacity building and evaluation strategy is espoused in the 'New Deal for Communities' and 'Neighbourhood Renewal Strategies' that have been promulgated in the United Kingdom in the past decade (Diamond and Liddle 2005). The PDNP is achieving a level of understanding commensurate with government expectations of partnership, innovation, creativity and entrepreneurship, coupled with a healthy respect for risk management and strategic focus. The public sector now needs to clarify the revised identity and vision with new objectives that reflect an open decision-making process for action plans. This new vision needs also to articulate the complex and often conflicting needs of stakeholders like the NFU and Natural England while maintaining democratic rights across boundaries (Taylor 2003).

Within this organizational and participatory framework and its complexity of players, the Park is seeking creatively to respond, within its developmental complexity, to the dimensions of accessibility that provide key challenges in the management and vision for national parks.

14 A ticket to national parks?
Tourism, railways and the establishment of national parks in Sweden

Sandra Wall Reinius

Introduction

The history of national parks in Sweden and the factors that contributed to the development of those areas can be analysed from various perspectives, such as policies for colonization, taxation, industrialism, exploration of nature, environmental philosophy, protection of nature, nationalism, and tourism (Ödman *et al.* 1982, Sörlin 1988, Hägerstrand 1991, Götmark and Nilsson 1992, Mels 1999, Nilsson Dahlström 2003). Such research is not purely 'academic' as an understanding of the historical dimensions of national parks is invaluable for appreciating the reasons for present-day conflicts between user interests and the environmental history of the region (Whited *et al.* 2005). This is particularly the case in relation to the historical process of establishing national parks, as it is relevant to understanding what the original motives and purpose were for nature reservation, how it was carried out, and by whom.

The point of departure for this chapter is the extent that commercial tourism interests, and particularly the railways, were a major factor in the establishment of national parks in the nineteenth and early twentieth centuries. This argument has been paramount, for example, in the works of Runte (1974a, 1974b, 1987, 1990), who argues that the western railways played a crucial role at the turn of the twentieth century for the establishment of national parks in the United States, including parks such as Yosemite and Yellowstone – an observation also supported by subsequent research in the US (e.g. Orsi 1985, Sellars 1997); while the significant role of railway companies in park development has also been emphasized in Australia (e.g. Hall 1992) and in comparative studies of the New World national parks of Australia, Canada, New Zealand and the US (Hall and Shultis 1991). The relationship between tourism and national parks acted to serve the interests of both park advocates and the railways. The railway companies promoted tourism in order to utilize and increase existing capacity, the construction of new lines and grow their profits, and the preservationists promoted tourism in order to gain public support for the national parks. However, the role of railways in the origins of national parks has not yet been addressed in detail in a Nordic context.

Nevertheless, there are significant parallels with the North American experience of park creation. For example, in the late nineteenth and early twentieth centuries many of the landscapes that later become designated as parks or as equivalent reserves had already started to find a place in expressions of national landscapes. For example, Romantic nationalism was a key element among Nordic artists, composers and writers. Perhaps nowhere was this clearer than in Finland where lakes and nature became clear cultural expressions of national identity and independence. Artists such as Eero Järnefelt and Pekka Halonen, composers like Jean Sibelius and writers such as Aleksis Kivi and Johan Ludvig Runeberg, all contributed to the development of an aesthetic appreciation of landscapes that are now synonymous with Finland's national heritage and its conservation in national parks (Häyrynen 1998, Lappalainen 2000). In addition, Saarinen (2005) maintains that the aesthetic values of nature for tourism were deeply involved in the process of the first Finnish national parks in the 1930s. As Finnish-born explorer Adolf Erik Nordenskiöld – sometimes referred to as the father of the Nordic national park movement – commented in 1880 with respect to establishing state-founded parks to preserve examples of the Finnish national landscape, 'If these days people are prepared to pay millions for paintings by the great masters, imaging what figures will be offered a hundred years hence for a true picture of the Fatherland ... Such pictures can still be found in most parts of the country, but they are steadily disappearing' (in Lappalainen 2000: 14).

However, the railways link to tourism and nature conservation is perhaps nowhere more pronounced than at Punkaharju in the southeast of Finland where the first conservation area in Finland, the Punkaharju State Forest, was declared in 1843. This was 40 years after a visit by Czar Alexander in 1803 when he ordered a ban on log cutting and the traditional means of fertilizing the lands by burning the forests. Runeberg, a poet who wrote the words of the national anthem 'Vårt Land' (Our Country), was a great early promoter of Punkaharju, 'who found it to be closer to his ideal of a beautiful and harmonic natural landscape ... Punkaharju was closer to a pastoral landscape, natural and yet idealized. In addition, it could be regarded as more Finnish: rapids and waterfalls existed nearly everywhere, but glacifluvial ridges surrounded by a lake landscape were something not many countries could boast' (Häyrynen 1998: 417). In a Finnish context Punkarju was not unique; however, its location gave it much better transport connections than other national landscapes, such as Koli, which did not receive any protection until 1923. Initially, these were a main road, but by 1906 this also included a railway station that served the substantial tourist traffic to the region that were using several hotels, including the Valtionhotellu, which is the oldest commercial tourist lodging house in Finland, built in 1845. Nevertheless, the development of the Finnish railway system to include Lapland, which was the area in which many of the first Finnish national parks were developed, was behind that of Sweden with the then northern terminus opening in

Rovaniemi in 1909 and the line from Rovaniemi to Kemijärvi opening as recently as 1934.

The development of the Swedish railway network and, in particular, the construction of railways in northern Sweden, runs parallels in time with the debate about the conservation of Swedish nature. The northbound railway line that reached the cities along the northeastern coast was completed in the beginning of the 1890s. At that time the development of the railways was regarded as symbolizing something more than transportation as 'it was a symbol of success and future' (Sörlin 2000: 17). Furthermore, the railway to northern Sweden was also a means of bringing both northern Sweden and Norway closer to the rest of the country (Oredsson 1969).

Through comparisons with the experiences of other park systems, this chapter describes the parallel processes of the development of national parks and railways in Swedish Lapland, which occurred a number of years ahead of Finnish Lapland, and examines whether there was a connection between the development of the railways and the creation of protected areas. Which were the motivations and justifications behind the establishment of the first national parks? What role did the railways and the tourists play in the early twentieth century for the establishment of national parks in northern Sweden? The result and analysis are based on an explorative literature study of historical documents on national parks, railways and tourism in northern Sweden. The main historical sources are private bills and government bills to the Swedish Parliament (1904a, 1904b, 1907, 1909a). Other useful historical sources have been Conwentz (1904), Starbäck (1909), Svenonius (1908), the Swedish Association for the Conservation of Nature (1910), and articles in the yearbooks from the Swedish Touring Club (1896–1910). Analysis has been focused on the statements, debates and proposals regarding the establishment of protected areas, and potential links to railways and tourism.

The study area, known as 'The Lapland Mountains', is located in the northwest subartic region in Sweden, in the county of Norrbotten, and stretches from Abisko to Kvikkjokk. Five of the 28 national parks in Sweden are situated in the study area, although only the national parks of Abisko, Stora Sjöfallet and Sarek, which were established in 1909, are part of this analysis. Significantly, in terms of understanding the history of national parks and tourism in a European context, these were also the first national parks to be established in Europe. The other two national parks were established in 1920 (Vadvetjåkka) and in 1963 (Padjelanta). Nature conservation together with Sami reindeer herding and tourism are the main land uses in the study area.

The time demarcation of this study is from the end of the nineteenth century to the first decades of the twentieth century, approximately 1880–1910. Table 14.1, provides a chronological overview over some of the main events referred to in this study.

Table 14.1 Chronological overview of national parks, tourism and railway development in the Lapland Mountains

National Parks	Tourism	Railways
Scientists start to discuss nature protection from 1880	Swedish Touring Club (STF) 1885	First construction fails in the 1880s
Parliament discussions mainly 1904–1909	Stora Sjöfallet Tourist cabin 1890	Kiruna-Narvik 'Malmbanan' opened in 1902, official inauguration in 1903
Established from 1909	Abisko Tourist Station 1902	Interior line, the northern part opened in 1911
	Saltoluokta Tourist Station 1913	

Setting the context: perspectives on nature protection

According to Saarinen (2005), the politics of nature and its spatial implications for landscapes are affected by ideas of nature and created images (see, also, Williams 2002, Brooks 2005). Different interpretations of what constitutes conservation, such as conflicts and disagreements between different forms of land use interests, can be traced back to differences in environmental philosophy and ideas of nature (Ariansen 1992). The most prominent view of nature in Western culture is derived from the anthropocentric Judeo-Christian tradition, which is predominantly interpreted as being founded on a belief that nature stands as the antithesis to culture and that humans should tame or dominate nature (Kellert 1995, Ewert and Stewart 2004). However, in a Nordic context, Riseth (2005) claims that the indigenous Sami thinking focuses more on the unity between nature and culture, which contrasts with nature conservation interests that hold a viewpoint rooted in the dualism between culture and nature.

Until the late eighteenth century, wild nature had no positive connotations in Western culture, but under the influence of romantic movements it changed in the nineteenth century to be regarded at least by the cultural elite as something like Eden; with the aesthetically championed wild areas requiring protection and conservation, not least for recreation opportunities (Johannisson 1984, Oelschlaeger 1991). Nevertheless, parallel to these romantic ideas was a focus on utilizing spaces for economic development. Runte (1973, 1974a, 1977) has examined the establishment of national parks in North America and found that national parks were only established if the areas were considered valueless for profitable lumbering, mining and agriculture. Areas that could not give annual economic output from the land or forest, termed 'worthless lands' or 'empty nature', could thus be set aside for conservation and thereby bring in revenues from tourism (Runte 1977). Similarly, in the Nordic context, Broberg and Johannisson (1986: 66) maintain that 'areas that do not coincide with economic interests' were of interest for nature protection.

At the end of the nineteenth century the Swedish nation and people began to change their relation to nature as the experience of direct contact and practical work in nature began to be replaced by a perception of nature as a scientific object (Ödman et al. 1982). Such changes ran parallel with the transition of Sweden from an agrarian and rural base to an urbanized and industrial society. Ödman et al. (1982) argue that this dissociation from nature as something negative and associated with work led to an increased importance of idealized and often romantic images of nature; thus, nature experiences gradually became more connected to the created aesthetic image than to the everyday experience of contact with 'wild nature' (see, also, Walter 1989). Not coincidentally, during the same period the new aestheticized nature became a key symbol of Swedish identity and nationalism, as in other Western countries. Ideas of protecting nature emerged at a time when these countries were industrializing and urbanizing, with changes in the socioeconomic situation and the environment, and the modern city threatened collective values and the national identity (Walter 1989, Hall 1992, Sörlin 1999, Hall and Page 2006). As in the United States, the reaction to increased industrialism and urbanism was fostered by the wealthy 'leisured' class, 'aristocrats', philosophers, artists and intellectuals, who encouraged nature conservation (Hall and Shultis 1991, Lövgren 1992, Sandell 1995, Neumann 1996, Mels 1999). According to Mels, rural unemployment, population growth, distressed areas and the end of the union with Norway (in 1905) contributed to a need for defining Swedish self-identity. The unstable social situation worried politicians and they expressed a necessity of finding unifying symbols for the people. The term 'national park' therefore underlined the patriotic importance of nature, as a symbol of the unity of the Swedish people and a supposed common experience of nature (Sörlin 1988, Mels 1999, Sundin 2001).

The ideas of nature – and the ideas of national parks – are also closely linked to power relations in a centre–periphery framework, as the centre has spatial control over the periphery (Riseth 2005, Sandell 2005). This is particularly important because of the way that many of the relatively high-quality natural areas in which national parks are created are in a relatively peripheral space with respect to the cosmopolitan cores in which, not only decisions with respect to park creation are made, but which are also often the intellectual homes of the national park idea (Hall 1992). In the Nordic countries, Lehtinen (2006) uses the expression 'postcolonial north' to describe the historical colonial pressure on Sami peoples from the scientific expeditions from the south, and full-scale exploitation of natural resources, later accompanied by the establishment of national parks, also introduced from the outside. The establishment of national parks in the Swedish Lapland Mountains was facilitated by large areas being owned by the state, and the absence of economic resistance against the national parks, unlike in many of the American cases of park creation (Runte 1987, Sellars 1997). Indeed, it is now generally acknowledged that the decision to establish parks in the north was undertaken with a lack of recognition that the mountains were part of the Sami cultural landscape. Instead, for the decision-makers in Stockholm the mountains were regarded as a natural landscape, without cultural heritage or human presence and activities, and the Sami people were not invited to participate in the land use debate (Mels 1999, Sandell 2005, Schough 2007). Such seeming cultural blindness with respect to the extent that the 'natural' lands that constituted the first national parks were actually part of a complex cultural and ecological landscape, has parallels with the attitudes and experiences of Australia, Canada, New Zealand and the United States (Hall 2000).

Scientists' and tourists' interests in the Lapland Mountains

As in North America and Australasia, scientific interests in the Lapland Mountains were developed in parallel with the tourist interests (see Chapters 3 and 7). The interior of northern Lapland came to be known to outsiders as a result of different travel expeditions from the mid-eighteenth century on. However, awareness of Lapland both in Sweden and elsewhere in Europe really started to become significant from the 1870s, when the Lapland Mountains were mapped and books from the scientific expeditions were published (Hamberg 1896). Svenonius (1908) wrote about Lapland as a tourist country and the mountains were described as pristine and magnificent with a monumental scenic beauty, which could offer tranquillity, meditation, adventure and wilderness, factors that were very important in attracting the first tourists (Vesterlund 1901). For example, Svenonius (1908: 278) described the small village of Kvikkjokk as a 'Nordic Paradise', although very few people could travel there, because the travel was expensive and time-consuming. In addition and again bearing some similarity to the American experience in the western states, the establishment of several small scientific research camps in the high mountains and their surroundings also served to attract the early mountain tourists.

During the second half of the nineteenth century alpine clubs were established in Europe. The national alpine club of Norway was founded in 1868 with the aim of promoting hiking in the mountains, and was the first organization in the world to use the word *tourist* in its name – the Norwegian Tourist Club. The Swedish Touring Club (STF) was founded in 1885, following the Norwegian example and the concerns of academics in Uppsala, who had interest in the Swedish high mountains as a recreation area (Sehlin 1986a, Sandell 1995). The construction of a railway (in 1882) crossing middle Sweden and the mountains of Jämtland contributed to the foundation of the STF, as the mountains became more accessible to visitors. According to Sillanpää (2002), the railway and the STF played a key role in the process of tourism development in Jämtland, where the club effectively promoted the mountains and their scenery and healthy air. The STF also wanted to develop and facilitate tourism more generally in Sweden, and under the slogan 'Know your country' the club formulated three main goals: to provide information, for example guide books; fieldwork, for example trail marking; and travel service, for example ticket selling (Carlquist 1986, Sehlin 1986b). The slogan had the intention to strengthen national unity through knowledge about, and love to, Sweden (STF 1903, Liljekvist 1906).

In order to facilitate the scientists and the tourists in the mountains of Lapland, the STF gave financial support to some scientists and established overnight cabins, boat connections and marked trails from the 1890s. Tourism in the Lapland Mountains first started in Stora Sjöfallet (the Great Waterfall), followed by the development of tourist cabins in the Lake Torne area – Abisko and Vassijaure (Anrick 1935). Since the Lake Torne area was also of great interest for scientists, the Natural Science Association in Stockholm bought one of the buildings from the railway company when the railway was constructed in the beginning of the twentieth century and the world's northernmost scientific research station rapidly became a centre for the Swedish scientific elite (Sörlin 1988).

Even at the beginning of the railway era, the Lapland Mountains were difficult to access for many people, due to the long distance and high costs. Even though the STF had the intention to promote tourism for all socio-economic classes, the early tourism was a strictly upper middle-class phenomenon (Forsslund 1915, Ödman *et al.* 1982, Thaning 1986, Sörlin 1988). Nevertheless, the club rapidly expanded in terms of hiking and skiing trails, boat connections, cabins, and information and publications about the Lapland Mountains, as well as members – from just below a hundred the year it was established to 25,000 at the turn of the century, and almost 50,000 members in 1909 (STF 1910).

National Park establishment in the Lapland Mountains

In Sweden, the ideas of nature protection came to be realized after a visit by the German Professor Conwentz in 1904 (Swedish Association for the Conservation of Nature 1910). Conwentz (1904) expressed his concern over the threat to nature from inappropriate development. However, his approach to conservation was quite strategic for its time; first he wanted to undertake inventories, followed by nature protection measures, and then he wanted to spread information in society about the protection. Moreover, according to Conwentz (1904: 34), organizations like the 'widespread and powerful' Swedish Touring Club, should promote conservation of the natural landscape. The ideas held by Conwentz had a direct influence on the Swedish Parliament member Starbäck, who in a proposal to the Swedish Parliament argued about which actions should be taken regarding nature protection. Starbäck was familiar with how nature was protected in the United States; above all, he referred to Yellowstone, but he was also aware of protection in

Prussia and Austria. Starbäck saw how economic and industrial interests exploited nature and in a romantic expression, reminiscent of the transcendentalist John Muir, argued that it was time for a reaction because people need nature for its 'original harmony' (Swedish Parliament 1904a, 1904b, Starbäck 1909).

In 1908, the year before the first national parks were established in Sweden, Svenonius wrote about 'the wonderful waterfall of Stora Sjöfallet':

> because of its harmonious beauty, it should be put foremost in the Nordic countries, . . . here you will never be disturbed by the railway or industrial establishments. An enormous area of several square kilometres will in all probability become a kind of *national park*, or an absolutely protected area with respect to fauna and flora.
>
> (Svenonius 1908: 280)

Svenonius (1908) wrote that those who were responsible for the tourism development in Sweden should in an effective way make the journey easier and cheaper for tourists to visit the proposed national park, but without disturbing nature. Some decades before Svenonius, Nordenskiöld had already proposed establishments of national parks (called *Riksparker*) in the Nordic countries. The central theme for Nordenskiöld had been the patriotic vision of nature and his initiative was motivated by an idea of creating natural areas where one could experience an outdoor museum (in much the same way that the artist George Catlin proposed a nation's park in the United States in the mid-nineteenth century). Even though many parts of the country were exploited by people, Nordenskiöld claimed that there still were areas where nature had been preserved in its original form, for example in the mountains, because of difficulties in access, and because the land and forest had no economic value (Swedish Association for the Conservation of Nature 1910).

The majority of the advocates of the idea of nature protection were academics from the universities in Uppsala and Lund, as well as artists and writers, and representatives from geographical and tourist organisations (Conwentz 1904). As a result, the Swedish Association for the Conservation of Nature (SNF) was founded in 1909. The SNF and STF cooperated very closely, and Wahlberg (1986: 231–2) has even claimed: 'what the STF wanted to show the people, the SNF wanted to preserve for the Swedish people'. Indeed, for several years the organizations even had the same person as chairman.

A review of parliamentary debates suggests that four factors appeared to primarily motivate the Swedish parliament with respect to nature conservation in the region: the importance for science; the comparison with other countries that already had established national parks; the unprofitable nature of the land (that is, no economic value from the forest); and the national parks' expected importance for tourists (Swedish Parliament 1904a, 1904b,

1907). The attraction of national parks as tourist sites was often mentioned, including with respect to the provision of accommodation for travellers as well as for managers (Swedish Parliament 1907). Furthermore, the Parliament bill from 1907 stated that it would be appropriate to charge a fee from the visitors to enter the parks, and that such a fee could be used for road constructions. In the same text the interior railway is mentioned: 'If the interior line will be constructed, it means that the area will come several daytrips closer to the rest of the world. . . . If that will be a reality, we can expect a flooding of tourists' (Swedish Parliament 1907: 69).

This interaction between nature protection, transportation and tourism can also be seen in the Government bill from 1909 where there was a statement regarding the government's involvement in tourist transportation, and that this could bring an income from these uninhabited areas, as well as income for the interior railway (Swedish Parliament 1909a). Starbäck (1909) believed that the national parks would attract tourists, but he regarded that as a disadvantage for nature, and therefore it must be associated with good management and guardianship. With respect to this, Starbäck underlined the importance of regulations and the first paragraphs in the *Swedish National Park Act 1909* concerned what was forbidden and what was permissible in the parks (Starbäck 1909, Swedish Parliament 1909b).

In June 1909, the Swedish Parliament decided that nine national parks were to be established. Starbäck (1909) regarded the chosen areas as representative types of Swedish nature. He also emphasized that the basis for selection was mainly scientific and that the parks were of the upmost importance for the development of science. According to Starbäck, the strongest motives for nature protection were: to stop exploitation of nature; to be able to follow the development of nature from a scientific perspective; the scenic beauty; and the love of nature. Three of the first national parks in Sweden were established in the mountains of Lapland: Abisko, Stora Sjöfallet and Sarek. Abisko National Park was first proposed by Svenonius, and the objective was to protect the wildlife from the tourists, but at the same time to provide tourists with recreational opportunities in one of the most beautiful areas along the northern railway (Swedish Parliament 1907). The closeness to the natural scientific research station was also mentioned in the report. When Starbäck wrote about Abisko, he described the rare flora in particular (Starbäck 1909). The national park of Stora Sjöfallet was first proposed as an enormous park, including the bordering Sarek. In the same way as Svenonius, Starbäck was fascinated by Stora Sjöfallet with its 'undreamt beauty', and the large glaciers of Sarek (Starbäck 1909).

The national park proposal met few opposing opinions, and it was accepted without objections in the Swedish Parliament (Swedish Parliament 1909b). Hägerstrand (1991) has given further details about the persons involved and the official trajectory of the establishment of national parks, but above all he emphasizes how information was collected and how individuals (particularly scientists), authorities and interest groups gave their opinions in the case. For example, Ödman and others (1982) and Rydén (2002) have maintained that Hamberg's doctorial dissertation from 1901 regarding the geology in Sarek and results from other geographical research in the area, were among the most important documents during the preparation of the issue of national park establishment. According to Hägerstrand (1991), not many members of the parliament spoke in the case, probably because they did not know much about it and because it was a new topic regarding nature. However, references to the Sami people and their interests were clearly missing in the national park debate. Since the Sami people in many cases were represented as part of nature or belonging to nature they were perceived more as a tourist attraction than an opponent in the national park process (Swedish Parliament 1907). In contrast, the writer Forsslund was of the opinion that the land belonged to the Sami people, and that other people were strangers in the mountains (Forsslund 1915). During the conversations with the Sami people, described in his book *Som gäst hos fjällfolket* (1914), Forsslund maintained that they had difficulties understanding the intention of the national parks, and they were particularly resistant towards the prohibition to shoot the bear, since it threatened the reindeer.

The construction of railways in the Lapland Mountains

Ever since iron ore was found in the seventeenth century in the mountains of Lapland, the problem has been to transport it to ports. The problem with transportation was thought to be solved by a British company, which began to construct a railway between the Swedish city of Luleå and the Norwegian city of Narvik in the 1880s, although this later went bankrupt. A connection between the Atlantic Ocean and the Bothnian Gulf was described as a 'vital question', not only for the county of Norrbotten, but also for the whole country, above all due to the incomes from the iron ore (Marius 1897). During the depression in the 1870s and 1880s, several private railway lines in Sweden ran into financial difficulties, and at the same time, interest increased in state-owned railway lines (Oredsson 1969, Andersson-Skog 1993). The struggle over the railway was the most debated issue – together with the union with Norway – in the Swedish Parliament during this period as the railway had significant political, economic and military dimensions. Oredsson (1969) maintained that the main argument for a state-owned railway between Kiruna and Narvik was to connect northern Sweden with the rest of the country in order to keep foreign (Russian) interests away. However, regardless of the reasons, the state purchased the parts of the railway that already had been built. The Luleå-Kiruna-Narvik railway line (*Malmbanan*, also called *Ofotenbanan*) opened in 1902 (with official inauguration the year after) and iron ore was shipped from Lapland ore fields to the Norwegian ice-free port of Narvik and to the Swedish port of Luleå and the Bothnian Gulf (Oredsson 1969).

In a text by Marius (1897) regarding railway lines and military interests, an interior railway line in a south–north direction from the southern part of the Swedish mountain range to the city of Gällivare in Lapland, was described, and in 1899 proposed to the Swedish Parliament. The interior line (*Inlandsbanan*) was supposed to make military transport faster and safer, and it could also be used for transport of natural resources. The interior line was constructed in parts (from the southern and the northern termini, respectively) over half a century; the whole railway line was opened to the public in 1937, but the northern part was opened in 1911. There were different motives behind the construction of the separate parts of the interior line. Selin (1996) has argued that the military arguments were stronger for the northern part of the railway line, whereas the commercial arguments were dominant for the southern part. Andersson-Skog (1993), however, has claimed that the military arguments were somewhat exaggerated, and that politicians in general were very positive towards railway construction, and that the goal was to include all Sweden in a national transport system. At the beginning of the twentieth century, according to Andersson-Skog, the passenger traffic on the Swedish railways was never an economic dominating factor behind the motives for the railway constructions. The railways in northern Sweden, for example, contributed to new communities, since the extraction of iron ore led to a dramatic change in economic and demographic development (Heckscher 1907).

Access for mountain tourists

With progressions in transport development it became easier for more travellers to reach the mountains, and it also encouraged the development of tourist facilities and services. The link between tourism and the railway was soon visible, and the STF followed the railway construction between Kiruna and Narvik. During the time of the railway construction and the time after it was completed, articles about the Lapland Mountains dominated the STF's yearbooks. Stadling (1900) stated that the Lake Torne area would one day become the county's most popular tourist destination. Similarly, Améen (1903) wrote that nothing could have greater importance for tourism than the development of railways.

Before the official opening of the railway line between Kiruna and Narvik, it opened up for passenger service and the first 'Lapplandsexpressen' went from Stockholm to Narvik (STF 1904). The cooperation between tourism and the railway was also visible as the Abisko tourist cabin was built in one of the railway workers' barracks (STF 1903). Tourist services in the Lapland Mountains therefore largely started with the railway-based and large-scale tourist station in Abisko. Svenonius (1908: 279) wrote that it was due to the railway that 'the landscape became a real tourist country. After the railway was constructed, Lapland has more visitors in a single year, than it used to receive in a decade.'

The same year that the interior railway line was opened in the north (in

1911), between Gällivare and Luspebryggan by the river Stora Lulevatten, the STF started a motorboat service on the river, reaching Saltoluokta, which became a tourist station. Forsslund (1914) was one of the few persons who criticized the occurrence of tourism, such as the construction of tourist cabins and other services, which made it easier for mass tourism, because it was a threat towards both nature and the Sami culture. In particular, the development of railways met resistance from Forsslund (1914, 1915) who opposed the railways as 'the untouched atmosphere of wilderness /. . ./ will be eliminated and destroyed through opening the gateways of "the outdoor museums"' (1915: 112), and around the railway stations and the tourist cabins, 'the Sami lose themselves' while trying to adapt to the new and foreign way of life (1915: 108).

Conclusion

This chapter has argued that patriotic and romantic attitudes towards nature as well as the scientific and touristic interests were at the heart of the process leading to the establishment of Swedish national parks. In the Swedish case the railways were not the driving force for national park establishment as they were in the case of North America and Australasia, at least in terms of a substantial alliance to lobby for park creation (Runte 1974a, 1974b). Perhaps one important explanation is the more commercialized railway policy in the American system, where the state played a somewhat passive role in contrast to the Swedish state, which built and owned the main railway network. Nevertheless, in North America and Australasia, as well as in Sweden, tourism provided the national parks with an economic justification for their existence, as the landscape was initially regarded as having no other economic value than as a tourist attraction (Runte 1987, Sellars 1997).

Tourism and the question of transportation was clearly an essential element of the documents and literature concerning the creation of the first national parks. Interestingly, the early mountain tourists and the early scientists were much the same persons during the period of this study; this means that it was to a large extent the same persons who were engaged in tourism as in the question of nature protection. The findings of this study as well as findings from Hall and Shultis (1991) reveal that a relatively small group of activists and influential men were able to persuade government that the establishment of a national park would be in the interest of the nation as a whole (see Chapter 7), although at the beginning of the twentieth century, very few tourists travelled to the Lapland Mountains, and the national parks in Lapland were more significant as a tourist dream or icon than as a real tourist destination. In addition, although the railway between Kiruna and Narvik was intended primarily for the transportation of iron ore, it came to play an essential role for the tourist organisation STF, and for the continuing tourism development in the region.

In understanding contemporary nature conservation issues, it is important to have the historical background, because disagreements and conflicts in current debates regarding protected areas can stem from the original designation process and the national park concept. It is possible to comprehend spatial disparities by analysing the relationship between powerful urban concentrations of demand and distant less powerful areas of supply. Ever since knowledge about the Lapland Mountains was spread, it is important to notice that it was produced by highly educated and wealthy Swedish (not Sami) men, most of them with their base in the southern parts of the country. Tourism and nature protection are therefore also to be understood within the centre–periphery relation, where the tourists and the ownership of tourist services, and the decisions about establishment of protected areas were from the centre; hence, the periphery was controlled by someone from the outside (Hinds 1979, Hall 1992). Moreover, the treatment of the Lapland Mountains as the aesthetic representation of an original nature set the parameters of environmental politics, and in fact even forms the basis of today's national park establishment. Contemporary tensions and conflicts between preservationists and local groups regarding protection of nature or utilization of nature, can be exemplified by the debate in the late 1980s, concerning the proposed creation of a new national park in the Lapland Mountains, which failed, due to massive resistance from local groups (Sandell 2005). Now, two decades later, the Swedish Environmental Protection Agency (2007) proposes establishment of additional and extended national parks in the Lapland Mountains. According to local newspapers and comments made on the official proposal, several local groups and organizations oppose restrictions on fishing, hunting and the use of snowmobiles, but also a more general resistance towards authorities and external interests, and a lack of local involvement has been expressed. This situation not only demonstrates that the national park concept continues to be contested, but that it can also continue to be understood within the context of the power relations between the centre and the periphery in terms of both nature conservation and aesthetic consumption.

15 'Protect, preserve, present'

The role of tourism in Swedish national parks

Peter Fredman and Klas Sandell

Introduction

Starting with Europe's first national parks in the Lapland Mountains in 1909, the area protected in Sweden has increased to over 10 per cent of the land with a heavily skewed distribution towards the mountain region in the north. The history of protected areas in Sweden involves many recurrent themes, but also a great deal of change and many international influences. For example, today we have the historically rooted theme of national park establishment locally perceived as the central authorities preventing local recreation traditions and development, simultaneously with the theme of protected areas as a motivation for international visitors and therefore an element in local tourism opportunities. The manifold nature of this frame of reference for the role of, and debate concerning, protected areas in Sweden is today reinforced by at least two very important global tendencies: (1) the increased number of different types of protected areas (e.g. the EU's Natura 2000, UNESCO's World Heritage and Man and Biosphere reserves) used parallel to more traditional instruments like national parks and nature reserves; and (2) the attempt to outdistance the 'fortress conservation' tradition and replace this with more of a bottom-up perspective involving local participation and regional development (Jones 1999, Attwell and Cotterill 2000, Brockington 2002, Gössling and Hultman 2006). In this diverse frame of reference of traditions, changes and ambitions, tourism is a key issue and since protected areas are located mainly in rural regions, tourism to such areas can play an important role in their socioeconomic development (Machlis and Field 2000).

From a tourism viewpoint, protected areas can be regarded as 'markers', that is, items that carry information about attractions (MacCannell 1976, Wall-Reinius and Fredman 2007), and protected area establishment may have positive effects on visitor numbers (Fredman *et al.* 2007). Paradoxically, however, when it comes to planning and management of protected areas, often too little focus is on recreation and tourism. But in order to consider the supply of outdoor recreation opportunities in protected areas, one must also know about the demand for such experiences. Therefore, in this chapter, besides giving an overview of the history, arguments and current situation in

Sweden with regard to national parks and tourism, data from a national population survey will be analysed with regard to current use and peoples' attitudes towards the 'role' of protected areas in Sweden. This will be done with special emphasis on the mountain region in the northwest where the proportion of protected areas (both national parks and nature reserves) is considerable higher than for other parts of Sweden.

Historical perspectives on protected areas, outdoor recreation and access in Sweden

Four main historical phases could be identified with regard to outdoor recreation and nature-based tourism in Sweden (Naturvårdsverket 1989, Sandell and Sörlin 2000): the end of the nineteenth century; the interwar period; the post Second World War period; and the end of the twentieth century.

Around the end of the nineteenth century it became more and more obvious that the growing industrial society with its railways, industries, urban areas and exploitation of natural resources also involved very many dramatic landscape changes. With inspiration from the US and Germany and for the sake of science and national identity, areas were protected from exploitation with the help of a nature conservation act in 1909. Related to such developments was a growing emphasis on recreation and outdoor education as a means for fostering youth and improving public health.

During the interwar period, and parallel to further industrialization and urbanization, involvement in tourism and outdoor recreation activities was broadened further. Less arduous places for tourism were emphasized as a complement to the previous focus on the high mountains and the small groups of better-off tourists. The establishment of Youth Hostels during the 1930s and a *Holiday Act* in 1938 (12 days of vacation per year) could be seen as examples of these ambitions. In addition, a more formal recognition of the Right of Public Access was an element of these social efforts. This 'right' had already been fundamental for leisure activities in rural landscapes, but simultaneously with larger numbers engaging in such activities it became a major element in official ambitions with regard to public health and the modern welfare society.

In Sweden, the *Right of Public Access* is laid down in common law and can be seen as the 'free space' between various restrictions, comprising mainly: (i) economic interests; (ii) local people's privacy; (iii) preservation; and (iv) the actual uses of, and changes in, the landscape. For example, camping for 24 hours or less is generally allowed and the traversing of land, lakes or rivers, swimming and lighting a fire are permitted, wherever the restrictions mentioned above are not violated. Even in the present day, although guidelines are provided by the Swedish Environmental Protection Agency, it is important to note that, to all intents and purposes, it is 'the landscape' that tells you what is or is not allowed, for example the way the land is being used may indicate how sensitive it is to people walking on it, and the weather tells you how safe it is to make a campfire. Therefore, the Right of Public Access is an important phenomenon, parallel to the designation of natural areas like national parks, with regard to outdoor recreation and nature-based tourism (Sandell 2006).

After the Second World War, economic welfare increased rapidly, encouraging further development of outdoor recreation, nature-based tourism and protected areas. Since the mid-twentieth century the amount of protected area in Sweden has increased tenfold, although this increase was mainly in the form of nature reserves rather than national parks. In addition, other forms of natural area management for conservation and recreation could be found as mentioned above. During the 1960s and 1970s, governmental efforts also increased with regard to tourism and recreation, using physical planning and various types of facilities to make it easier for the public to practise outdoor recreation. For example, in 1977 the Government took responsibility for the marked tracks in the mountain region that was established by the Swedish Tourist Association during the initial period in the late nineteenth century.

One hundred years after the first national parks were established, Swedish legislation declares, in line with the tradition, that these areas are to be representative biotopes, preserved in their natural state, but are also beautiful unique environments which should offer outdoor recreation experiences. More recent changes in Swedish environmental policy documents imply an increased recognition of social values, including economic values in, and around, park boundaries. 'Nature tourism and nature conservation should be developed for their mutual benefit. Swedish nature in general, and protected natural areas in particular, comprise an asset with great potential for future development' (Swedish Government Writ 2001/02: 173).

The writ concludes that Sweden has a long tradition of 'protecting' and 'preserving' environmental values, but when it comes to 'presenting' them, much more can be done, not the least in cooperation with local partners. This could be described as a shift from a more traditional national park perspective towards one of local participation and regional development where tourism holds a key position (Zachrisson *et.al.* 2006). A major reason for this shift in the talk about the role of national parks, in Sweden as well as internationally, is recurrent conflicts with local people arguing against protected areas in favour of local traditional use for consumptive as well as recreational purposes (Sandell 2005). The above-mentioned changes in Sweden also reflect many of the contemporary global trends of tourism in protected areas (Eagles 2007), for example, increased public participation, new forms of park management, a shift in population demographics, new information technology, increased demand from nature tourism businesses, as well as effects on visitation from energy prices and climate change.

Protected areas in Sweden

At the time of writing, Sweden has over 4,000 areas protected as national parks (28), nature reserves (>2,800) and conservation areas (>1,150), that together cover about 10.3 per cent of the country's land surface (Statistics Sweden 2007). This is less than the average proportion of land area protected globally (11.5 per cent) and less than the average of 14 per cent for European countries (Naturvårdsverket 2007). The national park system represents 1.5 per cent of the land area (approx. 700,000 hectares) or 14 per cent of the protected areas. In Sweden, a national park should represent a large continuous area of national interest. They are established on public land only in accordance with national legislation and IUCN criteria. Nature reserves are more flexible with respect to geographical scope, land ownership and natural resource use. Many of the national parks and nature reserves are included in the EU 'Natura 2000' network, and 14 areas in Sweden are listed in the UNESCO World Heritage List. In addition, Sweden also has a network of plant and wildlife sanctuaries of more local character. There are also two 'Man and Biosphere (MaB)' reserves and another handful of areas are being considered as potential future MaB reserves (Bladh and Sandell 2002).

From a geographical perspective, the areas protected in Sweden are heavily skewed towards the north. In the mountain region, stretching for over one thousand kilometres along the Swedish–Norwegian border, as much as one quarter of the land area is formally protected. While alpine environments may be more sensitive to disturbance and require larger continuous protected areas than other nature types, other explanations of this skewed distribution are a low recognition of the Samí peoples' land rights, fewer competing natural resource uses and a history of the region being a 'mental landscape' for recreational use (see Sahlberg *et al.* 1993, Nilsson 1999). The Swedish mountain region is also increasingly believed to have potential for various forms of tourism development. Almost a quarter of all adult Swedes visit the mountains in a single year for recreation and leisure (Heberlein *et al.* 2002). It provides good opportunities for outdoor recreation during both the winter and the summer seasons and, for many Swedes, the mountain region is regarded as a special place to visit (Fredman and Heberlein 2005).

As a direct consequence of the new environmental policy directions, the Swedish Environmental Protection Agency (Naturvårdsverket) developed an action programme with respect to protected area management (Naturvårdsverket 2004) for 2005–2015. This programme focuses on local participation, management planning, outdoor recreation and tourism, visitor information, nature protection, monitoring and evaluation. In this way protected areas in Sweden today are increasingly a key element in a broad course of action with local participation, regional development, tourism and a recognition of outdoor recreation benefits in terms of, for example, public health and environmental education.

More recently, the Swedish Environmental Protection Agency also published a draft of a new national park plan (Naturvårdsverket 2007), proposing 13 new parks and an extension of seven existing parks. In total, the proposal accounts for 1.2 million hectares of new national park area, most of which (75 per cent) today has the status of nature reserve. If the plan is implemented, it will imply a significant shift from nature reserves (down from 83 to 63 per cent) towards a higher proportion for national parks (up from 14 to 35 per cent), particularly in the mountain region.

However, despite the increased focus on recreational use of protected areas in Sweden recently, there is still a lack of systematic visitor monitoring to promote recreational development. In many areas, basic visitation number are unknown to managers, but in order to succeed in adequately maintaining and developing natural areas for outdoor recreation and tourism, it is crucial to collect relevant and accurate data on, not just visitor numbers, but also their characteristics, behaviour and attitudes (Kajala *et al.* 2007). For example, a survey among park managers estimated 2.2 million annual visits to the national park system (Naturvårdsverket 2007), but such numbers are for most areas, simply 'best guesses'. A study in Fulufjället National Park in 2003 estimated 52,000 summer visits (Fredman *et al.* 2007), but this is just one of few examples of comprehensive visitor counting.

Sweden has put considerable resources into area protection during the last decades, and more recently there has been an increased emphasis on their social and economic functions. However, the bulk of the protected area is found in the mountain region, far away from major population centres, and because of limited resources put into visitor monitoring there is little knowledge about actual use and the role these areas have for participation in outdoor recreation as well as peoples' attitudes toward their use – features that the rest of this chapter will shed further light on.

Data collection and analysis

Data for the study was collected in a national survey during the spring of 2004 and included a postal questionnaire sent to 12,483 individuals living in Sweden. The survey was part of a multidisciplinary research programme, 'Mountain Mistra', which aimed 'to develop scientifically based strategies for the management and long-term development of the [Swedish] mountain region's resources' (Price and Willebrand 2006: 303). Due to this regional context, the survey was over-sampled in favour of the population in the mountain region. Accordingly, 11,418 individuals were sampled from the four northern counties (Dalarna, Jämtland, Västerbotten and Norrbotten) and 1,067 individuals were sampled from the rest of Sweden. Each individual was contacted up to four times (one pre-contact card and two reminders). Addresses were obtained for individuals in the age group 16 to 65 years from the national census register. In total, there were 5,291 responses, with a response rate of 65 per cent in the four northernmost counties and 57 per cent for the rest of Sweden. Results presented in this chapter are based on weighted data to reflect a random sample of the total adult population in Sweden. For the purpose of our analyses, respondents are divided into three groups – respondents living in the 15 mountain municipalities (labelled as

'Mountain'); respondents living in the northern four counties of Sweden (labelled as 'North'); and respondents living outside the north of Sweden (labelled as 'Sweden').

Protected area visitation

Visitation to the nine Swedish mountain national parks was estimated from national survey data (see Table 15.1). In total, 6.5 per cent of the entire sample reported at least one visit to a mountain national park within a one-year timeframe. The second column of Table 15.1 is an estimate of the number of visitors to each national park, based on the proportions in the first column multiplied by the population of Swedes in the age frame of the survey. Some of the parks are obviously not attracting any large amounts of visitors (e.g. Vadvetjåkka, Töfsingdalen), while others are more prominent tourist destinations (e.g. Abisko, Stora Sjöfallet, Fulufjället). Among respondents living in the north of Sweden (i.e. the four northern counties Dalarna, Jämtland, Västerbotten and Norrbotten) 15.4 per cent have visited at least one national park in the mountain region, while the corresponding figure for people living elsewhere in Sweden is 5.3 per cent.

Because of limitations in the survey technique to measure visitation to specific areas, results will underestimate actual visitation since neither repeated visits, visitors outside the survey age frame (16–65 years), or foreign visitors are included. However, previous research (e.g. Lindhagen 1996) has shown that self-reported visitation often implies an overestimate of actual visitation up to 100 per cent. Using on-site visitor counters and surveys at Fulufjället National Park (Fredman *et al.* 2007), the number of visitors was estimated at 53,000 during the summer of 2003 (June – September), which is considerably lower compared to the figure in Table 15.1 (215,000). When comparing the figures one must also consider that the on-site estimates include all kinds of visitors (e.g. also foreign) and are limited to the summer season only (not entire year). The differences will most likely reflect methodological limitations.

It is not just numbers that matter; characterization of the national park visitors also provides important information about the demand for park visitation. Table 15.2 features descriptive statistics from a logistic regression using the dichotomous variable 'national park visitor (1) vs. non-visitor (0)' as dependent variable. Table 15.2 shows that the probability of being a mountain national park visitor *increases* if you live in the north of Sweden (i.e. closer to the parks), and if you are in a high-income category. The probability of being a national park visitor *decreases* if you or your parents grew up in a country outside Sweden (abroad), if you are female and if you do not have a university degree. Age does not explain whether you are a national park visitor or not. The tradition of nature-based tourism as initiated and to a large extent upheld by better-off members of society, as pointed out in the historical background above, is significant here.

Table 15.1 Estimated visitation to Sweden's nine mountain national parks

	Proportion of respondents (%)	Estimated number of visitors*
Any national park in the mountain region	6.5	377,000
Vadvetjåkka	0.4	23,000
Abisko	4.4	255,000
Stora Sjöfallet	4.0	232,000
Padjelanta	1.6	93,000
Sarek	2.2	128,000
Pieljekaise	1.6	93,000
Sånfjället	1.8	104,000
Töfsingdalen	0.6	35,000
Fulufjället	3.7	215,000

* Repeated visits not included. Figures based on a population of 5,800,000 in the age group 16–65 years (Statistics Sweden, 2007).

In addition to actual visitation, respondents were asked about the importance of protected areas to meet their demand of nature experiences, both in Sweden in general and within the mountain region. This question was more broadly framed and considered protected areas in general (not just national parks). Regarding protected areas in Sweden, 18 per cent of respondents report that these areas have no importance to meet their demand for nature experiences, 38 per cent say they have some importance, and 45 per cent that they have large importance (rather large, large or very large). Figures for protected areas in the mountain region specifically are very similar to these national figures.

The role of protected areas

Respondents were asked to mark on a five-point scale (completely disagree, partly disagree, don't know, partly agree, completely agree) their opinion on a

Table 15.2 Characterization of mountain region national park visitors

	Coefficient (B)	Standard error	Wald statistics (Sig.)	Estimated odds ratio (Exp(B))
Mountain resident	1.166	0.148	0.000	3.210
Childhood or parents abroad	−0.948	0.290	0.001	0.387
Female	−0.341	0.127	0.007	0.711
Age	0.001	0.005	0.912	1.001
University degree	−0.453	0.148	0.002	0.636
Income	0.181	0.048	0.000	1.198
Constant	−2.724	0.326	0.000	0.066

Model fit: −2 log likelihood 1945.0; Nagelkerke Rsq=6.2%

number of statements with respect to protected areas in general in Sweden (see Table 15.3). Eighty per cent or more of the respondents in all groups consider the areas as being 'important for biodiversity', 'pleasant places for excursions' and 'of great natural beauty' – which mirror the two main motives for area protection, that is, biodiversity and social values.

People living in the mountain region, with a high proportion of protected areas in their vicinity, generally consider them as pleasant places for excursions more than people elsewhere – which may simply reflect the availability of protected areas for recreation. However, pleasant places of great natural beauty will not suffice to support tourism when the areas are perceived to be difficult to access or unknown by about 40 and 50 per cent of the respondents, respectively. This result also indicates the limited role protected areas have for outdoor recreation opportunities in Sweden in general and the right of public access must be taken into consideration as a frame of reference.

Table 15.3 also shows that protected areas are not only associated with biological values and nature experiences, but to a large extent are considered culturally interesting. This result makes sense from the perspective that basically all nature in Sweden has some degree of cultural influence and the fact that some areas are protected as cultural heritage. These protected areas are

Table 15.3 Attitudes towards protected areas in general

		Disagree (%)	Don't know (%)	Agree (%)
Pleasant places for excursions	Mountain	6.2	13.8	80.0
	North	3.8	13.9	82.3
	Sweden	2.6	6.8	90.6
Important for biodiversity	Mountain	4.6	12.3	83.1
	North	2.8	13.1	84.0
	Sweden	0.4	11.0	88.6
'Dead hand' over communities	Mountain	36.5	44.4	19.1
	North	34.5	50.1	15.4
	Sweden	34.9	53.0	12.0
Difficult to access	Mountain	29.7	31.3	39.1
	North	26.9	32.3	40.9
	Sweden	27.2	32.5	40.3
Unknown	Mountain	22.2	31.7	46.1
	North	18.3	31.2	50.4
	Sweden	16.9	27.4	55.6
Of great natural beauty	Mountain	1.6	12.5	86.0
	North	2.8	12.0	85.3
	Sweden	0.9	6.2	92.8
Culturally interesting	Mountain	4.7	25.0	70.3
	North	5.4	24.4	70.1
	Sweden	4.4	20.4	75.2

neither considered to be a 'dead hand over communities' to any large extent, particularly among respondents outside the north of Sweden.

Now, considering what aspects are of importance when areas are protected, we find that approximately 90 per cent of the respondents agree with the statement that areas should be established primarily to protect nature (see Table 15.4). Biological arguments for protected areas (e.g. preserve nature from humans, protect species and ecosystems) are also considered to be the most important by a majority (see Table 15.5). People living in the mountain region are, however, somewhat less supportive of this view compared to respondents from other parts of the country. Looking at the social motives for protected areas we find a strong support (70–75 per cent) for designation to facilitate visitors in Table 15.4, with little variation between geographical regions. But when people are asked to consider the most important reason, only about 20 per cent think they should primarily produce opportunities for outdoor recreation and nature experiences (see Table 15.5). The argument that protected areas should primarily be reference areas for scientific research has support from 10 per cent of the Swedish population outside the mountain region and less than 7 per cent of the mountain population.

Looking at some current policy issues we find that a majority of all respondents support the view that the local population should have a large influence when areas are being protected, but less so for the general population (72 per cent) and people in the north of Sweden (76 per cent), compared to respondents living in the mountain region (84 per cent). We also find a larger support for commercial activities among mountain respondents, where such activities are most likely to have a positive economic impact, compared to people elsewhere in Sweden. Still, however, over 50 per cent of all respondents disagree with the statement that commercial activities should be encouraged. Table 15.5 also indicates that a significantly higher proportion of the

Table 15.4 Factors of importance when nature areas are protected

		Disagree (%)	Don't know (%)	Agree (%)
The local population should have a large influence	Mountain	9.2	6.8	83.9
	North	14.0	10.5	75.6
	Sweden	16.2	12.3	72.4
Areas should be designed to facilitate visitors	Mountain	15.7	9.8	74.6
	North	17.4	12.1	70.4
	Sweden	17.2	9.4	73.4
Commercial activities should be encouraged	Mountain	52.2	21.0	26.6
	North	57.6	21.1	21.0
	Sweden	63.3	19.9	16.6
Areas should be established primarily to protect nature	Mountain	6.1	6.5	87.4
	North	5.7	7.3	87.0
	Sweden	3.3	6.2	90.6

Table 15.5 Most important reason for protecting new nature reserves and national parks in the mountain region*

	Sweden (%)	North (%)	Mountain (%)
Preserve nature undisturbed by human beings	43.2	39.6	35.4
Protect endangered species and ecosystems	62.0	53.5	50.2
Produce opportunities for outdoor recreation and nature experiences	21.4	21.2	21.0
Reference areas for scientific research	10.1	8.1	6.7
I don't think more areas should be protected in the mountain region	3.4	10.4	15.7

* The table includes those respondents (approx. 25%) who left multiple answers.

people in the mountain region (16 per cent) and the north of Sweden (10 per cent) don't think more areas should be protected in the mountain region compared to the rest of Sweden (just over 3 per cent).

Discussion and conclusions

Sweden has a long tradition of both protecting nature and promoting outdoor recreation and tourism. In the current study, we find support for future protected areas to be designated to facilitate visitors, but less so for commercial activities. This result can probably partly be explained by the Swedish tradition of public access, which has made protected areas less important for the supply of outdoor recreation opportunities to the general public. In addition, commercial activities have also been forbidden (and still are) in several of the national parks. There is consequently no tradition of tourism operations in the parks, and due to limited promotion the general public may in some cases not even know if they are in a park or not. Given the shift in protected area policy in Sweden (and internationally), these things are likely to change in the future.

If the role of tourism in protected areas is to increase in the future, Sweden's protected areas need to be differently managed and better promoted. Current visitation numbers (here estimated for the mountain national parks) are low in an international perspective (Eagles and McCool 2002), and most likely far from any carrying capacity limits (e.g. Fredman and Hörnsten 2004). Improving information about, and access to, the protected areas should have priority. In the current study, parks with good accessibility also have the highest visitation. In a study of visitors to Fulufjället National Park (in the southern part of the mountain region) an on-site survey showed that the protected status had an effect on the decision to visit the area among 44 per cent of the visitors (Wall-Reinius and Fredman 2007). This study also showed that German visitors were more likely to visit the area because of the national park status than Swedish visitors, a result which has important marketing implications. However, given the results of the current study, commercial activities in protected areas should be developed with some care keeping the attitudes of the Swedish general public in mind.

The 'traditional' biological motive for protected areas (e.g. protecting nature and biodiversity) has strong support among the Swedish population. Even among people living in the mountain region, where a high proportion of the land is already protected, a majority want more areas to be protected and only about one in five think they are a 'dead hand' over the communities. Consequently, our results support a preservation perspective as the guiding principle also in the future, but good management practices should be able to balance the different functions of protected areas so that recreational use is not in conflict with preservation goals. If protected area management is to consider outdoor recreation opportunities more in the future, the potential to develop tourism opportunities will increase as well.

Part IV
Developing world
Beyond the eurocentric

16 National parks in Indonesia
An alien construct

Janet Cochrane

Introduction

The introductory sections to this book explain how attitudes to national parks vary across continents and across time. The different approaches do not depend on objective judgement based on ecological criteria, but are informed by political and cultural viewpoints, with the desire to protect tracts of land from damaging exploitation arising from attitudes shaped over centuries by cultural, scientific and religious influences.

The rapid economic development of European and New World countries in the nineteenth and early twentieth centuries stimulated interest in protecting natural resources, as realization grew that overexploitation could result in irreversible deterioration. As industrialized countries took the lead, an international system of protected areas was created around a Western (and in particular a North American) construct of wilderness. Towards the end of the colonial era and after former colonies gained independence, this model of national parks was applied to other societies, with mixed success.

In Indonesia, one of the most biodiverse countries in the world (National Development Planning Agency 1993, Stone 1994), a network of protected areas was established under donor-assisted programmes, which ran for around a decade from 1974 (Robinson and Sumardja 1990). The first five national parks were declared in 1980 with a further 11 in 1982 (Sumardja, Harsono and MacKinnon, 1984). The parks were selected according to sound ecological criteria and within the contemporary planning paradigm, but the cultural context was often neglected. The imposition of a Western model has resulted in a disjuncture between the conservation aims of national parks, framed within a rationalist learning paradigm, and the reality of their management in situations where a post-rationalist approach to conceptions of learning and development would be more appropriate. The result has often been conflicts over utilization of the parks' resources.

This chapter explores some of these conflicts and the reasons for them, illustrated by tourism examples. Starting with a brief examination of how protected areas were established in industrialized countries, the discussion will move to how the same model was applied in developing countries, specifically in Indonesia. Attitudes to national parks here will be discussed, along with tourism uses. Finally, more recent approaches to protected areas management, which centre on collaboration and shared learning will be explored; these show signs of being a more successful approach to reconciling the difficulties.

Protected areas in Europe and the New World

The earliest national parks were established in the late nineteenth century in North America, at a time when the frontier way of life was disappearing and there was increasing awareness that hunting and the conversion of forest lands to agriculture were affecting the balance of nature (Jepson and Whittaker 2002). Another driving factor was the romantic movement, which seized on wilderness as a source of inspiration and constructed majestic natural monuments and tracts of wilderness as symbols of national pride (Nash 1967). Spearheading the new interest was the inspirational writer and campaigner John Muir (1838–1914), who popularized the concept of wilderness, stressing the interconnectedness of everything in nature and moving it from a somewhat esoteric ideal to a policy embedded in the public consciousness (Oelschlaeger 1991; see, also, Chapter 7). Transportation systems improved, allowing people to explore the landscapes writers and artists enthused about. Camping in the wilderness – promoted by President Theodore Roosevelt – became popular. As a result of these streams of thought and activity, large areas of uninhabited land were set aside to be free of human exploitation.

In Europe a similar movement was taking place. The study of natural history had become popular as part of the nineteenth century's general thirst for scientific knowledge, and the romantic movement fuelled a nostalgia for pre-Industrial Revolution intimacy with nature. As travel systems improved and paid holidays became longer, the demand for greater access to the countryside increased (MacEwen and MacEwen 1982, Sheail 1995). Lobbying during the interwar period culminated in Britain, in the 1949 National Parks and Access to the Countryside Act under which 10 national parks were designated.

Across the industrialized world, then, there was a growth in appreciation of wilderness and of a sense of separation of people and nature, which cannot exist when humans have to grapple with their natural environment on a daily basis in order to survive. This 'landscape way of seeing' (Neumann 1995) and the idealized concept of wilderness moulded the early concept of national parks globally, with the result that anthropogenic landscapes were automatically excluded from protected areas networks, no matter how ancient and stable their ecosystems.

Despite shared cultural influences and an insistence on an ideal of untouched nature, North America and Western Europe produced very different models of national parks. This was because in contrast to North America, where parks were in state ownership and human influence was minimal or nonexistent, much European land was privately owned and had been in productive use for centuries (see Chapter 3). The protected areas necessarily included artificial landscapes and cultural features, with rights of public access and agreements on land management, agricultural and other commercial uses achieved by negotiation with landowners and tenants.

If the different historical and geographical circumstances on either side of the Atlantic gave rise to different outcomes in the style and substance of protected areas management, how much more divergence would arise where populations had not passed through a phase of industrialization and consequent loss of intimacy with the land? People have often lived for centuries off the natural resources of territories subsequently designated as reserves, and while the 'noble savage' ideal of humankind living in harmony with the environment is quickly dissipated by even a cursory understanding of the consequences of population growth and the market economy, it is arguable that both biodiversity and socioeconomic development aims would have been more successfully met if the cultural context had been better understood – or that if an externally derived model had to be implemented, then the European one of 'parks by negotiation' would have been more appropriate than the exclusionary, state-centric North American model.

In the early years of protected areas establishment in developing countries, managers were generally Europeans influenced by decades of evolving attitudes to nature and wilderness. As Schama (1995: 574) puts it: 'whether we scramble the slopes or ramble the woods, our Western sensibilities carry a bulging backpack of myth and recollection'. On the one hand, these sensibilities relied on an appreciation of wilderness as a source of wonder and inspiration, while on the other they rested on the desire to set large tracts of land – and their fauna – aside for hunting, as happened in Africa, where parks were established along the lines of the medieval and Victorian hunting traditions (Neave 1991, Adams 2004). In both cases, there was a sense of separation of people and nature, which cannot exist when humans have to grapple with their natural environment on a daily basis in order to survive, and which could only come about once the mechanisms of nature were sufficiently well understood for purposeful external management interventions to be applied. The selection of areas for protection and recommendations for management systems could not help but be informed by this cultural 'backpack', and increasingly by the 'ecosystem' view of biodiversity conservation. Management policy was situated within a rationalist paradigm, whereby knowledge is seen as objective, universal and transferable (McFarlane 2006). As administration of the protected areas was transferred into local hands, it was apparently assumed that the new managers would share both the consciousness of wilderness and the positivist sense of unchangeable and observable natural laws as the advisors, and that management training would impart sufficient technical knowledge to ensure fulfilment of conservation aims.

In the event, these assumptions often foundered on the complexity of local cultures and sociopolitical systems, underscored by nonrationalist beliefs and working practices. An examination of the way the national parks network was established in Indonesia by the colonial power – the Netherlands – and by subsequent foreign advisors will illustrate this point.

Twentieth-century park planning in Indonesia – a false start

In the first decades of the twentieth century the Dutch were influenced by the prevailing monumentalist view of landscape preservation, according protection to 'nature monuments' (Jepson and Whittaker 2002) such as Mount Bromo, in East Java, a volcanic crater at the centre of a dramatic landscape, which was given protected status in 1919 (FAO 1980). The Dutch had been present in the 'East Indies' since the early seventeenth century; the archipelago was by far the wealthiest of the Dutch colonies, and by the 1920s the settlers were able to cultivate their attachment to the country by viewing the land and the landscape with relative ease (Cribb 1995). As in other European colonies, there was as yet no sense that wildlife and forests might be finite unless carefully managed: hunting was a popular pastime, with the major species of Javan fauna hunted to near-extinction by the 1930s (Whitten *et al.* 1996). Numbers of the Javan tiger and rhinoceros fell so low that, aided by population pressure and consequent habitat change, the tiger went extinct in the early 1980s and the rhino survives only as a tiny population in a remote peninsula of the island (Whitten and Whitten 1992).

After proclaiming independence in 1945, the Indonesian government was too concerned with knitting together the new and fragmentary nation and establishing economic and social systems to worry about nature protection, and it was not until the 1970s that a modern programme of designing protected areas began, with technical assistance from overseas consultants paid for by donor agencies. The programme ran for several years and established a core network of national parks to cover representative ecosystems. According to the 1969 IUCN definition of national parks and the slightly modified version adopted in Indonesia, the parks were to have 'high recreation potential' and be 'of easy access to visitors' (FAO 1982: 6) – but the foreign advisors and their Indonesian counterparts were natural scientists with no knowledge of tourism. Neither did they have any anthropological expertise, even though all the parks had people living in or around them. In general, there was only a superficial understanding of the sociocultural and learning context in which their policies were unfolded.

The discussion now focuses momentarily on the Javanese rather than Indonesians in general, because the majority of government officials are from Java, the central island of Indonesia and the locus of administrative, financial and political power, and it is principally their world-view that has affected park administration. Rather than being informed by the rationalist narratives of Western planning dogma, Javanese management styles are more fluid, resting on avoidance of conflict and the preservation of an outward appearance of harmony, and on respect for a complex social system of power relations. An additional layer of complexity was added by the radical social

Figure 16.1 Mountainscape in Java, typical of the monumental scenery which often is included in national parks.

and economic changes the Javanese (and other Indonesians) were undergoing as they experienced the material fruits of economic growth for the first time. As in other countries, much of this growth was founded on exploitation of natural resources with, as yet, little concern for possible negative effects. In 1979, for instance, the senior environmentalist, Professor Otto Soemarwoto, commented of Indonesia that 'our societal values have "flipped" into another stability domain having the characteristics of carelessness for the well-being of our ecosystem, high consumption life-styles, weak or no social control, but also no science' (Soemarwoto 1979:19).

It was within this rapidly changing milieu that the plans for park management were written. The plans were produced to a formula and documented the flora, fauna and habitat of the proposed parks, briefly evaluated the threats to the area (mostly centring on resource use by local people) and listed the personnel, equipment and budget needed (for example, Blower *et al.* 1977, FAO 1980). While this resulted in some useful wildlife inventories, they did not recognize the livelihood needs of local inhabitants, local administrative capacity, national political will and resources for supporting protected areas, or tourism trends. In other words, the plans were written according to the contemporary 'top-down' planning paradigm and failed to engage with local cultural norms or take market forces into account.

In the case of local residents, a neat boundary was drawn round each new park, with inconveniently located villages designated as enclaves. Villagers were expected to remain inside the enclaves for their income-generating activities or were in some cases to be relocated, as at Komodo National Park (Hitchcock 1993, Borchers 2008). Local people were not consulted about the parks or even informed about them. Not surprisingly, acceptance of the new land regimes was poor, with attitudes ranging from fatalism to outright disregard. Infringement of the regulations was common because people were unaware of them or because enforcement was so weak that there was little incentive to obey them. In the early days of the programme the hierarchical traditions of Indonesian society (especially in Java) were still strong, with villagers acquiescing passively in the face of decisions imposed on them. Besides this, challenge to authority was impossible under the repressive Suharto government (1967–1998) – although since then, protests against national parks have become common, as in the case of continuing protests against Komodo (Borchers 2008), community disillusionment at Bunaken (Seivanen 2008), Wasur (Kompas 2003), Mt. Halimun (Tempo 2004) and at Mount Merapi, discussed below.

Bromo Tengger Semeru, in East Java, was one of Indonesia's earliest national parks, declared in 1982, yet even in the late 1990s, research here showed that local people were not familiar with the term '*taman nasional*' (the Indonesian translation of 'national park'), referring to it instead as 'PHPA's area', that is, the area under the jurisdiction of the government department with responsibility for the parks (at that time the Direktorat Jendral Perlindungan Hutan dan Pelestarian Alam – the Directorate General of Forest Protection and Nature Conservation, within the Ministry of Forestry). Furthermore, domestic tourists did not distinguish between *taman nasional* and other types of park (Cochrane 2003). The Indonesian word 'taman' in any case generally denotes urban parks or other heavily managed landscapes, and there has been no public information campaign to change this perception in the public consciousness.

If the label 'taman' has no connotations of wildness, the term 'wilderness' has no direct translation into Indonesian – and if it did, it would be unlikely to convey any positive meaning. For the majority of people – including the poorly educated rangers who are supposed to guard protected areas – forested places and high mountains still represent something to be feared, partly because many people were still engaged in a daily struggle to carve a livelihood from that wilderness, and partly because they are perceived in folk culture as places where wild beasts prowl and dangerous spirits are particularly powerful. While the existence of spirits has been neutralized in the West by a mechanistic understanding of the natural world, in rural parts of Asia such beliefs are still strong, and still matter because they influence daily decisions and attitudes.

Later, park planners also introduced zoning plans, but zoning, park or enclave boundaries were generally marked only on maps and rarely on the ground, and in any case, few maps existed. It is evident that use of maps as a planning and directional aid is a Western construct, as Indonesians rarely use them. At Bromo Tengger Semeru National Park in 1997, for instance, one of the main ranger stations had no map, while the head ranger at another station did not know exactly where the park boundaries were and had only a poor photocopy of a 1938 Dutch map. The situation was unchanged in 2005, when even the National Park Head Office still had no good maps.

Local capacity for administering the parks was affected not only by the lack of resources, but also by the cultural distance between the administrators and the original planners. Encumbered by their 'backpack' of notions of wilderness and ideal ecosystems, the advisors failed to recognize that their counterparts did not share their vision – or at least they failed to recognize the significance of this. Fundamentally, the local administrators shared in the prevailing sense of caution towards wilderness rather than the backwards looking or biocentric idealism of the advisors.

More prosaically, where opportunities to make money from the national parks exist, there is an overwhelming temptation for poorly paid rangers and other staff to do so. The enthusiastic way that personnel engage in accumulating tourism-derived income compares poorly with the apathy accorded their overt duties, as in the case of the head of an important ranger station at Mount Bromo. Judiciously marrying a local woman whose family owned property fronting on to the main street of the nearby village, he was able to develop one hotel here from the mid-1990s and another on national park land itself, a few years later (nominally, all developments on national park land are prohibited). He also managed to avoid being relocated to another posting, an arrangement normally imposed on senior staff in an effort to prevent abuses of power.

These factors call into question whether the imposition of a management system based on extraneous values can be successful. In this case, the protected areas system was based on norms held by its Western advisers rather than on indigenous ones. The philosophy of parks management in the West is based on decades of socially constructed attitudes to the world's wild places, and as Indonesian society does not yet share these attitudes, it is hardly surprising that the motivation for protecting natural resources is weak at both an individual and an institutional level. Having said this, there are now signs that developments in management regimes are beginning to achieve conservation aims, albeit in a different way and with a broader set of objectives than what was originally intended. The signs of change will be explored below, with the increasingly significant role of tourism in the parks outlined first, since this has been part of the drive towards a change in management structures.

Figure 16.2 Asian tourists climb to crater rim of Mount Bromo. Note sellers of endangered Javan eidelweiss in foreground. The plant is considered to bring luck, and is illegally picked and sold within the national park.

Tourism in Indonesian national parks

Tourism to Indonesian national parks was almost nonexistent at the time that the management plans for the first 'wave' of parks were drawn up around 1980. Any overnight facilities were primitive and poorly maintained, no tourist literature was available, and local tour operators knew nothing about them. At the time, government attention was still focused on the main revenue-generating sector of oil and gas, and international tourism arrivals hovered at a relatively low half a million per annum between 1978 and 1983. The only area given any substantial attention for tourism development was Bali, for which a master plan was produced and investments solicited from the 1970s onwards (SCETO 1971, Picard 1996). Domestic tourism was almost entirely undocumented.

The situation changed dramatically over the next decade. An economic recession and consequent decline in oil prices in the mid-1980s meant that forecasts for government budgets had to be drastically revised and other sources of income sought. Tourism was seized upon as an alternative source of foreign exchange and previously restrictive legislative structures were relaxed. Visa restrictions on visitors from major source markets were lifted, joint ventures with foreign investors were encouraged, the permits and licences needed to develop enterprises were simplified, the government department responsible for tourism was given higher status, and an archipelago-wide campaign of posters and slogans encouraged Indonesians to welcome their international visitors (the 'Seven Charms' campaign).

All this resulted in a sharp rise in international arrivals from 700,000 in 1984 to over 5 million in 1996. While the majority of tourism still took place to the coastal resorts of Bali and popular cultural destinations such as Central Java and South Sulawesi, more varied destinations also appeared on the tourist radar. This was partly because of government-sponsored promotional campaigns and partly because of trends affecting source markets. An affordable 'Visit Indonesia' air-pass was introduced, which allowed trips to distant islands that had previously been prohibitively expensive to reach (this was years before the expansion in low-cost air-carriers brought the cheap fares that the travelling public now takes for granted). At around the same time, a whole raft of new adventure tour operators had sprung up in important generating countries, such as Britain, Australia, New Zealand and Japan, meaning that the entrepreneurial framework now existed to send tour groups to destinations off the beaten track. These tours began to explore the natural and cultural riches of Indonesian national parks and remote villages. The earliest parks designated under the donor-assisted programme described above were generally the focus of such visits because accommodation facilities did at least exist there – even though they were primitive and tour participants had to be thoroughly briefed as to what to expect.

In addition to these organized international groups, other market sectors also took more interest in Indonesia's protected areas. A 'Backpacker Plus' category of better-off and more experienced independent travellers ventured to remote parks, and domestic tourists began appearing in larger numbers. The more accessible parks had since the 1970s hosted very large groups of students whose main activities were camping and hiking *en masse*, but other categories were now also arriving. One important market was domestic family groups, many of whom had become enriched by the growth of the Indonesian economy and could now afford a private car or a package tour. A school-teacher I met at Mount Bromo in 2005 recounted how she and other people from a city in West Java had arranged a tour of East Java and were then going on to Bali; it was the first time she had been outside her home province. After the Bali nightclub bombings in 2002, the domestic market became even more prominent as the Indonesian government turned to the home market to shore up a desperate industry deserted by international tourists; Hitchcock and Putra (2008) document managerial responses in Bali to the growing numbers of Indonesian tourists visiting the island post-2002.

A growing market for protected areas is also represented by an 'aspirational' group of young professionals who seek respite from the crowded cities and increasing pace of urban life. In addition, there are increasingly vocal groups in Indonesia who value their natural and cultural heritage (see, also, Picard 2008 on how the Balinese are revisioning their culture and Indrianto 2008 on attempts to use tourism to preserve historic buildings in Java), and taking an active interest in natural areas can be seen as part of this. Often visiting with just a small number of companions, these people explain that they are here because they enjoy the grandeur of nature, or because they are interested in specific nature-based activities. This trend is illustrated by the emergence of new tour service providers focusing on adventure and nature tours (see for example www.boodieadvindo.com, www.adventureindonesia.com, www.indonesia-trekking.com, which aim tours at both international and domestic markets). Having said this, few domestic visits to national parks are of the knowledge-seeking and contemplative type: activities such as white-water rafting and paragliding are common, off-road driving in 4-wheel-drive vehicles and trail bikes is popular, and some camping grounds have achieved notoriety as venues for taking drugs, away from nosy parents and authorities (author's doctoral research and subsequent observation).

Generally, the majority of domestic tourists treat national parks as pleasant places to relax with friends and family rather than wanting to engage in activities that lead self-consciously to greater physical or spiritual health. Studies elsewhere in Asia (for instance Backhaus 2003 on Malaysia and Nyiri 2006 on China) have made similar findings. Enjoyment of the fresh air and natural surroundings is sincerely felt, but enhanced by the collective gaze rather than an individual one: there is little search for a lonely, knowledge-based encounter with nature. It is anyway, extremely unusual for people to wander off alone in Indonesia. Lone treks are occasionally made to mountain summits and caves for the purposes of meditation and prayer; a purposeful choice is then made to be alone in liminal places such as caves and mountains where barriers between the corporeal and spirit world are particularly permeable. Otherwise, shunning the company of others is seen as socially deviant and the sign of a disturbed mind. There is little prospect here of seeing the emergence of wilderness purist visitors for whom 'experiencing solitude' and 'self-awareness/contemplation' are primary motivations (Higham 1997). Overall, however, the value for conservation of emerging domestic user-groups can be viewed as positive because it represents a shift in attitudes towards nature, as discussed in the next section.

Future prospects

Interest in environmental matters was strong during the three decades of the Suharto regime from student activists and other politicized groups, but they tended to focus on abuses of natural resource management and associated infringements of human rights rather than on greater protection for fauna and flora. For many, environmental protest was a form of displacement activity because open political criticism was forbidden. However, since the 1990s it has become evident that awareness of conserving biodiversity for its own sake and in some instances of the value of national parks is growing. Examples are the formation in 1994 of a group that campaigns for wild animal protection in Indonesia (sometimes through direct and risky action) and of a national bird conservation organization in 2002, while the organizer of regular birdwalks and an allied bird protection campaign in central Bali relates how trapping and hunting of birds in local villages has declined since the programme started in the 1970s (Mason 2007). In 2005, Agus Purnomo, a respected environmental policymaker and commentator, concluded that government agencies and NGOs were at last achieving some progress in conserving biodiversity (Purnomo and Lee 2005). To balance the protests against national parks there are also examples of community support for them, such as with Batang Gadis in Sumatra (Jakarta Post 2004) and Kayan Mentarang, which in 2002 was the subject of the first collaborative management agreement for any Indonesian national park (Eghenter 2004).

If awareness of conservation issues follows a similar pattern as in the West, it is possible that these moves will concretize into greater public support for protected areas. This could have several effects: parks may begin to figure so prominently in people's lives that territorial encroachment and overexploitation will receive public disapproval, and greater political will for their support may be forthcoming and translate into financial and human resources to manage them more effectively.

These positive indications of change are still rather tiny compared to the countervailing forces, however. Experience in other parts of Asia suggests that interest in the environment will remain a minority concern, at least in the medium term. In the first place, economic pressures dictate that countries will inevitably try to exploit all the natural resources at their disposal, even if these are irrevocably altered in the process, and this applies to the landscape of protected areas as much as to any other resource. There are numerous examples of how demand for leisure facilities takes precedence over biodiversity protection in protected areas: for Vietnam, Sofield (2008) relates how there are 53 hotels, nightclubs and karaoke bars inside the Tam Dao National Park; the World Heritage Site of Zhangjiajie National Park in China has received considerable criticism for overdevelopment and 'urbanization' (Zancai, Yi and Chen 2007); and 'copycatting' of sea-kayaking and elephant trekking tours in Phuket, Thailand, by unscrupulous operators has led to overcrowding in national parks and other natural areas (Shepherd

2002). These are, in effect, 'Tragedy of the Commons' scenarios as the true costs of using the parks and other natural areas are not internalized into tourism service prices.

Second, East and Southeast Asian constructs of nature militate against prioritizing environmental needs over human ones. Nature is there to serve the purposes of humankind, rather than having any intrinsic value. Thus, appreciation of national parks is mediated by facilities such as food stalls, concrete stairways – even elevators – and other 'signs' that these are places for a hedonistic style of nature enjoyment. The potentially dangerous elements of nature are tamed by being literally and figuratively fenced off, tidied up and heavily managed, in the process becoming vastly more popular with domestic tourists than more natural attractions (as described by Cheung, 2008, for the man-made Hong Kong Wetland Park). The third reason why in Indonesia wilderness has yet to achieve positive representation in the public imagination is that, as described above, venturing alone into the wilderness creates a vulnerability to other-worldly influences, which is frightening unless deliberately invoked.

Having said this, customary awe of a spirit dimension has not prevented Indonesians from clearing forests to make way for agriculture, industry and residences. It is possible that as the population becomes increasingly urbanized, the physical and psychological separation of humans from nature will have positive effects as anxieties about intangible dangers fade, in that it may serve to raise the existence value of rainforest, mountains and other natural habitats even as they become rarer.

New management methods for protected areas being implemented across Indonesia are in part predicated on the assumption that an indigenous lobby of support will, in time, appear. The new schemes typically take the form of co-management initiatives involving a range of societal partners, which generally include the government agencies formally charged with responsibility for the parks, private sector organizations, community groups and international NGOs. In that the NGOs provide funding and technical expertise the schemes echo previous exogenous interventions, but in keeping with the prevailing development discourse of sustainable livelihoods, there is now acceptance of the complexity and relevance of socio-environmental interdependencies.

These partnerships are in place in a number of places including well-known national parks such as Komodo, Bunaken and Rinjani. In some cases – for instance Rinjani – the partnership has won global recognition for its successful inclusionary policy, while in others – notably Komodo – there has been criticism that park conservation has been enforced, but local communities have not benefited (Borchers 2008). To blame this entirely on the management consortium is probably unfair, however: as with many protected areas, Komodo is in a remote part of the country where local people already experience distributional inequality and lack the financial and social capital to take advantage of tourism. A smaller example is in East Java, where a local NGO (Kaliandra Sejati) has won support from the Netherlands Committee of the

World Conservation Union to join forces with village organizations and the provincial forestry department to manage a Forest Park on the upper slopes of Mount Arjuna, one of Java's dramatic chain of volcanoes. The aim of using tourism to support management is expressed in the project's overall objective: 'to develop and promote responsible tourism in the Mt. Arjuna Tourism Area in order to conserve the Grand Forest Park of Mt. Arjuna while raising the standard of living of poorer local communities' (Kaliandra 2007).

In considering these encouraging schemes, however, we should not lose sight of the fact that the concept of a 'taman nasional' has gathered considerable negativity in Indonesia among many groups. At Mount Merapi, an iconic volcano in Central Java, a national park was to be declared in 2002, at a time when Indonesia was going through the painful social and political upheavals of early democracy. The community around Merapi felt that they had been excluded from the decision-making process and that their livelihood opportunities would be curtailed by the national park, and protested so energetically that plans were postponed (Down To Earth 2002, Sarono 2002). Since the park was finally declared in 2004, there have been continuing protests at national and international level (for example TILCEPA 2006). The declaration of Merapi as a national park can be seen as an example of the continuing tradition of hierarchical policymaking in Indonesia mentioned earlier, whereby decisions over resource use are taken at the apex of power and expected to be fatalistically accepted by those lower down.

Conclusion

What lessons can be learned, then, from the history and practice of protected areas management in Indonesia? It is only now that natural and traditional landscapes in Indonesia are under serious threat from industrial and tourism developments, along with customary practices associated with land use and the intimate connection of the peasantry with the land, that a movement to preserve them is stirring. Significantly, this is occurring among intellectuals, just as with the Romantic movement in eighteenth and nineteenth century Europe and North America, but there is no guarantee that national parks will become established as positive entities for all actors involved. Most domestic tourists enjoy parks as pleasant venues for leisure with friends and family, rather than appreciating the wilderness and biodiversity they protect, and although there is some evidence of more intimate forms of leisure use, there is no indication that this is founded on the same inspirational, preservationist wellspring as in the West. No studies of existence value have yet been carried out in Indonesia, and it would probably matter little to most domestic visitors and many foreign ones (bearing in mind that the major markets to Indonesia are from Asia) whether parts of the park outside the immediate landscape being enjoyed were covered in tea plantations or housing.

The previous imposition of conservation management practices generated by one society on the cultural framework of another has meant that protected areas are sometimes resented by the surrounding populations and inadequately supported even by those charged with responsibility for them. Indigenous people will certainly want to continue to benefit from farming or collecting resources in the parks, and will continue to view initiatives that impose new land-management regimes with suspicion. Government agencies may also have difficulty in accepting newer participatory and adaptive management paradigms, which run counter to the traditionally hierarchical approach to decision-making in Indonesia.

Overall, there can be little doubt that the cause of conservation in Indonesia and other developing countries was delayed by a wilderness focus, which gazed backwards and outwards to inappropriate models rather than focusing on the realities of what was close at hand; but it is easy to be wise with hindsight. At the time that Indonesia's national parks were first designated, the advisors were simply putting into practice contemporary received wisdom on managing protected areas. A narrow sectoral approach to resource management was not uncommon, the recognition that contextual social and political issues have to be taken into account had yet to be made, and tourism had not reached today's prominence. It has only been since development principles and practices have been iteratively tested and refined that techniques that may be more appropriate have been used. Co-management ventures firmly grounded in local circumstances seem to offer more promise of success than earlier, top-down interventions, but since these models have also been developed by exogenous agencies, it is too early to tell whether they can fulfil the elusive goal of protecting biodiversity while simultaneously providing livelihoods for indigenous people.

17 National parks in transition
Wuyishan Scenic Park in China

Xu Honggang and Zhang Chaozhi

Introduction

There is no unified system of national parks in China. Different ministries adopt different park systems. National natural reserves can belong either to National Environmental Protection agencies or to National Forestry, which controls all the forest parks. The National Construction Ministry sets up the National Scenic Park system. The geological park system is a newcomer and created by the National Land Ministry. Scenic parks are considered to be similar to the national parks definition created by the IUCN. The goal of the creation of Scenic Park stated in National Regulation of Scenic Park 2006 was to provide opportunities for the people to appreciate and enjoy the cultural and natural landscape. The first 44 national scenic parks were established in 1982. By 2004, 133 national scenic parks and 452 provincial scenic parks had been established, comprising a total area of about 1 per cent of China (Qiu 2007). By the same year, there were 226 national reserves established in China, about 8.86 per cent of the national land area (NEPA 2004).

The massive establishment of the national parks is closely related to this transition process from a centrally controlled system to a more decentralized management system, from an ideologically driven society to an economic-oriented society and from a closed to an open society. In this process, the local governments are given more incentives to promote regional development and also evaluated accordingly. As a result, the values of national brands, as a heritage for the society, or economic driver in the market, or the political achievements, are perceived easily. Most of the national parks were proposed by the local governments, either the prefecture level or provincial level governments, of the less developed regions and then approved by the National Ministries. The decision on which title the locals would apply for their parks is complicated. It not only depends on the nature of the resource, but also depends on the initial institutional arrangement of the land. Although these national parks provide opportunities for the preservation of valuable resources, they are also perceived to be, and are used as, the drivers for regional development through the market system via the attraction of tourists. Meanwhile, national regulations and policies are not clear and forceful on the ownership, management and evaluation. Therefore, national parks in China are potentially facing a more uncertain and complex situation than those in many other countries. Although it is well understood that social, economic and political factors ultimately determine the establishment, management and performance of national parks, limited empirical research has been conducted to systematically understand their complexity. This chapter selects the Wuyishan Scenic Park as a case study through which to explore the dynamics of this complexity and show a picture of national parks in this transition process.

The case of Wuyishan Scenic Park

Wuyishan City is located in northern Fujian Province, approximately halfway between Shanghai and Hong Kong. It is at the foot of the southeastern slope of Huanggang Mountain, on hilly land consisting of red gravel rocks. The city has a population of 208,000. Throughout the centuries, Wuyishan has been considered a remote and poor area, separated by mountains from the political and economic centres. Yet, Wuyishan has a landscape that has been valued for more than 12 centuries for its natural beauty, exceptional archaeological sites and famous Wulong Tea. It is a red stone geomorphologic area, famous for its 36 peaks. It includes a large area of subtropical rainforest and is home to some very rare species of plants and animals (Chinese Construction Ministry 1998). In addition, archaeological findings from the Minyue Kingdom (c 400 BC) add mystery to the place. Its fame was spread widely by a famous scholar, Zhuxi, the most important follower of Confucius during the Song Dynasty, who opened a school there for promoting his thoughts, called 'New-Confucius', when he retreated into Wuyishan. Many Chinese have read Zhuxi's extensive writing on the beauty of the landscape widely for over 1,200 years.

Concepts of conservation have a relatively long history in Wuyishan compared with other places in China. Even in the 1960s, when there were only several national reserves in China, the value of Wuyishan was recognized. The Wuyishan Management Bureau, directly supervised by the Fujian Provincial Government, was established to prevent large-scale destruction of the forestry and the historical sites in the mountains. Yet, the effectiveness of the conservation was poor.

Dramatic change to Wuyishan Park began in the 1980s when the potential for using natural and cultural reserves as tourist attractions to promote regional economic development was recognized. This was at a time when the decentralization process made it possible for the local government to control and manage these national parks. As a result, in 1979, Wuyishan Natural Reserve, of about 570 square kilometres, became one of the earliest of the five, key, natural reserves in China. A national scenic park, with a core area of 64 square kilometres, was granted by the National Construction Ministry in 1982. The Wuyishan Management Bureau managed it. This scenic park was close to the nature reserve but more accessible and attractive to mass tourism. Unlike the natural reserve system, whose purpose is conservation, the scenic park was to protect the landscape and make it accessible to the people. The scenic park is often regarded as a kind of cultural landscape in China. The cultural element in the scenic parks plays an important role in the identification and appraisal of their values. Scenic parks have also been the subject for poets over the centuries. Re-experiencing what has been written in the poems about the sights is still one of the most important motivations for domestic tourists. Right from the establishment of the scenic park, the possibility for using it as a tourism development tool was being considered.

When China was further integrated within the global tourism system, international branding and titles were perceived to be economically and politically valuable. The title of World Heritage Site was an ideal target for Wuyishan. In the early 1990s, the Wuyishan local government was among the pioneers to perceive its value and made substantial efforts for this to be achieved (Chinese Construction Ministry 1998). In 1999, Wuyishan was granted World Heritage Site status with a total protection area of 999.75 square kilometres. The heritage area includes the National Scenic Park, the nature reserve and a few exceptional archaeological sites, including the Han City established in the first century BC and a number of temples and study centres associated with Zhuxi.

The search for these titles and protection efforts was also motivated by the expectation that they would bring change, more specifically, development and modernity, to the locality. The establishment and management of the Wuyishan Scenic Park were fully integrated with the local regional development. It was a path-dependent way for the protection and regional development, although various tensions and conflicts started to grow soon after the selection of this path.

The institutional dynamics

Unlike other countries, the institutional structure of Wuyishan Scenic Park has undergone dramatic changes over the last 20 years, particularly as a result of decentralization. The changes also reveal a desire to balance the protection and the utilization of the park.

A dual structure system was initially adopted for the management of national parks. The central government, represented by the Construction Ministry, would only keep the right of final approval of the plans and monitor whether the construction was carried out according to the plans. The operational, financial and human resource management would be the responsibility of the local governments – either provincial governments, prefecture governments or the county governments, depending on the local situation. Different scenic parks may have a different management model (Xu 2003). However, recognizing the importance of Wuyishan as a heritage site, a special Wuyishan Management Bureau, directly under the control of the Hujian

Provincial Government, was established in 1978. In 1980, the Wuyishan Management Bureau was surprisingly shifted to be under the Jianyan Prefectural Government (now Nanpin Municipal), which was institutionally between Fujian Provincial Government and Chong'an County (Now Wuyishan City).

Decentralization also meant that the Jianyan Prefectural Government had to raise money for the protection. Between 1981 and 1985 it managed to lobby the Hujian Provincial Government to allocate 1 million Yuan for the restoration of Wuyi Palace (a heritage site at the entrance of the gate) walking trails and a pier for rafting. However, the money was insufficient for the maintenance, restoration and development of all the sites. Searching for economic and financial sources therefore became the outstanding issue. Jianyan Prefecture Government merged the Wuyishan Tourism Bureau and Wuyishan Management Bureau into the Wuyishan Management Committee. A state-owned company belonging to the Wuyishan Management Committee, called the Tenglong Tourism Company, was also set up. The Tenglong Tourism Company was responsible for the development of the tourism business and raising money for the protection and maintenance of the park. As a monopoly tourism company for running the business of Wuyishan Scenic Park, the Tenglong Tourism Company grew very fast. Up until 1998, the Tenglong Tourism Company had major shareholdings in some local tourism companies and turned into the biggest tourism business in the local region.

In 1989, Chong'an County made the decision to use tourism as the key economic development tool. The name of Chong'an was changed to Wuyishan City in order to make full use of the brand. In order to facilitate this strategy, in 1990, the Jianyan Prefecture Government decided to further decentralize management rights to Wuyishan City. The Wuyishan Scenic Park Management Committee was established under the Wuyishan City Government, although the Nanpin Municipal Government still had an advisory role in the management structure. In this way, the Nanpin Municipal Government had given up its benefits. As a result, the incentives for Wuyishan City Government to protect and utilize the Scenic Park increased. Immediately, Wuyishan City Government made the decision to apply for World Heritage Site status through the National Construction Ministry, although the financial pressure for the preparation was tremendous for such a poor and small city. A substantial amount of money was needed to finance the project, including infrastructure construction and environmental rehabilitation. The local agricultural bank allowed the Tenglong Company to borrow 29 million Yuan by mortgaging its old street and pier. It was the first time that a Chinese bank had given a loan in such a way. An additional loan was obtained by mortgaging the Tenglong Company's concession right of running rafting business inside the park and collecting the ticket money for the Wuyishan Scenic Park. Still, the Tenglong Company had not grown strong enough to cover all the necessary costs. As a result, a new joint venture was formed with another company from outside this region. This new venture was called Wuyishan Tourism Development Limited. Meanwhile, the Tenglong Company was re-engineered and changed into the Wuyishan Tourism Group Limited, which was the holding company of Wuyishan Tourism Development Limited. The re-engineering also separated the Wuyishan Tourism Group Limited, still a state-owned enterprise, from the Wuyishan Tourism Bureau when the central government began to forbid governments to run businesses.

When Wuyishan Tourism Development Limited was formed, it made efforts to become a public company in order to raise more money for local investment. However, due to changes in national policy on using the national natural and cultural resources in the asset evaluation, the process was delayed. For the locals, it was considered a great loss of an opportunity for development. Other similar places, such as Huangshan and the Emeishan Tourism Company, which went public one or two years before, had gained enough resources for investment and expansion.

Wuyishan Tourism Development Limited was given the concession of providing services related to the Scenic Park, including collecting the gate tickets and managing and maintaining the park, hotels and transportation. The company returns a certain portion of its income to the Scenic Park

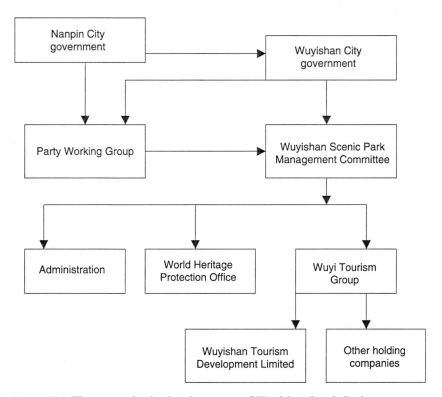

Figure 17.1 The current institutional structure of Wuyishan Scenic Park.

Management Committee, which would be submitted to the government as a resource fee. The local government budgets the resources for the Scenic Park for the coming year, based on the local financial availability and the income submitted by the Scenic Park Management Committee. The budget is mainly used for staff salaries and the planning, promotion and repayment of the loans. The major investment in the Scenic Park is from Wuyishan Tourism Development Limited.

Protection

The establishment of these parks restricts the large-scale destruction of these resources, which was frequently observed in the other places in a time when short-term economic gain was the target and when the popular strategy was 'eat the mountain when you are near the mountain, drink water when you are near the river'. The initial plan for the World Heritage Site was proposed only for the Wuyishan Scenic Park. When the proposal was examined by external experts invited by the provincial government, expansion of the area to include the national natural reserves and another historical site was suggested. The locals accepted the recommendation and more than 30 per cent of the total area is now under protection. Wuyishan became the largest World Heritage Site in China. It also explains why the World Heritage management office was located in the Wuyishan Scenic Park Management Committee. Moreover, the local government implemented a strict policy on investment in secondary industrial projects. The government selected tourism as the key industry, which is believed to be a 'non-smoking' and environmentally friendly industry to support their protection efforts.

In order to obtain these titles, various resource restoration and environmental rehabilitation projects were undertaken. Local regulations on the protection of the Scenic Park, the natural reserves, and the World Heritage sites were formulated and implemented in addition to the national regulations. These local regulations were pioneering in China. Accordingly, a series of protection plans for these sites was also formulated and approved by the Construction Ministry.

Major action on environmental rehabilitation, restoration of the historical sites, infrastructure building and the greening of the site were carried out through the World Heritage Site Project. The project was divided into two phases. The first phase was for the preparation for the application to UNSCO in 1998. The project had seven major components.

1. A resettlement plan was formulated for the people who resided inside the heritage place and in the buffer zone. The 140 square kilometres of developed area was restored and more than 2,000 people were resettled.
2. Seventy-kilometre-long lines of various networks, considered to be visual pollution in the Scenic Park, were removed and buried underground. Now no lines can be seen inside the Scenic Park.

3   A 15 kilometre ring road around the Scenic Park was built to enable the fencing of the Scenic Park. Meanwhile, the trails inside the park were rebuilt and maintained in good condition.
4   The infrastructure, including the parking lots, dustbins, the 22 flushing toilets and educational interpretation systems, was constructed.
5   Two hundred and twenty thousand square metres of green lands were developed.
6   The internal designs and interpretation of seven small museums and tourism centres were upgraded.
7   Inventories of the resources in the heritage sites were completed. Specialized plans were formulated for the protection of this heritage.

After obtaining the title of World Heritage Site, the second phase of the project was implemented. In 2000, a World Heritage monitoring centre was established in partnership with research institutes and universities with a major component of the project being to fence the Scenic Park. There was a key provincial road, with heavy traffic, which cut cross the Wuyishan Scenic Park. Action was taken to build a substitute road and abandon the existing one. Once the new road and the ring road of the Scenic Park were finished, the fence was possible. No vehicles, apart from environmentally friendly tour buses offered by the Company, would be allowed within the park. All these protection resources were made possible through tourism development. Table 17.1 is an overview of the financial situation of the Scenic Park. Among all these efforts, the reallocation of the residents took a substantial proportion of the financial resources. During 1998–1999, the reallocation of the local residents cost 100 million Yuan.

Potential economic tools

Traditionally, Wuyishan had been a place where economic development lagged behind the rest of Hujian Province, which itself is not well developed. Economic opportunities were few and traditionally its supporting industries

Table 17.1 The budget of the Wuyishan Scenic Park

Year	Construction costs (Million Yuan)	Tourists (10,000)	Ticket income (Million Yuan)	Tourism income (Million Yuan)	Expenditure on World Heritage projects (Million Yuan)
1996	3.97	30.18	4.47	10.10	
1997	8.25	35.62	11.93	11.85	
1998	39.41	34.86	14.48	11.37	32.3
1999	30.53	46.78	9.19	30.64	27.8
2000	39.84	47.78	26.28	32.92	20.00

Source: The Management Committee of Wuyishan Scenic Park 2001.

were tea plantations and rice growing, which is one of the reasons why the natural resources could be preserved and why the biodiversity was highly appreciated in the World Heritage appraisal. As the area opened up, resource-based tourism industry was evaluated as having comparative advantages and policies were formulated to support tourism as the key regional industry.

Among the tourism development models, mass tourism was selected naturally. Due to its fame, there is a high demand from domestic tourists for sightseeing in Wuyishan. For the locals, it is only through the mass tourism that investment can be attracted and modernity and quick changes for the region made possible. Few resources have been allocated to support the diversified products. For instance, a highway was built from the airport to the Scenic Park, but no public transportation is available. Today, over 90 per cent of the tourists are on package tours (Chen and Qi 2006). Yet, this strategy has been effective and Wuyishan has experienced a rapid growth in tourism. Within a short time, the service sector become the dominant industry in the local economic structure.

In 1992, the National State Council started a programme to set up 12 special tourism resort areas in China, which attempted to attract foreign investment and foreign tourists, following the model of Bali Nusa Dua Resort in Bali. Wuyishan National Resort at the border of the Scenic Park, covering an area of 12 square kilometres, is one of them. It was expected that the resort could rely on the brand of the Wuyishan Scenic Park to attract tourism investment and facilitate the urbanization process. It was also expected that the development of the resort could change the structure of the tourist market from domestic sightseeing tourists to international holidaymakers. Although the project did not successfully attract chain hotels and international holidaymakers and turn into another Nusa Dua, it did attract substantial small and medium-sized enterprises (SMEs) and migrants. In 1992, there were 1,860 residents; most were farmers relying on tea production and other agricultural products (The Management Committee of Wuyi Resort, 1995). By 2006, there were 140 hotels and over 14,000 beds, which can serve about 3.8 million persons per annum. In addition, there are over 400 retail shops, a golf club

Table 17.2 The tourists and tourism income of the Wuyishan Scenic Park

Item/Year	1998	1999	2000	2001	2002	2003	2004
Rafting income*	1826.46	2736.33	3857.91	5059.72	6025.57	6484.46	8374.74
Gate income*	1281.90	2262.19	3240.90	4731.02	5224.52	4282.77	5651.99
Total visitation**	34.53	46.71	51.72	65.77	70.05	59.23	73.55

* Thousand Yuan. ** Tourists (10,000).
Source: Wuyishan Tourism Bureau 2006.

and other entertainment centres. Overall, 6 square kilometres of land was developed and all the designated land was sold (The Management Committee of Wuyi Resort, 2006). This small resort town is more prosperous and busy than the city centre. Tourism functions as a major driver for urbanization. Moreover, the establishment of the resort makes it possible for the Scenic Park to forbid any accommodation inside the Park and avoid the common development patterns observed and criticized in other scenic parks.

Stakeholder conflicts

Although Wuyishan was considered a model place in China for reaching a balance of protection and the utilization of the resources, tensions have been accumulating among the stakeholders in the process of the protection and the development of the Scenic Park. The most outstanding one is the tension between the farmers and Wuyishan Tourism Company Limited.

Before the establishment of the Scenic Park, there were 5,800 people covering 1,306 households inside the Scenic Park. The mountain lands are collective land and belong to seven villages. The farmers live on tea plantations on these collectively owned lands. Dahongpao Tea – a kind of rock tea – grown there has been ranked for centuries as China's first-class tea. With the preservation and development of the Scenic Park, most of the local farmers were gradually persuaded to sell their houses and move out to the new settlements around the Park in the name of better protection. Their original houses were torn down and turned into green areas or rest spots for tourists. Every protection project always includes a resettlement plan. The first was in the early 1980s, when Wuyi Palace was restored; the second was in the late 1980s when Tianxin Temple was rehabilitated; the third and the fourth, the largest reallocation, were related to the World Heritage Application and Protection Project.

The Company did not buy the mountain lands from the farmers, apart

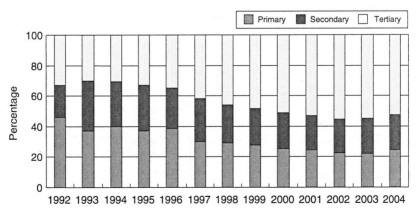

Figure 17.2 Value added as a percentage of GDP for Wuyishan Scenic Park.

from some key scenic spots visited by the tourists. There are various reasons why the tea plantation is an important component of the beautiful landscape. The farmers cannot live without these lands and also the Company cannot afford the cost of buying all the land. After the resettlement, farmers still have the legal right to the mountain land where they grow tea and other trees. Yet, the majority of them have to pay extra costs to travel to the tea plantation and the quality of tea was also affected because the farmers cannot take care of the plantations as often as before.

In order to facilitate the resettlement, some arrangements were made with the farmers. The new settlements were close to the entrance of the gate and near Wuyishan Resort to enable them to start some small businesses. A few also started tea tourism inside the Park. A rafting company was formed where the local farmers could find work. Yet, the farmers were not happy with the arrangements because they felt that they had not got a fair share of the costs and benefits from the Scenic Park development. The farmers already had a lot of complaints, but these remained dormant in the first phase of resettlement. The conflicts did not erupt until a fence project was going to be implemented in the second phase.

In the second phase of the project, the Scenic Park would be fenced and no vehicles, apart from the park bus, would be allowed within it. Also, with the fencing action, the rest of the farmers still located inside the park would be resettled. The plan seemed reasonable from a conservation perspective, as the Scenic Park would be better protected because traffic was substantially reduced. However, this decision lit the fire of conflict.

In the previous resettlement plans, although the farmers were reluctant to move, they finally complied, since the resettlements were arranged by the local government in the name of protection and the benefit of the city as a whole. However, since the Park was fenced by the Wuyishan Tourism Company Limited – the biggest winner who obtained the monopoly right of running tour buses inside the Park – the protests were taken up strongly by the local farmers. The provincial government became aware of the protests and made the decision to postpone the implementation. Also, with the development of tourism, farmers were exposed to the outside world and discovered that they were empowered financially, politically and knowledgably to protect themselves. They not only rejected the fence, but also began to claim the loss in the previous projects and other rights to the lands. On the other hand, the delay in the fencing damaged the Wuyishan Tourism Company Limited, which had invested 300 million yuan in the second phase of the World Heritage Project for environmental rehabilitation and improvement. These investments were loans and the benefits from the monopoly tours bus were expected to pay for the loans. Without this income, the Wuyishan Tourism Company Limited could only manage to pay back the interest, which increased its financial pressures.

Conflicts between the local government and the Wuyishan Tourism Company Limited also broke out. Before the formation of the joint venture, the local government had full control of the Tenglong Tourism Company and decided where it invested. Now the investment is more or less a decision by the Company, a joint venture formed with an outside company, which always searches for the best return on the investment no matter where it is located. The local government was reluctant to allow the Company to invest outside Wuyi City, because Wuyishan still lacks capital.

There are other conflicts associated with the use of the Scenic Park. The emerging conflicts are also observed between the tourism sector and other sectors. After a few years' tourism development, it has been gradually acknowledged that tourism is not a sustainable tool for regional development, especially for economic growth and financial accumulation. Wuyishan has lagged behind other regions in which the economic development depends on secondary industry, a more traditional system of development. It is perceived that protection of the resources and the overreliance on tourism has crowded out the financial and political resources for developing other sectors (Xu 2006). Now there are questions and criticisms of the development strategies. Support from the local government and the local residents for protection is declining.

A dynamic model of preservation and utilization of the national parks

The examination of Wuyishan has shown that within the social, economic and environmental contexts, the establishment and management of national parks in China is closely linked with the local population's search for development and economic growth. However, there is not a simple positive and negative answer to this linkage, which shows a more complex relationship between the two in the case of the Wuyishan Scenic Park. The process can be explained through the model in Figure 17.3 overleaf.

Initially, the values of the natural and cultural heritage are 're-discovered', explored and are highly appreciated. Yet, it is when these values can be branded with a national or international title that enthusiasm for protection is raised. Without the brands, the economic rents of this heritage are uncertain. These brands not only brought potential political gains, but more importantly, the potential investments and the mass tourists. The number of tourists increased substantially when Wuyishan obtained its World Heritage title (see Table 17.1). The growth of tourism reinforces the awareness of the economic value of the titles and accordingly, efforts for protection are reinforced. Of course, some of the actions are not for the resource protection but necessary for the development of tourism.

Furthermore, the increased commercial value of the heritage sites and the associated publicity increases the number of stakeholders in Wuyishan Scenic Park. The commitments of these stakeholders to watch and participate in the decision-making process of Wuyishan increased. Now, not only do the Wuyishan City Government, Nanpin Municipal Government, Hujian

Provincial Government and the National Construction Ministry participate in the decision-making process related to any major development in the Scenic Park, but the academics, mass media and NGOs all monitor and have a voice about the situation. Decisions and actions related to the Scenic Park become very slow and are examined carefully. Therefore, the overall protection efforts rise.

At the macro-level, the continuous and accumulated reports from the mass media and the scholars on the negative impacts on every scenic park inside China, due to heavy pressure for economic development, led the Chinese Construction Ministry to formally issue a regulation re-stressing the national scenic park as a national property with a noncommercial nature. No commercial companies are allowed to obtain and control ticket income. However, since a substantial part of motivation and resources for protection from the local government, Wuyishan Tourism Development Company and local residents, was for and from the economic benefits, the unmet expectations and different share of costs and benefits of the protection efforts can reduce the enthusiasm and support for the protection of national parks. Some of the key negative feedback is also perceived in Wuyishan National Scenic Park. This feedback includes the conflicts between the Scenic Park and the surrounding farmers (−3), the conflicts between the tourism sector and the crowding-out effect on other sectors (−2), and the conflicts between external investors and local SMEs (−1).

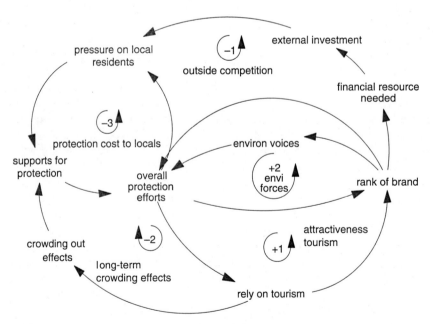

Figure 17.3 A feedback model of protection and utilization of Wuyishan Scenic Park.

Conclusion

Most of the national parks in China were established in the transition period from the planned to the market economy and from a centrally controlled to a locally driven system. This transition determines that the national parks are closely connected with local development. Although it is often the economic value of these national parks that assumes greater importance than other values in the decision-making process relating to the Scenic Park, it seems certain that the awareness of protection was raised and actions were taken. The need for development has led local people, especially officials, to search for all alternatives for development, including the one through the establishment of protected areas. The *Eleventh five year plan* of the Wuyishan Scenic Park Management Committee revealed this thinking. While the target of the Scenic Park is defined as the number of visitors, the actions were to increase the awareness of the need for the protection and maintenance of a good environment (The Management Committee of Wuyishan Scenic Park 2004).

There is no doubt that there is a tendency to overuse park resources. Yet, local government and local communities have shared the vision that they would adopt a more sustainable method of development, different from the other regions in which development strategies are formulated, without consideration of the negative impacts on the local environment. While they still attempt to search for development through resource-based tourism, it is agreed that the long-term development of this region depends on the protection of the Scenic Park.

Attempts have been made to explore and reach a balance between the conflicting goals. Some are successful and some not. The ecotourism model was selected as being suitable for development in the Wuyishan National Natural Reserve. A forestry plantation was also turned into a national forestry park where recreational and holiday resorts are allowed. Farmers and the Company are negotiating an acceptable way to settle their disputes and maybe they can reach a financial agreement. However, the way forward is not so obvious. They remain puzzled and discouraged by the ineffectiveness of obtaining development and catching up with the developed regions through tourism by selling the brands. When the National Regulation on Scenic Parks was issued in 2006, uncertainty about the return on the protection efforts increased. It is not now certain how Wuyishan City can readjust its social and economic structure when the new regulation is actually implemented.

18 'Full of rubberneck waggons and tourists'

The development of tourism in South Africa's national parks and protected areas

Jane Carruthers

> I understand that you have no stomach to see the place full of rubberneck waggons and tourists, but it was vulgarization or abolition, I suppose, and it was at that price only that the animals could be saved.
> (Publicist and artist Stratford Caldecott to Game Warden James Stevenson-Hamilton, 1926, quoted in Carruthers 1995a: 64)

Introduction

South Africa was proud to host the Fifth World Parks Congress in Durban in August 2003 – the first to be held on African soil. Every congress convened since the initial gathering in Seattle in 1962 has been associated with a particular theme and Durban was no exception. In surveying the changing topics over the past five decades it becomes manifest that national parks globally have focused their objectives on becoming ever-increasingly and ever-explicitly more human-centred. In Seattle, delegates worked around the concept of a 'more effective understanding of national parks' with the objective of promoting them worldwide, while in Yellowstone (1972) they directed the meeting towards 'a heritage for a better world' (Harroy 1972). By Bali (1982) the cry was 'parks for development' (*National Parks* 1982) and in Caracas (1992), 'parks for life' (Moosa and Morobe 2003: 247). By the end of the twentieth century, national parks and humanity were inextricably linked through sustainable utilization and development.

With its call for 'benefits beyond boundaries', the Congress in Durban was even more unambiguous in this regard. National parks were being instructed to deliver to society, and various ways in which that might be achieved were discussed. Environmental or ecological services, that is, the delivery to surrounding areas of 'goods', such as areas free of pollution, or the protection of healthy rivers and resilient ecosystems, seemed promising. Other, more tangible, benefits included allowing increased numbers of people to live within protected areas to continue with their traditional (or modern) economic lifestyles through sustainably utilizing resources, such as appropriate grazing practices and the collection of thatching grass, timber or medicinal herbs. But most importantly for the developing world, protected areas were also tasked to deliver direct economic benefits to the country at large. Perhaps not surprisingly, the most effective means of doing so was considered to be the development and nurturing of a strong and profitable tourism industry. This was not, of course, a new idea: it was the very reason why Yellowstone was declared a national park in 1872. In its modern form, this neoliberal approach to nature and its protection has been enthusiastically adopted by the South African national government and by provincial governments that manage their own national parks and other protected areas, although it is not without its critics (e.g. Büscher and Dressler 2007). Accepting such a policy requires that the economic benefit from national parks must be compared with that which would potentially be derived from other development schemes, such as urbanization, industry or agriculture. Thus, in the words of Hector Magome, a South African National Parks (SANParks) senior manager, '... in order to survive, SANParks must make money ...' (Magome 2003: viii).

As other responsible managers for SANParks have elaborated: 'This has resulted in what is often perceived to be a conflict between "conservation" and "tourism" as some magical balance is sought that would satisfy larger numbers of tourists without sacrificing the qualities of wilderness that attract people to national parks ... Indeed, in the case of national parks it is not possible to disentangle biodiversity conservation and human use, because without the one, you do not have the other ... Despite the reluctance of many people to associate nature conservation with money, the twin components of biodiversity conservation and human access within national parks are integrated through finance. While the biodiversity component is a consumer of funds, its enjoyment by people is a potential source of funds ...' (Fearnhead and Mabunda 2003: 186). Moreover, the connection between environmental protection and the tourist industry in South Africa is unambiguous because they are situated in the same national Cabinet portfolio, viz. the Department of Environmental Affairs and Tourism. Managing national parks is only one of the country's international conservation obligations; for there are others as well, many of them under this same ministry, including areas protected under UNESCO's World Heritage programme, Biosphere Reserves and Ramsar sites. Each of the nine provinces of South Africa also has departments that are responsible for environmental conservation and some of these departments incorporate other responsibilities too, for example economic development, agriculture, tourism or development planning. Some provincial governments own and manage protected areas as part of the civil service, others have devolved this responsibility to parastatal boards organized along lines similar to SANParks, such as the North West Parks and Tourism Board or KwaZulu-Natal's Ezemvelo Wildlife.

Tourism and South Africa

The tourism industry is extremely important to the South African economy. In all of Africa, South Africa has the highest number of international tourists, possibly because of its political stability, easy access, good roads and other infrastructure together with modern amenities. Growth has been exponential since apartheid ended. In 1990, 1 million foreign visitors arrived in South Africa: by 2006 there were 8.4 million (South African Tourism 2007(Q1): 2) and the number is expected to rise to around 30.5 million by 2020. Tourism contributes approximately 4.6 per cent to GDP and some estimates consider it higher, at around 6 per cent (ABSA 2002: 10). It is predicted that in the years up to 2010 it will grow annually by 5.5 per cent. Even by world standards, South Africa's tourist growth is one of the highest in the world (Van der Merwe, Saayman and Krugell 2004: 105–6). As is the case elsewhere, there are a number of sectors in the industry. In 2006, some 19.5 per cent of visitors came on 'general holiday', while the extremely large and fast-growing visitor numbers from other African countries reflect the expansion in business tourism (shopping, business travel and tourism combined) that totalled 34.8 per cent (South African Tourism 2007: 23). Tourism generated some R6.6 billion directly in 2006 (South African Tourism 2007: 18), a figure that does not include indirect economic gain. The local tourism industry is equally vibrant with some 42.2 per cent of the South African adult population being calculated as 'domestic travellers' (South African Tourism 2007: 44).

Many international and local tourists appreciate the wealth of eco-tourist destinations in South Africa of which there are more than 400 on state land alone, amounting to some 5.8 per cent (7,080,995 ha) of the total area of the country (Hall-Martin and Van der Merwe 2003: 46). The Kruger National Park (situated in the provinces of Limpopo and Mpumalanga) is a pre-eminent destination because it is the largest and best-known of the national parks. More tourists, however, visit the Cape Peninsula National Park because it includes Table Mountain in Cape Town, while another tourist magnet in close proximity is the Robben Island World Heritage Site where Nelson Mandela was imprisoned. South Africa's protected areas are very attractive to those who want to see the animals that are often referred to as the 'Big Five' (buffalo, lion, rhinoceros, elephant and leopard) as well as for birdwatchers and others interested in nature. It should, however, be noted that state protected areas are increasingly taking a smaller slice of the tourist cake as private protected areas increase in number and extent. It has been calculated that 80 per cent of nature conservation in South Africa takes place on private land, and some 13 per cent of the country's area is protected in this way, far more than the formal protected area estate (Hall-Martin and Van der Merwe 2003: 46, Van der Merwe, Saayman and Krugell 2004: 106). Tourists visit South Africa not only to view wildlife, but also to hunt it – 85 per cent of all Africa's trophies come from South Africa (ABSA 2002: 10–11).

South Africa: Political and environmental background

Any discussion about South Africa's protected areas and tourist history needs to be appreciated within the overarching historical and environmental context, particularly taking account of a new constitution and the coming to power of a black majority government in 1994 (with almost two-thirds of the vote). The bureaucratic landscape was reconfigured that year: nine provinces replaced the previous four, eleven official languages the previous two, new Cabinet portfolios were created and others rearranged, and the civil service was reorganized. South Africa is governed by a tripartite alliance (not a political party) consisting of the African National Congress (now merged with the National Party, the party of apartheid), the Congress of South African Trade Unions and the South African Communist Party. While numerous, other political parties are very small. Each of the nine provinces has a considerable measure of autonomy and its own premier and a full cabinet. There are two land regimes in the country: there is 'owned' land, either by the central and provincial governments as 'state land' or by private individuals and businesses, but there is also 'communal land'. Traditional leadership still exists and is powerful within government (the institution was shored up by apartheid) and brings with it a second legal system that operates alongside the modern one. Society itself is in a constant flux of sociopolitical and economic change, characterized by extremely high levels of crime, frequent and violent discontent against government's poor service delivery, declining labour needs, overwhelming poverty and the influx of many thousands of immigrants and refugees (particularly from Zimbabwe), which has brought a worrying degree of xenophobia. Almost half the population is unemployed and almost a quarter suffers from HIV/AIDS. A sharp decline in agriculture has put even more pressure onto the tourist industry to generate employment and revenue. From some 20 per cent of GDP in the 1920s, agriculture had shrunk to just 3.4 per cent in 2004 (Centre for Development and Enterprise 2005: 1–11).

South Africa's ecological wealth is unique; it is megadiverse (the third most biodiverse country in the world, after Indonesia and Brazil) containing at least ten biomes, a large number of 'hotspots', and an entire floral kingdom, the Cape fynbos. Despite its rich ecology, South Africa is a small country (1,219,912 square kilometres) with limited environmental resources. The western half is extremely arid and, with the exception of parts of the southwestern Cape and some of the east coast, there is an acute lack of arable land (only 13.7 per cent can be regarded as arable) (Centre for Development and Enterprise 2005: 10). In recent years the major form of diversification by commercial farmers (mostly whites) in the drier and savanna parts of the country has been into wildlife ranching. This has gained in popularity as a replacement for beef, dairy and sheep (wool) farming, because of lower subsidies (2.7 per cent of total output), rural insecurity, the need for less labour and government inducements to use the environment more sustainably through eco-tourism, trophy hunting and venison production. South Africa's black rural poor, on the other hand, rely on the communal lands to eke out a living from subsistence crops (mostly maize) and by keeping small herds of diverse domestic stock.

Defining national parks

In every discussion about 'national parks', the issue of definition raises its head. What is a national park? What kind of, or level of, management or governance regime defines it? Of what environmental or ecological attributes should it comprise? What is the optimum or minimum size, given that South Africa's largest national park under the aegis of the controlling organization, SANParks, is 1,948,528 ha and the smallest a mere 3,388 ha? (Hall-Martin and Van der Merwe 2003). Because of the variety of experiences in different countries, and the plethora of names, governance and management regimes applied to protected areas that can be included in the formal IUCN definition of 'national park' (Category II), the exercise is much like nailing jelly to the wall.

The South African experience may be instructive in this regard. The first area called a 'national park' was situated in the Drakensberg, a spectacular mountainous area of the province of KwaZulu-Natal, now a World Heritage Site (the Ukhahlamba-Drakensberg) and a proposed Transfrontier Conservation Area (the Maluti-Drakensberg). Natal was a separate colony until 1910, and thus a protected area proclaimed by the legislature of the Natal government was, in fact, 'national' in terms of being the highest level of government. It only became 'downgraded' (if that is the correct terminology) to 'provincial' when Natal became a province within the Union of South Africa in 1910. In 1903, some 8,000 ha of the Drakensberg near Giant's Castle was set aside for visitors to enjoy as a proclaimed game reserve, and, after various changes to the extent involved, in 1916, part of the upper Thukela River and the mountain peaks in that vicinity were officially declared the Natal National Park (Perry 1929: 9–10, Carruthers 1995a: 53, Zimmer 1998, Sycholt 2002: 32, Wright and Mazel 2007: 129). Within reach from Durban and Pietermaritzburg by rail, this magnificent area was conserved for recreational mountaineering and scenic appreciation, and a lodge and other visitor amenities were planned in those early days. In many respects as well as in name, the Natal National Park bears close comparison with the grand scenery of the American West, which comprised the first national parks in the United States. Given Natal's head start on use of the name 'national park' in South Africa, it is somewhat ironic to note that, for historical reasons, the province of KwaZulu-Natal is the only one of South Africa's nine provinces that does not contain any protected area administered by SANParks.

Further examples of the imprecision of the term 'national park' come from two other protected areas in South Africa. The Kruger National Park, fashioned from two game reserves in the Transvaal colony (also a province only after 1910), was given the title 'national park' in legislation passed by the Union government in 1926. At the time, a parastatal management body, the National Parks Board of Trustees (now transformed into SANParks) was founded to oversee it. However, the Dongola Wild Life Sanctuary was established by the same legislative assembly in 1947 (it was abolished two years later owing to political pressure) under a similar Board, but anomalously called a 'wild life sanctuary', rather than a 'national park' (Carruthers 2006). Throughout the 1920s and 1930s people who were seeking some kind of permanent protection for areas that were wildlife reserves, rather than mountains, geysers, or canyons, wrestled with the notion of what a national park entailed. Certainly, the idea of a level of legal permanence was attractive, but the word 'park' seemed inappropriate because it conveyed ideas of a public recreational playground – which is, of course, what the United States's national parks were initially intended to promote – not wildlife conservation. In those now well-known words, Yellowstone was to be a 'public park and pleasuring ground, for the benefit and enjoyment of the people'. In game reserves filled with dangerous wild animals, by contrast, tourists were difficult to accommodate and they certainly could not be permitted to enter on foot or walk around at will. Also problematic, however, were the appellations 'preserve' or 'reserve', because they seemed to connote exclusivity and medieval ideas of privilege. James Stevenson-Hamilton, South Africa's leading wildlife conservator and warden of the Kruger National Park area from 1902 to 1946, frequently came out in print against the phrase 'national park', although it has to be said that he was not averse to utilizing the term strategically for his own purposes, just as many others do today (Carruthers 2001). In 1905, he referred to his game reserves as 'game nurseries' (Stevenson-Hamilton 1905: 20–45) and later called for the Kruger National Park to be a 'National Faunal Sanctuary', 'National Wild Life Sanctuary' or an 'animal sanctuary'. As will be explained below, he rejected 'national park' because he wanted to encourage a tourist ideology of wilderness, of 'roughing it'. Unlike modern tourists who demand every luxurious amenity available, he considered a degree of physical discomfort to be integral to the experience of nature and, in his opinion this was not what the phrase 'national park' conveyed (Carruthers 1995a: 113–14).

Stevenson-Hamilton was, however, on a losing wicket because the word 'national', particularly after the First World War, was enormously powerful. In that era it became associated with the Treaty of Versailles, with the League of Nations and with Woodrow Wilson's and Jan Smuts's advocacy of 'nation states' being created from dying, discredited and defeated empires. In terms of its resonance with something 'good', it is appropriate to compare 'national' with 'democracy' on which, in the early twentieth century, the United States considered too that it had a monopoly and a mission to advance it worldwide. As George Orwell put it in his essay, 'Politics and the English language', 'It is almost universally felt that when we call a country democratic we are praising it: consequently the defenders of any kind of regime claim that it is a democracy, and fear that they might have to stop using the word if it were tied down to any one meaning' (Orwell 1950: 91).

It is in a similar manner that 'national park' has become a world famous brand, and it must be extremely rare indeed that products can be so very different and still be marketed under the same name. Because it is a desirable marketing tool, SANParks has tried to pressure government into reserving the words 'national parks' only for those 20 protected areas that are under its direct jurisdiction. Some of the provincial authorities that also own and manage 'national parks' have considered this predatory as well as prejudicial, because for them too, 'national park' is an advantageous tourist label. Moreover, it is one to which they are entitled because many of their protected areas fall well within the IUCN definition of a Category II protected area, called a 'national park'.

For the purposes of determining the nature of a national park, it is also important to recall that Yellowstone's conservation (1872) was predicated on two core principles: the first was its natural and geological wonders – its 'curiosities'; second, that it would be managed as a 'public park and pleasuring ground, for the benefit and enjoyment of the people'. In other words, it was a declared tourist destination. Moreover, it was only a scenic conservation effort, and there was no reference to wilderness, animal and plant life, ecosystem services or biodiversity, which are the values held in such high regard today (Nash 1967 and 1970).

Grove (1995) was correct to enquire whether nature conservation itself – that is, the protection of aspects of natural processes, rather than the protection of geological formations that the public could enjoy – was a North American invention at all. He concluded that it was an imperial legacy, and indeed, South Africa's experience owes more to its imperial connections than to the example of the United States.

In considering nomenclature, it might not be stretching the argument too far to suggest that rethinking the national park terminology in the first decade of the twenty-first century – as is taking place (see below) – may be related to the refiguring of the 'nation state', a form of government increasingly assailed by global forces, the fragmentation of former 'nations' into ethnic enclaves and the rise of powerful transnational ideologies. South Africa is a complex, diverse, multiracial, multiethnic, and multicultural country in which a shared vision of nationhood is still a work in progress after many centuries of racial division and ethnic violence. It is also as well to bear in mind that in South Africa the specific areas, management structures and objectives of South Africa's national parks are currently subject to the extremely strong political forces of land restitution and successful claims may well result in considerable adjustments. This has already been demonstrated in the case of the Makuleke in the northern portion of the Kruger National Park and the Khomani San in the Kalahari Gemsbok National Park (De Villiers 1999, Carruthers 2007).

At the World Parks Congress in Durban in 2003, the necessity for a new template for the international naming of protected areas was raised and is still being debated by stakeholders. Even before the Congress, however, South Africa drafted new legislation to provide for its protected areas under an updated and comprehensive law. This emerged as the National Environmental Management: Protected Areas Act No. 57 of 2003, amended the following year by the National Environmental Management: Protected Areas Amendment Act No. 31 of 2004. Its promulgation was preceded by fierce debate as to which organ of state would own the 'national park' brand, but pressure to maintain the status quo was intense and the legal framework of the protected area landscape was not dramatically altered by the new law, although some new terminology was introduced.

The historical context in which Yellowstone was first described as a 'national park' in 1872 was one from which international legislation or conventions were absent. Europe was still then in an imperial era, in which bilateral or regional treaties predominated. After the First World War this context changed with the establishment of the League of Nations, but its very ineffectiveness was partly responsible for the outbreak of the Second. The United Nations was founded as a far stronger and more representative global organization. One of its many subsidiary arms was the International Union for the Protection of Nature and Natural Resources, which emerged from a conference at Fontainebleau in 1948 (Jepson and Whittaker 2002).

In terms of the 1994 IUCN agreed definition (following on the first of 1969), the foundation of a national park is a 'protected area', viz. an 'Area of land and/or sea especially dedicated to the protection and maintenance of biological diversity, and of natural and associated cultural resources, and managed through legal or other effective means'. Those means are not elaborated upon further and it is thus not explicit that this has to be a national government or even any other formal level of a state.

A national park, viz. Category II, is defined as follows:

> Natural area of land and/or sea designated to (a) protect the ecological integrity of one or more ecosystems for present and future generations, (b) exclude exploitation or occupation inimical to the purposes of the area and (c) provide foundation for spiritual, scientific, educational, recreational, and visitor opportunities all of which must be environmentally and culturally compatible.
>
> (IUCN 1994)

In other words, protected areas themselves are defined both by management and governance regimes, national parks only by management (Phillips 2004, Mitchell 2007).

Whatever the history, and notwithstanding the debates over naming, the opinion that Dr Harold Eidsvik, the former Chairman of the World Commission of Protected Areas of IUCN, expressed when he visited South

Africa in 1996 in order to report on its national parks, remains critical. In his view, the key criterion that distinguishes a Category II protected area from the rest and thus makes it a true 'national park' is visitor access (Hall-Martin and Van der Merwe 2003: 47), an opinion that accords with Category II. Accordingly, tourism defines what a national park *is*.

SANParks owns just over half of South Africa's total protected area, but this can be misleading, because most of this comprises two enormous national parks, the Kruger National Park (1,948,528 ha) and the Kalahari Gemsbok National Park (960,029 ha). By contrast, some of the smaller national parks, such as Agulhas (8,528 ha), Bontebok (3,388 ha) or Wilderness (7,688 ha) are nothing more than large zoos or recreational areas. They, and perhaps others as well, but for different reasons, do not conform to other requirements of Category II of the IUCN system, but they are all open for recreational tourism. Many of the provincial game reserves and parks are very large and they too welcome tourists, indeed depend on them. In South Africa there is no law that prevents people from living in national parks or keeping domestic stock among wildlife (the Richtersveld National Park, 1994, was a precedent in this regard) but wherever dangerous wild animals have either been introduced or allowed to continue to exist, permanent human occupation is, naturally, not possible.

Tourism in national parks – case studies

The Kruger National Park

South Africa's national park system germinated during the First World War when white settler stock farmers cast covetous eyes at marginal pastureland that had been included in two game reserves, the Sabi (founded in 1898, extended in 1903 and that included many private farms) and the Singwitsi (founded in 1903) – the total area stretching from the Crocodile River in the south to the Luvhuvhu River in the north, approximately 19,000 square kilometres of the Transvaal. With the modernization of the South African economy after Union in 1910 and improvements in transport and disease control, voters were pressurizing the Transvaal provincial government to open the game reserve's eastern boundary to commercial agriculture. In order to evaluate the situation, a Commission of Enquiry was appointed.

The report of this Commission proved to be one of the foundational documents for South Africa's national parks and also for its eco-tourist industry. Although they did not visit the whole area, the commissioners were impressed with what they saw by way of scenery and fauna and they recommended that the objectives of these game reserves be reconsidered. They noted the 'uselessness of having these superb reserves merely for the preservation of the fauna' and considered that they should be transformed into 'a great national park where the natural and prehistoric conditions of our country can be preserved for all time' (Transvaal Province 1918).

A campaign was mounted to persuade politicians and the public to accept this national park philosophy. In 1926 – after a change of government and many delays and disappointments (Carruthers 1995a) – this came about, and the National Parks Act No. 56 of 1926 was passed unanimously by both Houses of Parliament. The issue that swung the argument and clinched the passage of the law (as it did in the United States in 1872) was that the national park would be visitor-friendly, which had not been the case with the game reserves. In agreeing to the establishment of a national park, white South Africans took the view that allowing citizens to look at wildlife was a government responsibility. It was also a legitimate – and financially viable – form of using state land. The law stopped short of defining a national park, but did provide objectives. In 1926, these were 'the propagation, protection and preservation therein of wild animal life, wild vegetation and objects of geological, historical or other scientific interest for the benefit, advantage and enjoyment of the inhabitants of the Union' (Section 1). By 1976, changes to the law had altered the objectives as follows: 'The object of the constitution of a park is the establishment, preservation and study therein of wild animal, marine and plant life and objects of geological, archaeological, historical, ethnological, oceanographic, educational and other scientific interest relating to the said life of the first-mentioned objects or to events in or the history of the park, in such a manner that the area which constitutes the park shall, as far as may be and for the benefit and enjoyment of visitors, be retained in its natural state' (South African National Parks Act No 57 of 1976, s 4).

As explained above, James Stevenson-Hamilton, warden of the Kruger National Park area from 1902 to 1946, grappled with the words 'national park'. What he visualized were extremely large nature reserves or 'sanctuaries' in which wildlife and natural processes could continue to occur, places which were not manipulated and which were expressly not a recreational 'park'. In this philosophy he differed from Piet Grobler, South Africa's Minister of Lands from 1924 to 1933 (the Cabinet position at that time responsible for national parks), who explicitly argued in favour of turning the Kruger National Park into 'a first-class holiday resort' (Union of South Africa 1926). Grobler's vision won the day: holiday resort it would be. As the minister had said in his persuasive speech to parliament, if 10,000 Americans visited each year, then more than a million pounds would be accumulated, 'a sum that should appeal to all South Africans' (Carruthers 1995b: 170–5).

It was as a consequence of this divergence of opinion that publicist and artist Stratford Caldecott wrote to Stevenson-Hamilton shortly after the Kruger National Park was established in 1926:

> I understand that you have no stomach to see the place full of rubberneck waggons and tourists, but it was vulgarization or abolition, I suppose, and it was at that price only that the animals could be saved.
>
> (quoted in Carruthers 1995a: 64)

After its establishment as a national park, tourism determined many of the management strategies of the Kruger National Park, as remains the case today. Within little more than a decade, the landscape was transformed by access roads, bridges and camps constructed for tourist convenience, for this was the era of the private motorcar. Because South Africa had become urbanized and the public had not been in a position to encounter wildlife for a generation – the vast herds of South African wildlife had all but disappeared by 1900 – the Kruger National Park was an instant tourist draw-card. Because of malaria and fierce thunderstorms in summer, only the southern portion of the park was opened to the public and only in winter when the threat of disease was less and torrential rain was less likely to destroy roads and bridges. The first camp, Pretoriuskop, was opened in 1927 and a game ranger's wife was amazed at the 'crowds of motor cars out this winter, as many as four in one day!' (Carruthers 1995a: 75). By 1929, there were 30 cars; the following year, 900. Investment in administrative staff, roads, bridges and accommodation and other tourist facilities could not keep abreast of the demand. There was no system of bookings (so visitor numbers were not known in advance), no formal opening and closing times, no qualified staff at the gates. Even regulations for appropriate tourist behaviour (such as driving at night, getting out of motor vehicles and carrying firearms for protection against wild animals) had to be initiated from scratch. At that time, a railway line ran through the national park, and the NPB initially considered allowing South African Railways to handle all the tourist facilities. However, most visitors came by motorcar, camped in their vehicles, or in tents or – the more fortunate – in huts, and took care of their own catering and ablution requirements. At some of the camps, bad behaviour was frequent: drunkenness and loud noise from gramophones and radios spoilt the wilderness experience for many. Caldecott had rightly warned Stevenson-Hamilton of 'vulgarization' and when visitors had left each year, Stevenson-Hamilton was relieved (Carruthers 1995a: 75).

The issue of 'national parks' being explicitly linked with 'national heritage' is a complex one in South Africa and a more detailed account is provided elsewhere (Carruthers 1994, Carruthers 1995a). Suffice it to say that in this country the overriding 'national heritage' celebrated by tourists was seeing wildlife and the idea of being among dangerous animals in their natural habitat. Thrilling accounts of encounters with, and sightings of, dangerous animals such as the very popular lion, and later elephant and rhinoceros, dominate the tourist literature. The heritage aspect, however, was restricted to white South Africans – black South Africans were not welcomed as tourists, although they formed the majority of those employed in menial positions (Carruthers 2001). The name 'Kruger' was used to inculcate a specific Afrikaner culture in the Kruger National Park, and rhetoric around the issue of political support for national parks and nature conservation predominates in the pre-1994 period (Carruthers 1995a, Carruthers 1995b)).

In considering 'heritage' as a tourist demand, one needs to take into account developments in the first – and topographically and climatically very different – national park in Natal. As outlined above, this was an area preserved for its scenery and geologically interesting high peaks and therefore had a specific attraction for mountaineers and walkers. Wright and Mazel (2007: 126–9) outline the increase in visitor numbers to the Drakensberg in the 1920s, 1930s and 1940s, but they give no statistics. It is, however, clear from the brief account that they provide that the tourists who visited the mountains were focused on rock climbing, trout fishing (introduced to South African rivers in the nineteenth century) and walking. In other words, physical activity, magnificent scenery and hotel accommodation, rather than driving around in cars, looking at savanna grassland and staying in rustic camps (such as was the case in the Kruger National Park) was the order of the day. By contrast, no specialized knowledge or requirements were needed to look at wildlife and no physical exertion was required. One merely drove through the bush on roads and looked out of the motorcar window. It seems therefore that the mountains catered for a different segment of the tourist market. Both, however, were 'national parks' and a 'heritage' for white South Africans then in denial that black South Africans might appreciate such a heritage in a different manner.

Although it is not the case today, at that time the Kruger Park warden had considerable power over all aspects of the park, and during the period Stevenson-Hamilton was in office, he tried to shape the tourism experience to conform to his vision of 'heritage'. He was determined to provide a wilderness experience for visitors to the Kruger National Park. In 1930, he was appalled by the suggestion that there might be 'dancing and gramophones at the rest camps' and he even dismissed the showing of instructive wildlife films and lantern slides as 'unnecessary entertainment' (Carruthers 1995a: 78).

Stevenson-Hamilton wanted to prevent the Kruger National Park from becoming a 'pleasuring ground' and he therefore deliberately refused to supply too many comforts. He favoured coir mattresses, rather than 'luxury' ones, he argued against the introduction of electricity or indeed against anything that would 'over-civilize' the park. But nothing he could do deterred the visitors and their numbers continued to soar: in 1940, there were more than 22,500 (Carruthers 1995a: 78). However, wardens who succeeded Stevenson-Hamilton when he retired in 1946 recognized the need to pamper visitors more, and in the 1950s far more consideration was given to comfort and convenience. New camps were established, old ones enlarged and revamped with better facilities, particularly by way of furniture, electricity, kitchens and ablution blocks.

As has been the case with many other protected areas throughout the world, the question of limiting numbers of tourists arose from time to time, as did ideas for the most appropriate administrative organization to manage them. In the early 1950s, the limit of 3,000 at any one time proved impossible to implement, and in order to control the burgeoning numbers more efficiently, the NPB took over the tourist operation. By far the majority of the visitors were local white South Africans, for whom an annual visit to the Kruger National Park became a regular pilgrimage. For the year ended March 1976 visitor numbers had topped 375,000, income was R2.7 million and turnover R4.8 million (Knobel 1979: 231).

The explosion of tourists to all protected areas in South Africa, whether provincial, national or private, after the Second World War requires more scholarly attention. One can, nonetheless, postulate a few of the factors responsible, including – as far as whites are concerned – an increase in population, greater prosperity and thus mobility and leisure time, urbanization, more state spending on infrastructure such as roads, and a growing international market because of improvements in air travel.

Until 1995, when SANParks took over the Cape Peninsula and Table Mountain as a 'national park', the Kruger National Park remained the country's premier protected area with an unequalled international reputation. None of South Africa's other parks – every one of which has its own history, special character and ecological significance – rivals them. Like the Kruger National Park, each of the country's 20 national parks was founded for a particular reason considered appropriate to 'national park' status at the time. In the 1930s, for example, the injunction to propagate, protect and preserve certain rare wild animal species inspired the creation of small national parks for this specific purpose. These included the Bontebok National Park in the Western Cape (1931), the Addo Elephant National Park in the Eastern Cape (1931), the Mountain Zebra National Park near Cradock in the Eastern Cape (1937) and the larger Kalahari Gemsbok National Park in the Northern Cape (1931). A generation later, national park principles had shifted from protecting certain species and habitats to identifying and conserving representative South African ecosystems. Among the first national parks established for this reason are Golden Gate Highlands National Park (1963), the Tsitsikamma National Park (1964) and the Augrabies Falls National Park (1966). More national parks have been declared since the ending of apartheid in 1994, viz. Cape Peninsula, Marakele, Namaqua and Vhembe-Dongola. This last-named is also a World Heritage Site and it includes some of the area that was within the Dongola Wild Life Sanctuary mentioned above. Since 1994 and the advent of a black majority government, there has been more emphasis on celebrating African cultural heritage in national parks, but this aspect of park management has still a long way to go before it can begin to compete with the popularity of wildlife viewing among tourists (Mbenga 2003). Not only do cultural sites have to be excavated or interpreted by specialists and scholars, but also almost the entire philosophy, research, conservation managements – and tourist publicity – of SANParks and of South Africa's other national parks and protected areas is concentrated on the natural environment and specifically on large mammals.

Visitors want to see large wild animals and their requirements have not only shaped tourist facilities, but also national park management. However, they like wild animals to be seen in appropriate landscapes (even when they are reclaimed farmland) and when the growing numbers of elephant in the Kruger Park began to flatten large trees and alter riverine views, the issue of culling elephant numbers again (this was halted in 1994) arose. Surrounding the elephant culling debate are threats of a tourist boycott (from international tourists in particular) if culling is done, and there are threats of a large decline in tourist numbers if elephant populations continue to expand because the landscape is being detrimentally altered aesthetically by its over-utilization by elephants. There is a fear that Kruger might begin to resemble Tsavo (Scholes and Mennell forthcoming). It will be interesting to follow how this plays out in the future, given that more than 75 per cent of SANParks visitors are local South Africans. Moreover, in terms of international visitors, Germans (many of whom are hunters) dominate with 35 per cent, the Netherlands and the United Kingdom both hover around 11 per cent, while visitors from the United States – the stronghold of the Humane Society of the United States and the International Federation for Animal Welfare who spearhead the campaign against culling – amount to some 4.8 per cent (Phillips 2008).

There are many new tourism initiatives in South Africa's national parks, all of which benefit from a historical perspective and from the association with 'national parks'. For example, when the Kruger National Park was established on state land in 1926, there were many private farms on its southwestern boundary. At first, owners of these properties tried to farm them, but the climate and soils did not suit either agriculture or pastoralism. For decades, most were used for intermittent recreational sport hunting by owners and their friends. From the 1970s and 1980s, however, some of these properties were converted into private game reserves and they have been extremely successful financially, offering wildlife experiences impossible to duplicate in the government's portion of the protected area estate. Among these were luxury accommodation and catering, fewer rules, dusk, night and dawn drives in open vehicles with qualified game rangers at hand to act as interpreters, to lead guided walks and, thanks to radio contact between 'camps', even to almost guarantee sight of all 'Big Five' (Rattray 1986). Needless to say, this top-end of the tourist market is the most profitable and very high prices (usually quoted in US dollars) are the order of the day. At first reluctant to compete in this sector, SANParks has now been obliged to do so in order to provide those 'benefits beyond boundaries' that have previously been alluded to and many fences between the Kruger National Park and their adjoining private properties have been removed.

There are many benefits that protected areas can deliver; among them are sociopolitical, environmental and educational outreaches. However, it is principally on tourism that many expectations depend. South Africa's idea of a 'nation' was explicitly racial in character from the outset and the history of the majority of national parks reflects the restrictions on black South Africans who were discouraged from visiting protected areas through racist legislation, excluded from structures of governance or management and only employed in menial capacities. Moreover, some Africans in some places were evicted from protected areas, a matter which is being corrected through the land restitution process. The domestic market still dominates national park tourism and, as mentioned above, in respect of SANParks protected areas, the ratio is around 75 per cent South African. As a total of that domestic market, it is pleasing to record that black visitors now comprise more than 20 per cent, having climbed steadily since 1994 (Phillips 2008).

However, for historical reasons many black South Africans, particularly the poor, still regard protected areas as playgrounds for the rich, luxuries that the country can ill afford to support with taxpayers' money. Demands on central and provincial governments are enormous and although biodiversity conservation is a national priority, so too – and often more urgent – is the provision of housing, water, sanitation, schools, health services, electricity supply and other infrastructure and employment creation. One area in which the state has decided not to compete, but rather to cooperate and manage for profit, is eco-tourism in the protected areas it controls. Although what follows is relevant directly to SANParks, the general outlines apply also to many provincial protected areas of which the extremely popular Madikwe in North West Province is a prime example (Davies 2000).

The Department of Environmental Affairs and Tourism has taken the decision that tourist amenities do not fall within government's core functions and thus public/private partnerships (which must have a substantial Black Economic Empowerment dimension) have come into play and altered the pattern and variety of tourist experience. In parks that have the potential to cater for the high-end of the tourist market, a 'commercialization for conservation' strategy has been developed in which SANParks grants concessions to private operators in exchange for a commission. Thus the tourism part of the enterprise is profitably outsourced, but not, of course, the scientific research or conservation management. Not only does such a policy generate an income to that particular national park, it allows SANParks to cross-subsidize those parks that are nationally more significant for reasons of biodiversity conservation than they are in generating tourist revenue. This is not to suggest that the local, lower-end of the tourist market is being ignored, because the decision has been taken that in the short and medium term

SANParks will continue to manage 'rest camps', which offer budget to middle-level accommodation. This policy has not been without its critics. In 1998, the report of the Investigation into the Institutional Arrangements for Nature Conservation in South Africa concluded that 'Nature conservation as such can never be self-supporting. This is axiomatic . . . It is therefore short-sighted and fallacious to expect a protected area to be economically self-sufficient' (Kumleben *et al.* 1998, quoted in Fearnhead and Mabunda 2003: 189).

Private funding partnerships have also enhanced the tourist experience through initiatives to enlarge the currently established protected areas and also to establish Transfrontier Conservation Areas that straddle international boundaries, and a number of these 'Peace Parks' have either been established or are in this process (Cumming 1999, Sandwith *et al.* 2001, Pabst 2002, Hall-Martin, Braack and Mabunda 2003).

Pilanesberg National Park

The second case study that well illustrates the tourist history of South Africa's national parks is the Pilanesberg National Park, situated fairly close to the town of Rustenburg in what is today the North West Province. This popular protected area was founded for political reasons and for tourism, and owes its origins to the 'homelands' policy of the Afrikaner nationalist government of the 1960s. In this period, the National Party government carved South Africa up into 'homelands', places in which Africans were forced to live according to their ethnic background. The policy was 'legitimated' by giving these homelands independence, calling them 'national states', with the accoutrements of 'nationalism' by way of government officials, a national capital, state flag and the like. The Pilanesberg National Park was a way of promoting one of apartheid's 'independent states' and easing it into a place of international respectability (Carruthers 1997).

In 1969, when self-governing and independent homelands were being mooted as apartheid blueprints for the future, the idea of some kind of wildlife tourist attraction in the unique Pilanesberg volcanic crater originated from the extremely conservative Potchefstroom University for Christian Higher Education. At first, opposition to the scheme was intense from the local Pilane clan and the initiative collapsed. However, the idea re-emerged when the Sun City hotel, casino and entertainment complex began nearby some years later, and a game reserve was considered an attractive complement. At the time, gambling was prohibited within 'white' South Africa, but permitted in the 'homelands' (Brett 1989: 111–12).

By no stretch of the imagination could the Pilanesberg crater at that time have been regarded as 'pristine' nature or an intact ecological system. It has been used in pre-colonial times by stock-farmers and in the early colonial period the land had been taken over by white settlers. Ownership changed hands too, when by the 1913 and 1936 Native Trust and Land Act legislation the area reverted to African ownership, and white farmers were expropriated by the state. Both groups were intent on eking out a living, not on protecting nature, and when development occurred, it was generally unsightly. Roads, dams and other structures were unsuitably sited; invasive exotic plant species had taken hold and soil erosion was considerable. But mindful of the benefits that tourist revenue would bring to this depressed part of the country, and contribute to the viability of the homeland, named Bophuthatswana, people were evicted from some 50,000 ha of the crater and surrounding hills and the national park took shape.

The Pilanesberg National Park came into being – as did the Kruger and other national parks – when paramilitary wildlife management and anti-human ecology were powerful in national park dogma. Initial signs were not propitious. Not only did local people oppose it, but also capital to begin land reclamation and game introduction projects was not forthcoming from the Bophuthatswana homeland. In the event, the South African Nature Foundation (today the South African arm of WWF, the Worldwide Fund for Nature) came to the rescue with R2 million and created a national park. Planning was thorough, with tourism, management and wildlife consultants used for the process. Tourists were the target, and wildlife the attraction. Many difficulties had to be overcome. Opposition from local communities continued, early management by the National Parks Board of Bophuthatswana was incompetent, even corrupt. Wildlife had to be translocated into the area and scientists argued about the wisdom of reintroducing wildlife species from other areas of the subcontinent and possible problems resulting from mixing gene pools (Brett 1989: 112–14).

However, over time, experienced wildlife rangers principally from Zimbabwe and Natal were employed, land reclamation procedures began to take effect and wildlife populations settled. Tourism – at first from the luxury casino but subsequently sightseers from the Witwatersrand

> independent of that hotel, as well as international visitors – has increased. This has swollen the coffers of the park, and since 1994 and the abolition of homelands, has spread tangible benefits well beyond its boundaries to people of what is now the North West Province. Community conservation in South Africa began, it may be justifiably argued, in this national park, which was the first to benefit neighbouring people and to begin education and other environmental schemes (Magome and Collinson 1998, Ringdahl 2003). Currently, there is a land restitution claim on the Pilanesberg National Park (as there are on parts of the Kruger and other South African national parks), instigated in 2002 by the current Chief Pilane of the Bakgatla-ba-Kgafela and it is not yet resolved. Should it be successful, the proceeds of the lucrative tourist industry in the national park may well accrue to the Pilane clan, rather than to the coffers of the province.

Conclusion

It is extremely doubtful whether the vast majority of tourists are aware of the fine distinctions between 'national parks' and other protected areas. In terms of planetary health, it might well be argued that the IUCN category of wilderness is the one that is globally underrepresented (South Africa has no wilderness areas) and national parks the most numerous. Tussles over who 'owns' national parks can be fierce, but they generally occur among managers, perhaps determined to protect their status and their revenue streams. It is unlikely, for example, that the number of national parks in Australia – or even in the United States – meets every one of the current requirements of the IUCN Category II definition. Certainly in South Africa, many, even those administered by SANParks, do not. Bontebok, Mountain Zebra and Knysna, for example, are not integrated and intact ecosystems. On the other hand, some provincial national parks, of which Pilanesberg is one, allow hunting under strict conditions, about which an argument can be raised (but also countered) as to whether this is a visitor opportunity compatible environmentally or whether it is a form of exploitation inimical to the area's purposes. On the other hand, protected areas such as the Ukhahlamba-Drakensberg World Heritage Site meet every aspect of the definition, and some of the provincial game reserves in KwaZulu-Natal do so too. What is certain, however, is that no protected area once called a 'national park' will willingly ever renounce the name. It is too entrenched as a desirable product in the international tourist vocabulary.

Part V
Beyond nature

19 National parks as cultural landscapes
Indigenous peoples, conservation and tourism

Heather Zeppel

Introduction

Globally, 50 per cent of all national parks and protected areas are on indigenous lands, with 85 per cent of parks in Latin America and Africa declared on indigenous territories (Kemf 1993, Amend and Amend 1995, Colchester 1996, Mackay 2002). This chapter reviews the history of indigenous exclusion and removal from national parks regarded as 'wilderness areas', starting with Yellowstone, and the relationship to tourism. By the 1980s, there was emerging recognition of national parks as cultural landscapes, inhabited by local people, with legal recognition of indigenous land rights and the advent of co-managed or jointly managed parks. Some key national parks that are now managed with indigenous groups in Australia, New Zealand, South Africa, Canada and the US are critically examined. These co-managed national parks represent the cultural identity of affiliated indigenous nations along with other political, economic and environmental values of national parks as symbolic landscapes and conservation or tourism icons representing the identity of modern nation-states. There is a focus in this chapter on governance structures allowing for indigenous inclusion in park management, cultural interpretation, cultural values and conflicts in parks and the involvement of indigenous groups in tourism and conservation within these co-managed national parks. Regaining legal control over tribal lands within parks and the reassertion of indigenous cultural identity in protected areas will increasingly define the management of conservation and tourism in these parks.

Indigenous peoples and national parks

Indigenous peoples are defined as:

> tribal peoples in independent countries who are regarded as indigenous on account of their descent from the populations which inhabited the country (who) retain some or all of their own social, economic, cultural and political institutions ... and whose status is regulated wholly or

partially by their own customs and traditions or by special laws or regulations.

(IUCN 2004, based on ILO Convention 169)

Other terms used for indigenous people are Aboriginal, Native or Tribal peoples, First Nations (Canada), or Fourth World peoples in colonized countries. There are estimated to be around 400 million indigenous people, comprising 5,000 tribal groups (Weaver 2006).

Globally, indigenous peoples inhabit 20 per cent of land, in areas of high biodiversity, compared to the 6 per cent of the world declared as protected areas (WWF 2005). 'Indigenous peoples comprise 5 per cent of the world's population but embody 80 per cent of the world's cultural diversity. They are estimated to occupy 20 per cent of the world's land surface but nurture 80 per cent of the world's biodiversity on ancestral lands and territories' (UN 2002: 2–3). The remaining forest areas or global 200 eco-regions with the highest biodiversity are linked with surviving indigenous groups in Asia, Africa, the Americas and Oceania (Nature Conservancy, 1996, Oviedo *et al.* 2000, Weber *et al.* 2000, WWF 2000). More national parks are being declared in these indigenous land areas.

Indigenous land practices and cultural knowledge have thus ensured the conservation of global biodiversity in areas declared as national parks or protected areas by colonial rulers or national governments. However, over 12 million people – mainly hunter-gatherers and pastoralists – have been removed from their ancestral lands to make way for protected areas, conservation and tourism. They are affected by poverty, limits on resource use and land degradation, with few benefits from tourism (MacKay 2002, 2007, African Initiative 2003, Colchester 2001, 2003a, b, Lasimbang 2004, Mackay and Caruso 2004, Hill 2004, Stevens 1997). In many regions such as Africa, protected areas deny indigenous rights, or involvement in conservation (Negi and Nautiyal 2003, Nelson and Hossack 2003, Lasimbang 2004). Since the 1980s, there has been growing legal assertion of indigenous rights to own or occupy ancestral lands and use natural resources now within national parks (MacKay 2002). The World Conservation Union (IUCN) has only since 2000 devised guidelines to involve indigenous communities in co-managing national parks, other protected areas and community conservation areas (Beltran 2000, Borrini-Feyerabend *et al.* 2004, Bushell and Eagles 2007). These IUCN guidelines focus on securing indigenous land rights in legislation together with policies for co-managed protected areas and also support for community conservation and resource management (Carino 2004, Grieg-Gran and Mulliken 2004, Hill 2004, UNESCO 2005). The IUCN Commission on Environmental, Economic and Social Policy (CEESP) has provided professional guidance and input on these indigenous issues through its Theme on Indigenous and Local Communities, Equity and Protected Areas (TILCEPA), a Co-management Working Group and other working groups on governance of protected areas, and also conservation and human rights. The World Summit on Sustainable Development in 2002 and the 5th World Parks Congress in 2003 also included resolutions on the rights of indigenous peoples and conservation of biodiversity in national parks (FPP 2003). Some 200 indigenous delegates actively participated in workshops at the 5th World Parks Congress. The *Durban Accord and Action Plan* from this Congress recognized the rights of indigenous peoples in regard to biodiversity conservation with a specific recommendation on indigenous peoples and protected areas (*WCPA News* 2003, Brosius 2004, Larsen and Oviedo 2005). Indigenous communities were also included in recommendations on community conserved areas, cultural and spiritual values, tourism, governance, mobile indigenous peoples, and co-management of protected areas. Tourism in protected areas was to support indigenous communities and contribute to their livelihood through economic opportunities; recognize indigenous customs and values; include traditional knowledge and protect sacred sites (IUCN 2004). Specific management goals for indigenous peoples and national parks to be implemented by 2010 were:

Key Target 8: All existing and future protected areas shall be managed and established in full compliance with the rights of indigenous, mobile peoples and local communities.
Key Target 9: Protected areas shall have representatives chosen by indigenous peoples and local communities in their management proportionate to their rights and interests.
Key Target 10: Participatory mechanisms for the restitution of indigenous peoples' traditional lands and territories that were incorporated in protected areas without their free and informed consent established and implemented by 2010 (Mulongoy and Chape 2004: 22–23).

The *WCPA Strategic Plan 2005–2012* sets a target of 2012 to achieve these measures, with an objective of 25 indigenous/local communities involved in protected area governance (WCPA 2005). However, according to Mackay (2007), indigenous rights in South American protected areas are being legally affirmed through the court system and by land claims rather than by governments or conservation organizations, despite the policy directives of the IUCN on indigenous peoples in parks.

The 2007 UN ratification of the *Declaration on the Rights of Indigenous Peoples* and national laws on indigenous land rights will further expedite this process of involving indigenous peoples in parks. In many countries, government policies on protected areas and indigenous lands are now linked with legal recognition of indigenous land tenure and resource use rights (Zeppel 2007). These national policies are shaped by international conventions on protected areas; biodiversity conservation, indigenous rights and cultural heritage (see Table 19.1). The development of co-managed national parks with indigenous communities since the mid-1980s reflects the strong links between global initiatives on biodiversity, legal assertion of indigenous rights,

Table 19.1 Biodiversity conservation, indigenous rights and co-managed national parks

Biodiversity Conservation	Indigenous Rights	Co-managed National Parks
1980s		
Biosphere Reserves World Heritage Areas World Parks Congress (1982)	UN Working Group on Indigenous Populations (1982) ILO Convention No. 169 on IP (1989) UN *Draft Declaration on the Rights of IP* (1989/90)	Gurig (1981), Kakadu (1984), Uluru (1985), Nitmiluk (1989) *Australia* Gwaii Haanas (1986) *Canada*
1990s		
GEF established (1991) UN *Convention on Biological Diversity* (1992) World Parks Congress (1992) World Conservation Congress (1996) IP Biodiversity Network (1997) Latin American Congress on National Parks and PA (1997) Ramsar Wetlands and IP (1999)	World Bank Policy on IP (1991) UN *International Year for the World's IP* (1993) UN *Decade of the World's IP* (1995–2004) Caracas Declaration on PA and Human Futures (1992) WWF/IUCN Principles on IP and PA (1996) WCC Resolution 1.53 on IP and PA (1996) IP of Africa Coordinating Committee (1998) Latin American Declaration on IP and PA (1997) Minority Rights Group International (1999) IWGIA Arusha Resolutions on IP in Africa (1999)	Booderee (1995), Witjira (1995), Mutawintji (1998), IPA (1998>) *Australia* Kluane (1992) *Canada* Richtersveld (1991) *South Africa* Kaa-lya del Gran Chaco (1995), Isiboro Secure (1997) *Bolivia* Aoraki/Mt Cook (1998) *NZ*
2000s		
TILCEPA set up by IUCN (2000) World Summit on SD (2002) World Parks Congress (2003) World Conservation Congress (2004) Protected Areas and IP (IUCN) (2000 and 2004) Conservation and IP (CBD) (2004)	UN Permanent Forum on Indigenous Issues (2000) Dana Declaration on Mobile IP and Conservation (2002) Durban Accord and Action Plan – Rights of IP (2003) Business for Social Responsibility Rights of IP (2003) World Social Forum includes IP (2005) UN *2nd Decade of the World's IP* (2005–2014) Inter-American Development Bank Policy on IP (2006) UN ratified *Declaration on the Rights of IP* (2007)	Arakwal (2001), Biamanga and Gulaga (2006) *Australia* Kgalagadi (2002) *South Africa* Alto Fragua-Indiwasi (2002), Cahuinari (2001) *Colombia* Lanin (2002) *Argentina* Kayan Mentarang (2002) *Indonesia*

Notes: IP=Indigenous peoples; ILO=International Labour Organisation; GEF=Global Environment Facility; SD=Sustainable development; IPA=Indigenous Protected Area; PA=Protected Area; IWGIA=International Work Group on Indigenous Affairs; NZ=New Zealand; TILCEPA=Theme on Indigenous and Local Communities, Equity, and Protected Areas; CBD=Convention on Biological Diversity.

participatory conservation and ecotourism (Ghimire and Pimbert 1997, Zeppel 1998, 2006, 2007, Weber *et al.* 2000, Alcorn 2001, Mark 2001, MacKay 2002, IUCN 2004, Johnston, A. 2006, Poirer 2007).

Indigenous issues in protected areas

Since the early 1990s, several key books have critically examined the situation of indigenous peoples living in or near protected areas (West and Brechin 1991, Kemf 1993, Furze *et al.* 1996, Stevens 1997, Brechin *et al.* 2003, Colchester 2003b, Igoe 2004), especially in Africa, Asia, and Latin America (IWGIA 1998, Colchester and Erni 2000, Nelson and Hossack 2003). Other books have focused on community-based conservation (Notzke 1994, Western and Wright 1994, Ghimire and Pimbert 1997), and community partnerships with parks (Stolton and Dudley 1999) to more recent indigenous and local involvement in protected area governance (Jaireth and Smyth 2003, Child 2004). Other books about indigenous peoples and tourism include chapters about indigenous involvement in conservation and protected areas (Ryan and Aicken 2005, Johnston 2006, Notzke 2006, Butler and Hinch 2007, Gale and Buultjens 2007). Chapters about indigenous peoples have been included in books or reports about managing protected areas (Borrini-Feyerabend 1999, Oviedo and Brown 1999, Stolton and Oviedo 2004, Larsen and Oviedo, 2005, Scherl 2005, Kothari 2006a, b, Lennon 2006); conserving biodiversity (Tongson and Dino 2004); park values (Acha 2003, English and Lee 2003); indigenous resistance (Adams 2005); heritage tourism (Carr 2008); ecotourism (Cochrane 2007; Zeppel 2007); and tourism in national parks and protected areas (Hall 2000, Scherl and Edwards 2007, Wellings 2007). Key themes are indigenous rights, park partnerships and providing benefits to local communities from conservation and tourism. Since 2000, special issues of the IUCN's *Parks* journal, and *Policy Matters* newsletter, included articles about indigenous participation in conservation; park partnerships; local communities and protected areas; community empowerment and community conserved areas; poverty; history and culture; and human rights. Other reports also review the rights of indigenous peoples in biodiversity conservation and co-managing protected areas (Wall 1999, Clay *et al.* 2000, Larsen 2000, Alcorn 2001, Colchester 2003b, Borrini-Feyerabend *et al.* 2004, Hill 2004, Pannell 2006, Redford and Painter 2006, Bauer 2007, Poirer 2007).

Inhabited national parks

Preserved for its natural features and scenery and seen as uninhabited, Yellowstone was in fact used for hunting, fishing and living areas by Native American people for 11,000 years. The Shoshone and Bannock people used the area for hunting and fishing with one group residing year round. Fire was also used to shape the ecosystem with the Shoshone still setting fires in Yellowstone in 1886. Following the Battle of Little Big Horn (1876) and subsequent military campaigns, the US government removed all Native American people in the area to nearby reservations and their hunting trips ceased in Yellowstone. In 1877, Nez Perce Indians fleeing from US troops attacked a hotel and camp in the park, killed two visitors and abducted others (Whittlesey 1995). Some 300 Shoshone people were subsequently killed in battles with park authorities (Kemf 1993, Stevens 1997, Keller and Turek 1998). Many Native American groups considered Yellowstone a sacred place but these cultural links, mythic legends and tribal stories are not interpreted to park visitors (Whittlesey 2002). The Crow people also legally controlled a small part of Yellowstone Park as part of their reservation land in Montana, up until 1882. In California, the Yosemite National Park area was the home of the Miwok Indians. After the area was declared as a state reserve in 1864, some Miwok people returned to work in hotels and as national parks staff. Other remnant Miwok settlements were evicted from inside the park (Keller and Turek 1998). The Miwok people have resided in or near Yosemite for over 150 years and still annually gather acorns in the park (Stevens 1997).

In 1931, the Navajo Nation and US Parks Service jointly declared Canyon de Chelly National Monument in Arizona. The Navajo retained land ownership and residence under an 1868 treaty, along with rights to hunt, graze sheep, and grow crops. Canyon de Chelly received 881,000 visitors in 2004, with Navajo people hired as tour guides and comprising 80 per cent of park staff (Wall 1999, Tuxill 2006). Since 1936, Western or Timbisha Shoshone people have resided on 40 acres inside Death Valley National Park in California. In 2000, the *Timbisha Shoshone Homleland Act* established a reservation of 7,600 acres, including 314 acres inside the park (Stevens 1997, MacKay 2002, Mitchell *et al.* 2002). In 1908, when the Grand Canyon was declared a National Monument, Havasupai Indian people were left on a reservation of 518 acres. In 1975, *the Grand Canyon Enlargement Act* transferred 185,000 acres of parkland to the Havasupai Indian Reservation with a further 95,000 acres inside the park set aside for traditional use by Havasupai Indians including grazing, in the first recognition of native land use in a US park though with any future claims extinguished (Hough 1991, MacKay 2002). Marketed as Grand Canyon West (and much closer to the tourist centre of Las Vegas), the Havasupai charged entrance and camping fees at these areas, with 20,000 annual visitors staying at these facilities (Wall 1999). Glacier National Park in Montana comprises land ceded through a treaty by Blackfeet Indians, who do not pay park entrance fees, gather wild plants by permit, and hold ceremonies in the park. In South Dakota, the northern portion of the Pine Ridge Reservation became part of Badlands National Park in 1978. Oglala Sioux people are allowed to hunt, gather wild plants, herd cattle and grow crops in this park. Half of the park entrance fee goes to the Oglala Sioux Parks and Recreation Authority for use in park development and facilities, while tribal rangers patrol the area. In 2002, Oglala activists occupied the park, demanding that the land be returned (Burnham 2000, Igoe 2004).

Throughout Africa, India and Asia indigenous groups were also forcibly removed from areas declared as national parks or nature reserves. Maasai people and their herds of cattle were excluded from all the savannah national parks and game reserves of Kenya and Tanzania, with the exception of Ngorongoro Conservation Area (Berger 1996, Igoe 2004). Other indigenous groups were forcibly removed from the national parks of Tsavo, Lake Rukana and Sibiloi in Kenya; Kidepo in Uganda; Myika in Malawi; Kalahari Gemsbok in South Africa; Dumoga-Bone in Sulawesi, Indonesia; and Rara in Nepal (Colchester 1997, Stevens 1997). In South America, though, local people have continued residing in or using natural resources inside 86 per cent of protected areas while in Central America indigenous people reside in 30 per cent of protected areas (Kemf 1993, Amend and Amend 1995). In Costa Rica, Aboriginal reserves are included in protected lands (28 per cent of all land), along with national parks and wildlife reserves (Eagles and McCool 2002). In Asia, about 16,000 Dayak people live within or around Kayan Mentaring National Park in Kalimantan, Borneo (Borrini-Feyerabend *et al.* 2004).

The right of indigenous peoples living inside or near protected areas is a key issue in terms of conservation, park management and involvement in tourism (West and Brechin 1991, Terborgh and Peres 2002). According to Scherl (2005), key aspects for indigenous peoples affected by parks include livelihood security, cultural and spiritual integrity, psychological well-being, governance, and educational and economic opportunities in park management and tourism. There are four stages in the indigenous co-management of national parks and tourism, ranging from exclusion of indigenous people, then conflict and negotiation about indigenous access and use of parks, to park development (Table 19.2). Parks subject to indigenous land claims are in the conflict stage and moving towards co-management. Indigenous involvement in cultural interpretation and tourism is part of this process.

Tourism, indigenous peoples and national parks

Few books or chapters on tourism in national parks address the involvement of indigenous peoples (Hall 2000, Scherl and Edwards 2007), most focusing

Table 19.2 Four stages of indigenous co-management of national parks and tourism

1: Exclusion/Removal: Ejection of People, Denied Use of Natural Resources, Informal Park Services
2: Conflict/Contestation: Indigenous Rights, Land Claims, Cultural Conflicts, Indigenous Tours
3: Negotiation/Co-Management: Agreements, Leases, Settlements, Resource Use, Interpretation
4: Development/Consolidation: Park Board, New Enterprises, Joint Ventures, Tourism Ownership

on management issues in protected areas (Eagles and McCool 2002, Lockwood *et al.* 2006). According to Hall (2000), the mountain areas and indigenous peoples of early national parks such as Yosemite were regarded as 'worthless', with the land and native people not seen as having any economic value. These national parks were declared for recreation and tourism, with indigenous people then expelled or resettled on other marginal land. With the growing popularity of nature-based tourism in rural areas, more US national parks were declared, some including areas of land on indigenous reservations. The Mesa Verde National Park in Colorado, with Anasazi ruins and cliff dwellings, was carved out of a Ute Indian reservation. The US Congress annexed Ute tribal land with a further coerced land swap adding more Ute land with Anasazi ruins to the park. The Ute tribe now operates helicopter tours over the park, with Navajo people operating craft stalls and fried bread stands in the park. Mesa Verde has 650,000 visitors annually, while the adjacent Ute Mountain Tribal Park receives 2,000 visitors (Igoe 2004).

In Australia, the area around Uluṟu (then known as Ayers Rock – see Chapter 5) was excised from the Petermann Aboriginal Reserve in 1958 by the federal government as a national park and only returned to Aṉangu Aboriginal ownership in 1985 (Power 2002). The Uluṟu park board has a majority of Aṉangu Aboriginal people, who receive 25 per cent of park entrance fees, operate Aṉangu Aboriginal Tours, work as rangers and have part-ownership in a tourist resort. Hall (2000) reviewed the co-management and tourist promotion of Uluṟu-Kata Tjuta National Park, with visitor education about Aṉangu Aboriginal culture a key approach (i.e. signs, brochures, cultural centre, tour operator handbook, guided walks). He noted there was less Aboriginal control over visitor numbers or tourists climbing Uluṟu, regarded as culturally inappropriate, and images of Uluṟu used in tourist marketing (Mercer 1998, Digance 2003; see also Chapters 5 and 10). Aṉangu people have twice closed the climbing point on Uluṟu, as a sign of respect over the deaths of traditional owners.

In a similar manner, New Zealand has Maori people on conservation boards for national parks, with Maori cultural and environmental values interpreted to park visitors such as the spiritual significance of Aoraki/Mount Cook as a *taonga* (treasure) for *Ngai Tahu* Maori people (Hall 1996, 2000, Carr 2001, 2004, 2008, McIntyre *et al.* 2001). However, again, there was no indigenous control over tourist numbers, concessionaires or climbers on Aoraki/Mt Cook. Hall (2000) noted that the recognition of some national parks as cultural landscapes with joint management and interpretation of indigenous heritage values were not aspirations shared by all commercial tour operators.

Eagles and McCool (2002) mentioned the forced removal of Native Americans from US national parks, and recent recognition of indigenous park ownership and use rights in Australia and Canada. Their final section on future trends for park tourism did not mention indigenous people, other than as a minor caption for a photograph of a restored Indian village in

Canada. This caption referred to policy debates on the role of indigenous peoples in park management, shaping park values, and park services.

Scherl and Edwards (2007) analysed the benefits of cultural and nature tourism opportunities for indigenous peoples living in or near protected areas in developing countries. Their focus was on tourism as a means of poverty alleviation and sustainable development for local people living around parks. Tourism in parks needed to address human rights and social justice in conservation, with limitations on indigenous use of land and natural resources or other viable activities in national parks. They identified three management models for park tourism enterprises such as community-based (i.e. lodges, guiding, concessions, visitor services), partnerships (i.e. government, NGO, private sector), and joint ventures (i.e. shared ownership, management and profits of lodges on community land). The Iwokrama Canopy Walkway in Guyana was an indigenous tourism enterprise in a nature reserve. They recommended that indigenous people be active partners and managers of park tourism ventures; however, apart from Iwokrama, there was no review of indigenous tourism in national parks.

Some indigenous peoples support the declaration of national parks to prevent other land uses. In 1887, in New Zealand, the Maori chief Te Heu Heu Tukino gifted the volcanic mountain area of Tongariro to the nation as a national park. This gift was to prevent land subdivision and sale, assert Maori territorial ownership and retain the *mana* or spiritual power of the mountain. Local Maori people now want greater involvement in the management of Tongariro Park and its two ski resorts (Te Heuheu 1995, Green 1996, Hall 2000). A century later in Bolivia, the Kaa-lya del Gran Chaco National Park became in 1995 the first park in the Americas declared and run by indigenous peoples, with a 10-year agreement to co-manage the park. Compensations funds (US$3.7 million) from a gas pipeline and highway funded indigenous park guards and 41 per cent of the Kaa-lya park budget, with recent funding for ecotourism projects (Winer 2003).

There is a long history of indigenous involvement with tourism and national parks, despite their exclusion or forced removal from these protected areas. This includes indigenous people working in hotels and as national parks staff in Yosemite, and as tour guides, transport providers or vendors selling food or crafts at other parks. Recent political, legal, economic and social changes have increased the profile and participation of indigenous peoples in cooperatively managed national parks and tourism enterprises (Scherl and Edwards 2007, Carr 2008). At Kgalagadi Transfrontier Park in South Africa, the indigenous San people (or Bushmen) and the Meier community were each awarded 50 per cent shares in a lodge constructed in the park, the right to build and operate community entrance gates and commercial rights to develop eco-tourism ventures in adjacent land areas (Holden 2007). Other key drivers are western recognition of 'natural' areas as cultural landscapes, tourist interest in indigenous culture, national parks on indigenous lands, and consulting diverse stakeholders in park management. Hence, strategies for park management and tourism now also address indigenous rights, interests and employment or enterprises in co-managed national parks.

Conservation and national parks

The ideal of national parks as uninhabited 'wilderness' areas, deriving from the United States, has influenced the global declaration and management of protected areas. This 'Yellowstone model' of national parks is based on a protectionist conservation paradigm that prohibits human settlement and bans the consumption, use or sale of natural resources (see Table 19.3). This Western ideal of nature protection led to the removal of indigenous and other resident local people from areas declared as national parks and the prohibition on local use of natural resources within park boundaries (Stevens 1997, Adams 2004, 2005). This model of national parks as pristine natural areas devoid of people, however, conflicts with indigenous land claims, human rights, cultural preservation and rural development needs (Stevens 1997, West and Brechin 1991, Mowforth and Munt 2003, Poirer 2007, Riseth 2007). This Western social construction of national parks led to a model of 'fortress conservation' (Brockington 2001, Igoe 2004) that excluded local people from park management and tourism. In 1978, the IUCN redefined national parks as natural and scenic areas used for scientific, educational and recreational purposes, where commercial extractive uses were not allowed. The 1994 IUCN definition of a national park still focused on ecosystem protection and recreation, excluding human exploitation or occupation that was adverse or detrimental to nature conservation, while providing for compatible spiritual, scientific, educational and recreational visitor opportunities. It also recognized both natural and associated cultural resources in parks and spiritual links with natural areas. In 1995, the US National Park Service established a liaison office to consult with Indian groups about park management, followed by Parks Canada in 1999 (Weaver 2006). Since the early 1990s there has been a changing paradigm of protected areas moving towards participatory or community-based conservation, involving local people in resource use, park management and tourism (Adams and Mulligan 2003, Lockwood *et al.* 2006). This was driven by the growing global declaration of national parks since 1980, particularly in developing counties (Mowforth and Munt 2003), and recognition of local people still residing in protected areas, especially in Latin America, Asia and Africa (West and Brechin 1991). This 'Uluru model' of national parks is thus based on participatory conservation, with co-management and indigenous use of natural resources (see Table 19.3).

National parks as cultural landscapes

There is increasing Western recognition that natural areas and national parks are a cultural landscape, modified by human actions and activities and shaped by cultural perceptions of the environment (see Chapter 5). The

Table 19.3 Paradigms of protected area management

Protectionist Conservation (1872–present) 'Yellowstone Model'	Participatory Conservation (1982–present) 'Uluru Model'
Government managed, national asset	Co-managed, community heritage
Local people removed and excluded from parks	Local people living in or included in parks
Preservation of nature and wildlife	Sustainable protection of nature and wildlife
No local use of natural resources	Managed local use of natural resources
Nature separated from culture	Nature linked with culture and community
Natural and environmental values dominant	Natural, cultural and spiritual values recognized
Manage land, wildlife and natural resources	Manage relationships, rights and resources
Community ignored and disempowered	Community engagement and empowerment
No revenue sharing with local communities	Revenue sharing with nearby communities
Local people not involved in parks tourism	Local people engaged/work in parks tourism
Non-local rangers and parks staff	Local people work as rangers and parks staff
Top down management, few links to communities	Shared governance, equity, community development
Park management within boundaries, 'islands'	Park management beyond boundaries, linkages
Park used only by rangers, scientists and tourists	Park also used by local/indigenous people
Isolated/individual nature conservation units	Integrated conservation development projects
Protected areas part of dominant/colonial space	Protected areas as shared intercultural spaces

Source: Expanded from Phillips 2003, Borrini-Feyerabend *et al.* 2004, Lockwood and Kothari 2006.

Caracas Resolution and Action Plan from the 4th World Parks Congress in 1992 recognized national parks as indigenous cultural landscapes linked with cultural identity, spirituality and subsistence use of resources. Recommendation 6 from this Plan was about respect for indigenous knowledge, recognizing customary tenure and land use, and indigenous participation in park planning. In this approach, indigenous cultural traditions and knowledge were now recognized as a key resource for community-based conservation

(Kleymeyer 1994). The US National Parks Service defined a cultural landscape as: '. . . a geographic area, including both cultural and natural resources and the wildlife or domestic animals therein, associated with a historic event, activity or person, or exhibiting other cultural or aesthetic values' (NPS 1994, cited in Lennon 2006: 455). Ethnographic landscapes include subsistence and ceremonial grounds, sacred religious sites and settlements of Indian groups. Canyon de Chelly, with cliff dwellings, pit houses, petroglyphs and sacred landforms, has cultural, historical and spiritual significance for the Navajo Nation and other Indian tribes (Tuxill 2006).

For indigenous people, culturally significant landscapes also have symbolic or spiritual meanings associated with specific places (Taylor 2000, McAvoy 2002, Acha 2003, Hay-Edie 2003, Carr 2004, 2008). 'An Aboriginal cultural landscape is a place valued by an Aboriginal group (or groups) because of their long and complex relationship with that land. It expresses their unity with the natural and spiritual environment. It embodies their traditional knowledge of spirits, places, land uses, and ecology' (Buggey, 1999, cited in Lee 2000: 6, Neufeld, 2005, Parks Canada 2007a). In 1992, the World Heritage Convention recognized a new category of associative cultural landscapes based on 'the powerful religious, artistic, or cultural associations of the natural element' (UNESCO 2005), such as indigenous spiritual beliefs linked with landscape features. In 1993, Tongariro National Park in New Zealand was the first World Heritage Area listed as an associative cultural landscape, based on the spiritual significance of this mountain area for the local *Ngati Tuwharetoa* Maori people (Te Heuheu 1995). Uluru-Kata Tjuta National Park in Australia was also relisted in 1994 for its cultural and spiritual significance to Anangu Aboriginal people (Layton and Titchen 1995). The recognition of Tongariro, Uluru and other national parks as indigenous cultural landscapes involves the integration of indigenous people and heritage values in their operation and public presentation.

Cultural and spiritual values of parks

Cultural landscapes include both tangible sites (e.g. monuments, ruins, tools, archaeological remains) and intangible spiritual beliefs associated with natural places. There are a range of intangible or nonmaterial values associated with national parks, such as recreation, education and science, with a more recent focus on spiritual, cultural and identity values in parks (see Table 19.4). Intangible values are those that enrich 'the intellectual, psychological, emotional, spiritual, cultural and/or creative aspects of human existence and well being' (WCPA, 2000, cited in Harmon 2003: 55). Cultural values connect people 'in meaningful ways to the environment', spiritual values 'inspire humans to relate with reverence to the sacredness of nature', while identity values 'link people to their landscape through myth, legend or history' (Harmon 2003: 56). Lockwood (2006) listed culture, identity, spiritual and social well-being and bequest as part of community values for parks, while individual values for parks included satisfaction, health and spiritual well-being but not identity or meaning. According to the Millennium Ecosystem Assessment (2003, cited in Scherl *et al.* 2004), cultural services such as religious values, tourism, education and cultural heritage are part of the ecosystem services of protected areas, along with provisioning, regulating and support services provided by the natural environment. Within indigenous societies, culture, nature, spirituality and personal identity are interlinked and indigenous 'cultural-identity values are often transcribed (either figuratively or literally) into an ancestral landscape' (Harmon 2003: 59). That is, tangible cultural heritage sites and spiritual beliefs about creator beings imbue indigenous cultural landscapes with meaning and identity. Interpretation in co-managed national parks highlights the ongoing spiritual, cultural, ecological and historic connections between indigenous peoples and natural landscapes (Pfister 2000, Carr 2004). Cultural interpretation at Aoraki/Mount Cook National Park in New Zealand presents Maori creation beliefs and spiritual links, with traditional use of plants and *pounamu* (greenstone) (Carr 2001, 2004).

Pfister (2000) describes the written and verbal interpretation by Canada's Nisga'a First Nations people from the Raven Clan about the sacred significance of coastal mountains where 2,000 Nisga'a people were killed by a volcanic eruption at Nisga'a Memorial Lava Bed Provincial Park in British Columbia. It is the first jointly managed state park, where Nisga'a park guides accompany visitors to restricted areas. Park signs convey Nisga'a values, beliefs and legends about the cultural significance of the lava bed. This strategy protects the cultural meaning and heritage values of Nisga'a sacred sites and stories, by presenting the lava bed as a cultural landscape that reaffirms the presence and identity of First Nations people (Murtha 1996). The Nisga'a people are also negotiating rights to other traditional land areas. Land claims progress indigenous cultural interpretation of sites in public parks.

Table 19.4 Values of protected areas

Tangible Values (Material)	Intangible Values (Nonmaterial)
Conservation values (ecosystem services+)	Recreational and Therapeutic values
Economic values (tourism revenue)	Spiritual and Cultural values
Land values*	Artistic and Aesthetic values
Infrastructure values* (buildings, roads, utilities)	Educational and Scientific values
	Peace values (equity and social justice)
	Existence values and Identity values

Source: English and Lee 2003, Harmon 2003, Harmon and Putney 2003, * = tangible values for parks added by the author

+ *Note:* Ecosystem Services include 1) Provisioning, 2) Regulating, 3) Cultural, and 4) Supporting services (Millenium Ecosystem Assessment 2003).

Cultural conflict in parks

Indigenous cultural and spiritual beliefs about sacred natural sites, however, can create conflicts with recreational users of national parks. This includes tourists climbing Uluṟu (Ayers Rock) in the Northern Territory (Brown 1999, Head 2000, Weaver 2001, Digance 2003; see, also, Chapter 10) and Mt Warning (Wollumbin) in northern New South Wales (Gale and Buultjens 2007); both considered sacred sites. The Uluṟu climb was closed for 20 days in 2001 as a sign of respect for a deceased traditional owner, angering tourism operators (Weaver and Lawton 2002). The construction of walkways, barriers and signs at sacred rock art sites in the Keep River National Park offended the Miriuwung people who are custodians of the sites (Mulvaney 1999). In New Zealand, Ngai Tahu Maori people revere Aoraki/Mt Cook and in 1998 climbers were asked to show their respect by stopping just below the main summit and by not leaving litter, food or human waste (McIntyre *et al.* 2001; Weaver 2001, Carr 2004). At Devils Tower National Monument (Wyoming, US), rock climbers came into conflict with American Indians who performed sacred ceremonies in June for the summer solstice. A US court endorsed a voluntary climbing ban on Devils Tower in June, with an 84 per cent reduction in climbers since 1995 in this month and park interpretation of Indian religious values for Devils Tower (Ruppert 1994, Hanson and Moore 1999, Linge 2000, Dustin and Schneider 2001, Dustin *et al.* 2002, Harkin 2002, McAvoy 2002, Taylor and Geffen 2003, 2004). Niaitstakis Mountain in Glacier National Park, Montana, is sacred to Blackfoot Indian people who hold vision quests at this site. Conflicts between climbers and Indians performing ceremonies at this mountain led to the closure of access roads and hiking trails (Reeves 1994).

Purpose of national parks

The recognition and documentation of indigenous cultural and spiritual values in natural landscapes, since the mid-1980s, has also affected western perceptions about the purpose and identity of national parks and protected areas (see Table 19.5). Governments declare national parks for conservation, tourism and scientific purposes with these protected landscapes reflecting national pride and identity. Non-indigenous cultural and identity values for parks comprise secular indicators such as biodiversity and scenic amenity. Park visitors and local people have a strong personal affinity with protected areas. Indigenous groups, however, see these protected areas as cultural landscapes and homelands that embody personal, cultural and community identity (McAvoy 2002, Neufeld 2002, 2005, English and Lee 2003, Harmon 2003, Hay-Edie 2003, McAvoy *et al.* 2003, Prosper 2007, Carr 2008). These two world-views are reflected in government approaches to conservation based on strict nature protection and indigenous claims or cultural conflicts about the ownership and co-management of national parks. A recent social-ecological

Table 19.5 World views on purpose and identity of national parks

National Government	Indigenous Group
National park	Indigenous land
National identity	Indigenous identity
Natural landscape	Cultural landscape
Wild land/wilderness	Home land
Protected area	Community area
Scenic preservation	Subsistence resources
Conservation	Caring for country
Biodiversity	Biocultural diversity
Scientific knowledge	Traditional knowledge
Recreational, educational and economic benefits	Spiritual, cultural and community benefits

Source: Author.

approach to conservation and management of protected areas, however, places parks within landscapes, links local people to parks, acknowledges indigenous social and cultural values and recognizes the rights and needs of local communities (Lockwood *et al.* 2006, Carr 2007, 2008). This new approach is particularly evident in co-managed national parks.

Co-managed national parks and indigenous tourism

Indigenous peoples are involved in the co-management of selected national parks with government agencies in Australia, Canada, New Zealand, South Africa, Bolivia, Colombia, Argentina and Indonesia (see Tables 19.1, 19.6 and 19.7). Canada has negotiated cooperative park management with Aboriginal groups since the 1970s as part of Treaty settlements. There are now 13 jointly managed national parks in Canada (Johnston, J. 2006), mainly in the Inuit territory of Nunavut, other northern Arctic areas and the west coast of British Columbia. To resolve land claims, South Africa has also negotiated new contractual national parks with San and Khoikhoi people at Ai-Ais/Richtersveld and Kgalagadi Transfrontier Parks, with community members on a park management board (Reid 2006, Holden 2007). In 1991, Richtersveld was the first national park in South Africa to acknowledge ownership by the indigenous Nama people. A joint management plan was signed in 2002 with preference given to Nama for park positions and a lease fee of US$32,000 (Magome and Murombedzi 2003, Dutton and Archer 2004). The legal assertion of indigenous rights and land claims over protected areas, treaty negotiations, changes to government laws and regulations, and indigenous park policies, have driven this trend. Indigenous groups are involved in co-management, park planning and other activities only where they have successfully defended a land claim or treaty settlement (Peckett 1998).

Table 19.6 Indigenous co-managed national parks in South Africa, South America and Indonesia

National Park, Country, Year of co-management	Indigenous People(s)	Management Stakeholders	Key Park Policy
Richtersveld NP, South Africa (1991)	Nama IP (San and Khoikhoi)	Nama community (5, including chairperson) South African NP (4)	SANP agreement (24 years, 50c/ha/year) Management Committee
Kaa-Iya del Gran Chaco NP, Bolivia (1995)	Isoseno (Izoceno)-Guarani IP 1st PA in Americas declared and run by IP 10-year agreement to co-administer park in 1996	CABI (IPO), Ministry of Sustainable Development and Planning, WCS, SERNAP, municipalities, Chiquitanos and Ayoreo groups, Izozog women	*Agrarian Reform Act* 1993 $3.7 million in compensation from gas pipeline and highway funded park guards and 41% of park budget Kaa-Iya Foundation set up in 2002 – funds for ecotourism
Isiboro Secure NP, Bolivia (1997)	Moxenos (Mojeoo), Yuracares and Tismanes (Chiman) IP	Ministry for Sustainable Development and Environment, TIPNIS	General Regulations of Protected Areas July 1997
Lanin NP, Argentina (2002)	Mapuche people 7 Mapuche communities	Nationa Park Service, Neuquen Mapuche Confederation	May 2000 co-management committee established (50/50 government and Mapuche)
Alto Fragua-Indiwasi NP, Colombia (2002)	Ingano IP – 'House of the Sun' sacred place – 1st park where IP control declaration, design and management of PA based on Ingano Life Plan	Colombian government, Indigenous Ingano Councils, National Parks System of Colombia (UAESPNC), Amazon Conservation Team	Policy for Social Participation in Conservation (1998) MoU for park management and 50/50 park board of 8

Cahuinari National Natural Park, Colombia (2001)	Bora-Mirana IP Park overlaps Mirana indigenous reservation Formal agreements	Bora-Mirana IP and National Natural Parks System (SPNN), Minister of the Environment, NGOs	Policy for Social Participation in Conservation (1998) Joint management plan
Sarstoon-Temash NP, Toledo, Belize (1997)	Kekchi Maya and Garifuna IP	Belize Forestry Department, Kekchi Maya and Garifuna, EcoLogic NGO	Park committee (2 members from each of 5 villages) Park management strategy
Kayan Mentarang NP, Kalimantan, Indonesia (2002)	Kayan Dayak people, Central Borneo	FoMMA (Alliance of the Indigenous People of Kayan Mentarang National Park), Forest Protection and Nature Conservation Agency, district government	Park Policy Board (*Dewan Penentu Kebijakan*) established in April 2002 by Ministry of Forestry decree

Sources: Borrini-Feyerabend 1999, Clay *et al.* 2000, Winer 2003, Borrini-Feyerabend 2004, Scherl *et al.* 2004, Tongson and Dino 2004, Lockwood 2006, Reid 2006, Redford and Painter 2006, Tongson and McShane 2006.

Notes: PA=Protected Area, IP=Indigenous People, CABI=Capitania del Alto y Bajo Izozog (Bolivia), SERNAP=Servicio Nacional de Areas Protegidas (Bolivia), TIPNIS=Subcentral Indigenous Territory and National Park Isiboro Secure (Bolivia), IPO=Indigenous Peoples Organisation, WCS=Wildlife Conservation Society, MoU=Memorandum of understanding, NGO=Non government organisation.

Co-managed protected areas (CMPAs) are defined as 'government-designated protected areas where decision-making power, responsibility and accountability are shared between governmental agencies and other stakeholders, in particular the indigenous peoples . . . . that depend on that area culturally and/or for their livelihoods' (Borrini-Feyerabend et al. 2004: 32). Collaborative management involves park agency consultation with other stakeholders, while in joint management, multiple stakeholders sit on a park management board with decision-making authority. Co-management (or joint management) of parks involves recognition of cultural differences, power sharing, negotiated decision-making, and equitable sharing of benefits (Borrini-Feyerabend et al. 2004). Co-managed national parks where indigenous peoples comprise a majority on a park board require the government conservation agency to negotiate and develop new agreements on park management, resource use, interpretation and tourism. These new park agreements with indigenous peoples address subsistence use of natural resources (i.e. hunting, fishing, gathering), living areas, ceremonial use, naming/renaming land features, park employment, training and tourism. Agreements for co-managed national parks in Canada and Alaska mainly focus on subsistence hunting and indigenous resource use rights, with less interest in providing access roads and tourist facilities (Nelson 2000). In these parks, indigenous people also work as rangers or in eco-tourism. The next section focuses on some co-managed national parks with indigenous tourism agreements.

Canada

In Canada, indigenous land claims to national parks include: Nunavut; Yukon (Kluane); Northwest Territories; Ontario (Pukaskwa, Bruce Peninsula and Point Pelee); Manitoba (Riding Mountain); Alberta (Banff and Wood Buffalo); and British Columbia (BC) (Gwaii Haanas and Pacific Rim). Indigenous hunting rights are recognized in northern parks, while Parks Canada retains ownership of national parks (East 1991, Morrison 1993, 1997, Notzke 1994, Sneed 1997). Tourism use of Gwaii Haanas and Aulavik parks has been reviewed (Shackley 1998, McVetty and Deakin 2007) along with Nuu-cha-nulth First Nations attachment to park lands in BC (McAvoy, McDonald and Carlson 2003). In 1986, the Haida Nation and Parks Canada jointly established the Gwaii Haanas National Park Reserve and Haida Heritage Site in the southern portion of the Queen Charlotte Islands off the west coast of BC. Haida people and wilderness activists initiated this park declaration, with Haida traditional lands under pressure from logging. The 1998 South Moresby Agreement committed CA$106 million to developing the national park and compensating forestry operators. A 1993 Gwaii Haanas agreement with the Haida Nation established a joint management board and Haida harvesting rights with an unresolved Haida land claim (Morrison 1997, Parks Canada 2007b). The park protected forests and old Haida village sites with carved totem poles, interpreted by Haida Watchmen since the early

1980s. Park entry fees help to support the Haida Watchmen, located at five cultural sites from May to September. The local economy shifted from logging to tourism with Haida people comprising 50 per cent of park staff (Borrini-Feyerabend et al. 2004). The Qay'llnagaay Heritage Centre has opened in the main town of Skidegate, with tourism mainly controlled by non-Haida people and two Haida tourism operators in 2001. Protection, control and access to land were higher priorities for the Haida Nation than tourism (Blangy and Martin 2002a, b). At the Pacific Rim National Park, on the west coast of Vancouver Island, a final agreement was signed in 2003 with the Maa-nulth First Nations. It included provisions for co-management and the addition of tribal areas to Pacific Rim National Park. Park signs now present Maa-nulth tribal culture and history along the 2.5 kilometre Nuu-chah-nulth Trail opened in 2003. First Nations people provide visitor services along the West Coast Trail and Nuu-chah-nulth Trail in trail maintenance, as trail guides and cultural interpreters (Indian and Northern Affairs Canada 2006).

New Zealand

The *Ngai Tahu Claims Settlement Act 1998* was an agreement negotiated between the South Island Ngai Tahu Maori tribe and the New Zealand (NZ) government. It addressed Ngai Tahu claims under the 1840 Treaty of Waitangi and covered 65 per cent of conservation lands in the South Island. The Ngai Tahu settlement included full title to Aoraki/Mt Cook National Park, gifted back to NZ, areas of the Te Wahi Pounamu South Island World Heritage Area, and title to other nature reserves. Some 90 names were changed to dual Maori/English place names in recognition of Ngai Tahu land ownership (MacKay 2002). A cash settlement of NZ$170 million was used by the Ngai Tahu to acquire commercial businesses including mainstream nature tourism enterprises. Ventures operating in national parks include Franz Josef Glacier Guides, Hollyford Valley Walks, Aqua Taxi and Kaiteriteri Kayaks in Abel Tasman. In 2006, 10 Ngai Tahu-owned tourism ventures generated revenue of NZ$35 million and earnings of NZ$3.3 million (Te Runanga o Ngai Tahu 2006). Ngai Tahu had sought sole responsibility to control tourism and guiding concessions in national parks (Hall 1996). At Abel Tasman, Wakatu Incorporation purchased Ocean River Kayaks and Abel Tasman Kayaks in 2003, and the Barn Backpackers, forming a joint venture with Ngai Tahu Tourism in 2007. There are plans to develop a cultural resort on 15 ha of Maori-owned beachfront land in Marahau, the main gateway town to Abel Tasman National Park with 150,000 visitors annually (Wakatu, 2008). The NZ government has resisted Maori land claims to protected areas, although the *Conservation Act 1987* recognizes principles in the Treaty of Waitangi (Hall 1996, McIntyre et al. 2001). Through the 1800s and early 1900s, Maori land was confiscated, coerced, compulsorily acquired and purchased inexpensively by the NZ government in areas redesignated as national parks (Coombes and Hill 2005). There are few formal agreements with Maori for co-management of national parks, with discussions proceeding with Maori Trust Boards for Mt Egmont-Taranaki, Whanganui (River), Tongariro and Te Urewera on the North Island (Manuera, Te Heuheu and Prime 1992, Green 1996, Taipea *et al.* 1997, Federation of Maori Authorities 2003, Coombes and Hill 2005, Ifopo 2007). Some Maori regard co-management as diverting Treaty claims to national park lands (Coombes and Hill 2005). Conservation lands also have limited economic viability for local Maori groups (Ell 2000). Maori tourism ventures though, reinforce cultural identity, claims to tribal sovereignty and land use.

Australia

In Australia, most states and territories (except Tasmania) have formal joint management arrangements between government park agencies and Aboriginal groups for selected national parks (Smyth 2001, Wearing and Huyskens 2001, Poirer 2007) (see Table 19.7). Agreements with Aboriginal groups claiming national parks under land rights or native title legislation include an annual park rental fee, a portion of park entry fees, park maintenance contracts, employment as rangers, support for Aboriginal-owned or joint venture tourism enterprises, cultural centres and interpretation. The first jointly managed park was Gurig (1981), followed by Kakadu (1984), Uluru (1985) and Nitmiluk (1989), all in the Northern Territory and all with Aboriginal involvement in tourism. At Uluru and Kakadu, traditional owners work as rangers, operate cultural or wildlife tours, invest in tourist lodges, receive an indexed annual rental payment of AUD$150,000 and 25 per cent of park entry fees. A new tourism strategy for Kakadu promotes local Aboriginal businesses and tourism joint ventures, along with marketing Aboriginal cultural heritage (Morse *et al.* 2005, Welling, 2007). At Gurig, traditional owners also receive safari hunting royalties and resort land rental fees (Foster 1997), while Jawoyn people at Nitmiluk National Park have full ownership of Nitmiluk Tours, operating boat tours, kayak rentals and a campground in Katherine Gorge, and also receive half of park revenue. The Northern Territory now has 39 jointly managed protected areas, mainly comprising Aboriginal freehold land leased back to the government as a park or reserve. Jointly managed parks in other states include Booderee (1995, ACT), Witjira (1995, SA) and Mutawintji (1998, NSW). State control and management of national parks in Australia means there is varied land ownership and lease-back arrangement for indigenous co-managed parks. In New South Wales in 1996, three national parks were specified for Aboriginal ownership and a 30-year lease period: Mungo, Jervis Bay and Mootwingee (Mutawintji), followed by Arakwal (2001), Kinchega (2002), Biamanga and Gulaga (2006), and Worimi (2007) (DECC, 2007). In 1983, Aboriginal people blockaded the entrance to Mootwingee Historic Site in western NSW, demanding that a campground be relocated, public access to sacred sites be banned, walking trails realigned, accredited tour guides at rock art sites and a

Table 19.7 Aboriginal joint management of national parks in Australia

National Park, Year,* Indigenous Group(s)	Management Board Structure	Framework Arrangements	Aboriginal Tourism
Uluru-Kata Tjuta National Park, NT (Cwth) 1985 Anangu (Pitjantjatjara, Yankunjatjara) IP	Traditional owners 6 Director of NPWS 1 Tourism Minister 1 Environment Minister 1 Arid Zone ecologist 1	Granted Aboriginal freehold, 99-year lease to Cwth, Board management plan, Annual fee	Anangu Tours Maruku Arts and Crafts Ayers Rock Resort, Yulara Annual rent of $150,000 25% of park entry fee ($25)
Kakadu National Park, NT (Cwth) 1984 Gagadju, Jawoyn IP	Traditional owners 10 Director of NPWS 1 Parks Australia 1 Conservation expert 1 Tourism expert 1	Combined Aboriginal freehold and Crown land, 99-year lease to Cwth, Board management plan, Annual fee ($7,200, 1978), 50–60% Aboriginal staff	Yellow Waters Cruise Guluyambi Cruise Wildlife and cultural tours Gagadju Lodge, Jabiru Annual rent of $150,000 (1991), 25% of park entry fees (abolished in 2004)
Booderee National Park (Jervis Bay, ACT) (Cwth) 1995 Jerrinja IP	Traditional owners 6 Director of NPWS 1 Territories office 1 Conservation expert 1 Tourism expert 1	Granted Aboriginal freehold, Annual fee to Wreck Bay Community Council, 99-year lease to Cwth, Board management plan, Park renamed	Annual rent % of park entry/use fees Park service contacts (cleaning, road maintenance, operate park entry station) Cultural Centre
Garig Gunak Barlu NP (Gurig National Park), NT 1981 Agalda, Muran, Madjunbalmi IP	Traditional owners (including chairman) 4 NT government reps 4	Granted Aboriginal freehold, Perpetual lease to NT, TO chair casting vote, Board management plan, Annual fee, Park renamed	Annual rent of $20,000 (1981) Park entry fees Safari hunting royalties Resort land rental payments

(*Continued Overleaf*)

Table 19.7 Continued.

National Park, Year,* Indigenous Group(s)	Management Board Structure		Framework Arrangements	Aboriginal Tourism
Nitmiluk National Park (Katherine Gorge), NT 1989 Jawoyn IP	Traditional owners NT parks officers Resident appointed by local mayor	8 4 1	Claimed Aboriginal freehold, 99-year lease to NT, Board management plan, Annual fee, Renamed	Nitmiluk Tours (100%) (boat tour, canoe hire, campground, kiosk) Annual rent of $100,000 50% of park revenue
Mutawintji National Park (Mootwingee NP), NSW 1998 Wiimpatja (Paakantji) IP	Traditional owners Land Council Shire Council Director NPWS Conservation group Park neighbours	8 1 1 1 1 1	Aboriginal freehold vested in Mutawintji Land Council; 30-year lease, Board management plan, Annual fee paid to Board and must be spent in park, Park renamed	Annual rent of $275,000 Mutawintji Heritage Tours NPWS Discovery rangers
Witjira National Park, SA 1995	Traditional owners Dept. Environment Regional committee Aboriginal affairs	4 1 1 1	Lease of park to TOs, SA government retains land ownership, TO chair of Board, No annual fee, Name	
Gulaga National Park, NSW 2006 Yuin IP	Traditional owners Aboriginal Councils Shire Council NPWS	9 2 1 1	Aboriginal freehold vested in Merrimans and Wagonga Land Council; 30-year lease, Board management plan, Renamed	Annual rent of $210,000 Umbarra Cultural Tours Umbarra Cultural Centre

Sources: Based on Smyth 2001, Mundraby 2005.

Notes: Cwth=Commonwealth, NPWS=National Parks and Wildlife Service, NT=Northern Territory, ACT=Australian Commonwealth Territory, SA=South Australia, NP=National Park, IP=Indigenous People, TO=Traditional Owner

* National Parks with formal joint management arrangements between government agencies and Aboriginal groups. Based on year of park hand-back and/or joint management agreement with Traditional Owners.

Mutawintji Culture Centre to present Aboriginal history. The area was closed from 1983 to 1989 to implement these measures, with cultural training for non-Aboriginal guides and operators since 1991, an Aboriginal ranger employed in 1993 and local Aboriginal people contracted to provide guided tours. In 1998, it was the first park in NSW returned to traditional owners with the name changed to Mutawintji (Sutton 1999). In Western Australia, Aboriginal people are involved in the management of Purnululu and Karijini National Parks, with a cultural centre and tourist camp in the latter park owned by Aboriginal groups. In Victoria, Aboriginal groups are involved in managing the Grampians/Gariwerd National Park, operating Brambuk Culture Centre and a backpacker lodge. In 2007, Aboriginal native title to Mt Eccles National Park was recognized and it will now be co-managed. A key feature of most jointly managed national parks is developing Aboriginal cultural interpretation and tourism ventures. The restoration of Aboriginal names for national parks and key natural features also occurs, recognizing park areas as indigenous cultural landscapes (e.g. Uluṟu).

Conclusions

This chapter has reviewed some key national parks co-managed with indigenous groups in Australia, New Zealand, South Africa, Canada and the US. Globally, 50 to 85 per cent of national parks have been declared on indigenous lands including Yellowstone and Yosemite in the US. There is a long history of indigenous exclusion and forced removal from these parks, regarded as 'wilderness' areas, with few benefits for indigenous groups from conservation or park tourism. The legal assertion of indigenous rights since the 1980s has seen increasing land claims and treaty settlements for national parks. This involves government negotiation about indigenous co-management of selected national parks, subsistence use of natural resources and indigenous involvement in cultural interpretation and tourism opportunities. There are also cultural conflicts between indigenous groups and some visitors over recreational use of sacred sites in national parks. From the 1990s, there was growing recognition of parks as indigenous cultural landscapes, of cultural and spiritual values in natural areas and involving local people in park management through participatory conservation. Through the twenty-first century, indigenous land claims over parks and the reassertion of indigenous cultural identity in these areas will increasingly define the management of conservation and tourism in many national parks.

20 National Mall and Memorial Parks

Past, present and future

Margaret Daniels, Laurlyn Harmon, Min Park and Russell Brayley

Introduction

The National Mall & Memorial Parks (National Mall) includes icons such as the Washington Monument, Lincoln Memorial, Jefferson Memorial, Franklin Delano Roosevelt Memorial, Constitution Gardens, Korean War Veterans Memorial, Vietnam Veterans Memorial and World War II Memorial. A unit of the National Park Service (NPS), the National Mall contains some of the oldest protected park lands in the United States, which provide relaxation and recreation opportunities for tourists and residents (NPS 2007d). In addition, the park includes significant cultural resources and downtown visitor destinations such as Ford's Theatre National Historic Site, the African American Civil War Memorial and Pennsylvania Avenue from the US Capitol to the White House. Museums on or near the National Mall, such as those operated by the Smithsonian Institution as well as the National Gallery of Art, National Archives and the US Holocaust Memorial Museum, are separate entities from the National Mall.

Taken together, the monuments, memorials and natural resources of the National Mall comprise an enduring symbol that provides an inspiring setting for national memorials and a backdrop for the legislative and executive branches of the US Government. Enjoyed by over 25 million national and international visitors every year, the National Mall is a primary location for public gatherings such as demonstrations, rallies and festivals. Annually, the park receives over 3,000 applications for public gatherings, resulting in more than 14,000 event-days (NPS 2007a).

The wear and tear of concentrated activities on the National Mall has left its mark. Lawn areas are worn, soils are compacted and irrigation systems are outdated. Budgets for maintenance and upkeep are strained. The NPS is currently conducting research that will provide a management framework for the National Mall over the next half century. The authors of this case study are collaborating with the NPS to provide data analysis for this research endeavour, entitled the National Mall Plan (NPS 2007b). The purpose of this case study is to provide a comprehensive overview of the history, ongoing challenges and future vision of the National Mall. The first section of the

Figure 20.1 Spring visitors to the National Mall.

chapter details the historical origins of the National Mall, including intent and development. The second section focuses on current public perceptions of the National Mall, based on national feedback. The final section summarizes the primary management recommendations under consideration for the National Mall Plan. This chapter illustrates that national parks are a work in progress and adequate resources and public input are needed to balance the rivalling perceptions of their use and management.

Past: the history of the National Mall

Over 200 years ago, the idea of a purpose-built capital city for the United States was conceived. The original plan used design strategies from prominent European cities to provide the nation with a capital that was geographically advantageous and connected the various units of government (Penczer 2007). The National Mall, envisioned as a greenspace area, was a significant component of that plan. Originally known as the public grounds of the city, and sometimes referred to as our Nation's front yard, the National Mall has experienced a variety of planning stages, construction, development, and several significant renovations since its conception.

Pierre (Peter) Charles L'Enfant's plan of 1791 for Washington was the first federally accepted plan that guided design of the National Mall and was founded on formal design principles including axial lines, vistas, terminal focal points and rectilinear shapes. The Senate Park Commission Plan of 1901–1902, also known as the McMillan or Burnham Plan, guided the majority of new developments and renovations from 1901 to the present and included the original vision of tree-lined walkways and open public grounds.

The National Mall was developed in the late 1840s at the same time as the Washington Monument and Smithsonian Castle, the first of many Smithsonian buildings. In L'Enfant's original design, available land extended to approximately the current location of the Washington Monument and a drainage canal connecting Tiber Creek to the Potomac River ran along the entirety of the northern edge of the mall crossing south in front of today's Capitol location. The US Capitol was completed in 1829. Some of the first improvements to the National Mall included canal renovations and the addition of the Botanical Gardens in 1842.

The intrusion of the Civil War stopped any work at the time, including building of the Washington Monument, which had begun in 1848. During the Civil War, a variety of temporary buildings were added to the Mall, for example, temporary hospitals, armories and officers' quarters (Penczer 2007). After the war, with the subsequent increasing stability and prosperity of the nation, efforts turned again to the design of the Mall.

The Washington Monument, completed 1884 and the tallest masonry structure in the world, was constructed slightly off from the east/west axis. The reasons for this are unclear as some scholars suggest it was due to the instability of the subsurface soil, while others suggest it was a design decision (Scott 2002). Regardless, the Monument remains a key focal point of the National Mall and is visited by over 800,000 people annually (NPS 2008).

During the initial development of the National Mall, the Commissioner of Public Buildings controlled its planning and, during this first management period, the Mall was one of several projects under its directorate, which included maintaining the President's home and all nearby public buildings. However, in 1867 management was transferred to the Army Corps of Engineers (Penczer 2007).

Andrew Jackson Downing, a follower of the Capability Brown informal landscape design movement, began the first significant landscaping renovations in the 1850s. Departing from L'Enfant's formal design, Downing added dense plantings and winding paths, which stood for the next 30 years until the introduction of the Senate Park Commission Plan (Longstreth 2002). Prior to Downing's untimely death in 1852, only the grounds near the Smithsonian Castle were landscaped. After his death, the design implementation was fragmented, due to the lack of strong proponents of the new design, which departed significantly from L'Enfant's.

During the 1890s, residents of the area and politicians realized that the Mall, with its dense foliage and fragmented design, had evolved into a public area used by the city's less desirable clientele, for example, thieves and prostitutes. Senator James McMillan, interested in improving the Mall area, pulled together an elite team of architects and designers who brought back the formal-inspired design of L'Enfant (National Coalition to Save Our Mall 2008). The City Beautiful movement of the 1890s, which was an attempt to bring environmental aesthetics to urban areas, was significant to the development of this plan (Hines 2002). Guided by the Chair of the Senate Park Commission, Daniel Burnham, the resulting Senate Park Commission Plan of 1901 reaffirmed L'Enfant's original formal plans with some changes. Rather than open vistas to the Potomac River, the Lincoln Memorial was proposed as a terminus and necessitated filling the swamplands adjacent to the Potomac. Completed in 1922, the Lincoln Memorial also is a focal point of today's National Mall.

In 1926, the National Capital Park and Planning Commission, later modified to the National Capital Planning Commission, was created for the purpose of formalizing the improvement process to the National Mall. In 1933, the responsibility for management of the Mall was transferred to the NPS (Luria 2005). A primary focus of the Commission was to implement the Senate Park Commission Plan to its fullest, as informed by landscape architects committed to the conservation of national parks such as Frederick Law Olmsted, Jr. (NPS 1990, Streatfield 2002). As part of the plan, trees intruding on the formal design were removed to the strenuous objection from many citizens. In addition, the Jefferson Memorial was added in the Tidal Basin, much to the chagrin of critics who believed the Japanese Cherry Trees, a post First World War gift from Japan, would be irreparably harmed. However, the support of the administration at the time resulted in the chosen site and the memorial was completed in 1943.

It was not until the introduction of the Skidmore, Owings, and Merrill Plan (the SOM Plan) in the 1960s that the National Mall experienced more dramatic changes. The plan architects revitalized the ideas and formality of the L'Enfant and Senate Commission Park Plans in the design. As part of that plan, and as a result of an agreement between the National Gallery of Art and Department of Interior, the design of the National Sculpture Gardens was presented. Directly across the mall, the Hirshhorn Museum and Sculpture Garden was completed in 1974 (Streetfield 2002). And in 1981, Maya Lin's simple but moving entry to the Vietnam Veteran's War Memorial competition was accepted.

More recent revitalization efforts included the Washington Monument and the US Botanical Gardens. In the 1990s, noted architect Michael Graves was commissioned to design a temporary scaffolding, hiding the renovation work on the Monument, which started in 1998 and was completed in 2000. The deteriorating structure became the framework for a temporary piece of artwork throughout the entirety of its refurbishing. Additional revitalization efforts targeted the Botanical Garden conservatory, which was restored under the guidance of Alan Hantman, Architect of the Capital in 2001, and the addition of the National Gardens, designed by EDAW, Inc.

Recent additions to the Mall include the Holocaust Memorial Museum in

1993, the National Museum of the American Indian in 2004, and the highly controversial World War II Memorial at the east end of the Reflecting Pool in 2004 (National Coalition to Save Our Mall 2008), the last of which has been simultaneously praised and faulted for the prominent design. Future plans include a Vietnam Veterans Memorial Centre, Martin Luther King Jr. Memorial, and the National Museum of African American History and Culture (NPS 2007c). In order to reduce the likelihood of excessive development on the National Mall, Congress reinforced the Commemorative Works Act in 2003 and continues to support the goals of the National Capital Planning Commission.

The overarching prominence of the National Mall has remained over the last 200 years; however, the specific purposes of the Mall are dynamic. Changing politics, economics and significant historical events facilitate people's evolving perception of the National Mall's purpose and priorities. Even now, the management issues confronting the NPS, caretakers of the National Mall, are continually shifting, particularly from a development focus to a preservation-oriented focus. While there are similarities in challenges facing the Mall, there are also new concerns and the management of this prominent urban park will continue to evolve accordingly. Unlike some large-scale urban parks, the National Mall continues to be an evolving design, albeit framed within its foundational plans. It continues to be a symbol of an open, active, and accessible urban park, which influences city planning across the nation.

'From the Capitol to Fourteenth Street, the broad, linear space creates a sense of expansiveness commensurate with the bold visions that shaped it, a space that may be said to be emblematic of the vast reaches of the country itself' (Longstreth 2002:14). The Mall exemplifies both formal design, for example the lines of majestic elms running parallel to the long, rectangular reflecting pool, and informal design, for example the curvilinear Tidal Basin area filled with cherry trees in which one feels almost completely removed from the adjacent urbanity. The cohesiveness of the Mall's design belies the fact that not one, but several designs have guided its development over the years. Coupled with those designs have been a plethora of ideas for additions and renovations to the Mall, for example more monuments, less vehicular traffic, more walkways and more amenities. The Commission of Fine Arts and the NPS have worked tirelessly to retain the character of the Mall while meeting the needs of a changing society. As Longstreth so eloquently states, 'after two centuries of development, the Mall does not bespeak those numerous changes (of society); a sense of continuity in the whole is the overriding force ... and is perhaps the one reason so many people have long considered the Mall so remarkable a place' (2002:16). This remarkable place is currently facing many difficulties, due to overuse and constrained budgets. The second part of the chapter is dedicated to summarizing the public's perception of current and pressing issues that need to be addressed by the NPS.

Present: public feedback regarding the current state of the National Mall

Because the National Mall is many things to many people, it is vital that NPS planners receive substantive feedback from the public on what they want to see and experience when they visit this centre of heritage and national identity. Accordingly, as part of a larger study to gauge current public sentiment regarding the National Mall, the NPS posted eight open-ended scoping questions for public comment on the National Mall Plan website. The public was encouraged to comment by means of the associated website, but email and mail responses were also collected. The eight questions were as follows:

1 What is most important to you about the National Mall?
2 What, if any, improvements to the appearance of the National Mall are needed?
3 What types and amount of facilities do visitors need? Where should they be located?
4 What should visitor facilities and sidewalk furnishings look like, or what character should they have?
5 What programs, activities, educational and recreational opportunities do you want on the National Mall?
6 What kinds of information would help you get around more easily?
7 What kinds of events and recreational opportunities do you feel can be accommodated in addition to First Amendment demonstrations and open public access?
8 Do you have any other comments you would like to share about the National Mall?

Respondents did not answer in any defined pattern. Some systematically addressed each of the items, while most simply posted their thoughts and feelings about the National Mall and changes they would like to see made.

The public comment period extended from 1 November 2006 until 16 March 2007. During this period, 4,833 respondents posted comments on the National Mall Plan website, while an additional 183 pieces of correspondence were sent via email or letter, for a total of 5,016 pieces of correspondence. Of these, 4,039 pieces of correspondence were identified as a replication of agency stakeholder form letter. The remaining 977 pieces of correspondence were unique contributions and therefore comprised the data set for this case study.

Four researchers each independently reviewed and coded a sample of the data in order to identify groupings, themes and categories. Each piece of correspondence was assessed for distinct comments, and thus any given submission could be segmented to separate the variety of ideas therein. From the 977 unique pieces of correspondence, a total of 2,901 comments were extracted and coded.

This section of the chapter will offer an overview of the primary groupings and themes that emerged from the public comments, with sample quotes offered to support the findings. All 50 states and the District of Columbia were represented in the study. The six primary groupings include: 1) Vision of the National Mall; 2) Landscape Design; 3) Facilities; 4) Services; 5) Activities; and 6) Regulation. Each of the groupings is supported by themes and direct quotes from the respondents. Table 20.1 summarizes the six primary groupings by number of comments made per thematic area, offering a sense of the issues that were most relevant to respondents. The direct quotes provided in this chapter are representative of the variety of feedback offered and are presented exactly as written by the respondents, with the exception of changes made to correct spelling errors. The groupings and themes are not

Table 20.1 Summary of 6 groupings and 18 themes

Grouping/Theme	Count		% of Total	
I. Facilities				
1. Bathrooms	275		9.48	
2. Monuments / Memorials	212		7.31	
3. Walkways	153		5.27	
4. Furniture	133		4.58	
5. Water	75		2.58	
6. Lighting	33		1.14	
7. Museums	28		0.97	
		909		31.33
II. Services				
8. Information	294		10.13	
9. Transportation	261		9.00	
10. Food	163		5.62	
11. Retail	13		0.45	
		731		25.20
III. Landscape				
12. Greenspace	460		15.86	
13. Sustainable practices	45		1.55	
		505		17.41
IV. Activities				
14. Events	196		6.76	
15. Recreation	114		3.93	
16. Entertainment	23		0.79	
		333		11.48
V. Vision				
17. Purpose of National Mall	328		11.31	
		328		11.31
VI. Regulation				
18. Safety and security	95		3.27	
		95		3.27
Total		2,901		100.00

intended to be mutually exclusive; instead, coding was based on the primary indicator of the comment.

The five themes that emerged most prominently in the public comments were: the protection of greenspace (15.86 per cent of total comments), the purpose of the National Mall (11.31 per cent), the need for improved information services (10.13 per cent), the desire for more bathroom facilities (9.48 per cent) and concerns or suggestions regarding transportation (9.00 per cent). For organizational clarity, however, the themes will be discussed in the order presented in Table 20.1, which is based on the total count by grouping.

Grouping I: Facilities

Grouping I was specific to facilities and contained seven themes: bathrooms, monuments and memorials, walkways, furniture, water, lighting and museums. Each of these themes will be discussed in turn.

Bathrooms

Public feedback regarding bathrooms on the National Mall was frequent and strongly opinionated. Visitors consistently voiced the need for more facilities, with comments such as, 'You desperately need more public bathrooms' and 'As a parent of a young child who frequently visits the Mall, it would be nice not to have to search for a toilet.' The maintenance of the bathrooms was also commonly noted, for example, 'Fix the bathrooms! Heavens, there was a urinal covered up with a plastic trash bag under the Lincoln Memorial! This should not be' and 'The third world bathrooms at Constitution Garden and Hanes Point need to be razed. At present they are better than nothing – but not much!'

Monuments and memorials

Comments pertaining to monuments and memorials reflected suggestions regarding the current number, future commemoration, as well as the character, placement, maintenance and public access. Over half of the comments in this theme reflected a desire to see limits placed on further construction. Commentary such as the following was consistent: 'One of the things that I hope does not happen on the Mall is more memorials. We have many heroes and events to honor but any more memorials and it will look like a jumble sale lot. Enough.' Other responses reflected on the feelings associated with the monuments and memorials: 'The monuments are beautiful, bold and commemorate famous people in U.S. history along with the wars and honoring those who served our country.' Points regarding the monuments and memorials were often opposing, as seen with the following two entries: 'What impressed me was the World War II Memorial. This may be because it was so new, but it reminded me of my parent's America' versus 'The World War II Memorial is grossly overdone, imposing and ridiculous.'

Walkways

National Mall walkways were discussed primarily in terms of material used and accessibility. The current use of gravel came up as a point of contention, with two schools of thought. Many visitors feel the gravel is aesthetically displeasing and limits accessibility, as noted by one visitor, 'Do something to get rid of the gritty, dusty, sandy walkways. Inlaid bricks with historic designs would be easier to walk on, eliminate dirty air and remove possibility of muddy quagmires.' Those with mobility issues were particularly concerned about access, 'Please consider the challenges of moving crowds, baby carriages, wheelchairs, others with mobility or sensory limitations.' However, others appreciate the gravel, as it restricts certain types of use, 'I would strongly suggest to keep the walking paths in its current format. Don't do anything that would encourage wheeled traffic through the Mall area such as skateboard, rollerblades, bicycles, etc.'

Furniture

Feedback regarding park furniture focused primarily on the request for more benches, 'Place some benches along the walkways so that pedestrians can stop and rest, or just sit and watch.' Other commentary was specific to design elements, with suggestions related to style, maintenance and integration, 'Furnishings should complement, but not compete with the various architectural styles surrounding or the natural environment. Simple, clean lines.'

Water

Water features that emerged during analysis included drinking fountains, decorative fountains, cooling stations and large areas such as reflecting pools and the Tidal Basin. Visitors made it clear that there are not enough water fountains and the current ones are often not operable, 'More operational, clean-looking water fountains. Dehydration is a REAL danger on the Mall in the summer!' Other water features were discussed in terms of the need for more, 'The last time that I visited it was hot and I remember wishing that there was a fountain to cool off by' or the maintenance of what is currently there, 'The reflecting pools are in desperate need of a good cleaning.'

Lighting

Visitors' primary concerns with lighting dealt with personal safety at night, with either a focus on feelings of fear, 'One of the biggest improvements that needs to be made is on lighting the Mall up at night. It's so dark at night that the space could be seen as dangerous', or navigation concerns, 'Crossing the Mall area was very dark and there were ruts and very uneven surfaces to deal with. There definitely needs to be more lighting.'

Museums

Public comments regarding museums included those that were specific to Smithsonian museums as well as those stated in the broader context of the National Mall, essentially suggesting that many visitors do not readily separate their experiences on the National Mall from those within the adjacent museums. Primarily, respondents noted the desire to see services on the National Mall similar to those readily available in the museums: 'The Mall is a large area, and about the only place where I've found where you can get more than a hot dog is at one of the museums.' Respondents also used the discussion of museums as a forum to consider maintenance of the entire area: 'I know there has been talk of having a nominal fee to get into the museums and possibly the Washington Monument. I don't think people would mind paying a few dollars if it meant that money would be spent on upkeep and improvements to the museums and monuments as well as the mall area.'

Grouping II: Services

Grouping II focused on feedback particular to services and contained four themes: information, transportation, food and retail. Visitors had strong opinions regarding service improvements.

Information

The information theme involved a wide spectrum of suggestions for improvement, with signage, maps, requests for specific types of content, the desire for a visitor centre, technology directories, guided tours and accessibility establishing the associated categories. Information requests specific to wayfinding were most frequent, with ideas such as, 'Install a map at each end of each block on each side of the Mall with a "you are here" and one of those standard NPS maps with the entire Mall' commonly set forth. Visitors also felt strongly about the nature of the information, 'The National Park Service should incorporate stronger contextual messages about the Mall Memorials so the public not only understands the importance of the memorials themselves, but also the context of why they are among the select group so honored.'

Transportation

The transportation theme was a compilation of issues pertaining to parking, on-site mobility, public transit options, bicycling, sightseeing and tour bus services. While parking was a concern for many, there was also a strong sentiment against building additional parking areas, unless they are underground. As noted by one visitor, 'The Mall is supposed to be a park, not a parking lot.' Mobility is also a key issue for many visitors, in particular those with specialized needs, 'All of those major monuments at the other end of the Mall are out of reach of many families and elderly people who are unable to walk the mile and a half.'

Food

Visitors who commented on food were particularly opinionated, with ideas focused on the general desire for more food on the Mall, as well as suggestions for venues, types of food desired, the location and pricing. While food is readily available in the museums surrounding the National Mall, food access on the Mall itself is limited, 'The problem with the Mall is that it is a desert! You could starve to death there: there is nothing to eat! There is nothing to drink.' The nature of the food that is available was also of significant concern, 'Refreshment stands that offer something more than the crap served at the few stalls currently on the Mall, and for slightly more reasonable prices.'

Retail

Public feedback regarding retail refers to Mall-based locations with gifts, souvenirs and convenience items, as well as bookstores that are located within NPS areas. Public consensus did not suggest the desire for more retail availability on the National Mall; rather, visitors were more interested in removing or replacing what retail is currently present, 'Get rid of the ramshackle vendor huts. They junk up the Mall. In particular the area in front of the Lincoln Memorial. They are enough to make Abe weep. If you want vendors, let them operate within the existing Mall buildings.'

Grouping III: Landscape

Grouping III contained two themes: greenspace and sustainable practices. Greenspace was the theme that resulted in the most comments across all groupings. Sustainable practices often reflected back to the need for recycling and other conservation methods.

Greenspace

Comments on the topic of greenspace addressed issues relating to grass and soil conditions, other vegetation, general aesthetics and apparent maintenance of the landscape. The stressed appearance of the lawns (due to heavy pedestrian use and limited watering) was generally deplored, and recommendations most frequently included closing areas to heavy use, redirecting pedestrian traffic, regular re-seeding, and more landscape-friendly programming. Representative comments such as, 'I have always found the threadbare appearance of the lawns to be a detraction', emphasize the impact of the landscape's aesthetic quality on the visitors' experiences. Many visitors to the National Mall felt that 'It makes a bad first impression when the first thing you notice is how messy the entire area is'. The poor quality of the resource was often attributed to maintenance deficiencies, although some commentary was also devoted to marvelling at the ability of the NPS to rehabilitate and revive the area after major events or peak season usage. Suggested improvements to the aesthetics of the landscape usually included replacing or planting trees, and adding shrubs and flowers.

Sustainable practices

Conservation and recycling were highly valued as priorities for management of the National Mall. As expressed by one commentator, it was generally agreed that 'a "green" mall would certainly be a credit to our country'. The National Mall, it was suggested, should be a national example of environmental sustainability.

Grouping IV: Activities

Grouping IV contained three themes: events, recreation and entertainment. The use of the open National Mall space for these activities was often considered in terms of priorities and competing demands.

Events

Receiving the majority of visitor comments was the annual Smithsonian Folklife Festival, a major regular event held on the National Mall. Hailed as 'an unmatched opportunity to celebrate a diversity of cultures', it is conversely assailed as encouraging 'widespread destruction of massive parts of the lawn.' Other events such as demonstrations, national celebrations and public entertainment were seen as important to the mission or purpose of the National Mall, but it was frequently recommended that clear standards be developed and enforced, and some management support services be made available such as central space for event operations.

Recreation

Open recreational activities such as picnicking were viewed as appropriate for the area, but organized league play or recreation that requires significant infrastructure development was considered to be a much lower priority than free play activities. Local sports leagues, for example, were considered by many to be inappropriately facilitated on the National Mall. Indeed, unstructured, casual use of the park setting was more valued by visitors who 'appreciate the open space to run, bike, people-watch, make friends, and be exposed to new ideas.'

Entertainment

Feedback suggested elements that would add life to the area, such as 'street singers, jugglers, magicians, orators or other spontaneous entertainment', most of which are currently prohibited or restricted. Entertainment was largely seen as noncontroversial in nature and deemed as adding value to the overall experience.

Grouping V: Vision

Grouping IV comments related to the purpose of the National Mall and included a single theme: purpose of the National Mall. This purpose was viewed in a myriad of ways, including physical, cultural/historical, emotional and social.

Purpose of the National Mall

As a physical resource, most visitors saw the National Mall as unique open space that 'provides perspective and breathing room for everything around it'. Its openness was reflective of the expansive geography of the nation, and the American appreciation of natural space. The National Mall was also 'seen as a measure of the country' with respect to history and culture, and 'our front yard', which is 'a reflection of our self-esteem and national conscience.' To many, the National Mall is a peaceful refuge for introspection and contemplation, but can also be a lively national 'town center' or 'gathering place where we celebrate our communities and families.'

Grouping VI: Regulation

Comments included in Grouping VI were concerned mostly with safety and security issues. Visitors to the National Mall were very concerned about their personal safety and felt that the police/security presence in the immediate area was insufficient. Even when present, the security officers were not effectively uniformed, or performed 'false cosmetic security – looking into old ladies' and moms' purses'. Physical security measures such as jersey barriers and snow fences were considered to be unsightly, ineffective and annoying. A typical comment on this subject was, 'I do sincerely understand the need for safety but – this looks tacky and make-do.' Suggested improvements such as 'Use ornamental barricades rather than simple concrete ones', and make the security barricades 'discreet and porous to pedestrians, but capable of stopping vehicles' were commonly advanced.

Future: The National Mall Plan

At the time of writing the NPS is in the process of creating a 50-year vision plan to improve existing conditions on the National Mall. The status of The National Mall Plan is detailed in the final section of this chapter. Alternative management plans were developed through a combination of public feedback, stakeholder suggestions and internal expert review. In additional to the public comment study summarized in the second part of the chapter, alternatives were informed by open public meetings, closed meetings with national experts and agency partners who presented ideas for preserving, improving and maintaining the National Mall, and suggestions from NPS management and staff.

The NPS is already making progress on issues of public comments that are amenable to immediate change. Pedestrian wayfinding and signs are designed to be improved and coordinated with the city's wayfinding system. Many actions are being taken to improve turf conditions including establishing a regular turf aeration and renovation schedule, launching a pilot project to demonstrate different levels of treatment methods and recording soil data, labour, and machinery used, placing new sod, and installing fencing to protect sod. Also, improvements and solutions are being provided or pursued in the areas of recycling/solid waste management, bicycle parking, memorial lighting, water feature operations, vehicle circulation and interpretive visitor transit system. These common actions are in progress regardless of alternatives.

Four alternative plans were proposed to get a consensus on the development of a preferred alternative: no-action alternative, alternative A, alternative B and alternative C. Among them, the no-action alternative serves as the baseline against which to measure the action alternatives because it outlines existing conditions. All alternatives were designed to include main missions of the National Mall: 1) preserving historic resources; 2) providing space for constitutionally based civic activities, national celebrations, and public enjoyment; and 3) exemplifying the best of sustainable urban ecological practices. However, each alternative maximizes opportunities for one of the main missions to be reflected in the redesign and direction of the National Mall Plan while combining the actions to meet multiple needs. In each alternative, basic planning concepts were established by facilities and services such as preservation, landscape, public access and wayfinding, events, visitor facilities, commercial visitor services and recreation. Additionally, major planning actions were proposed for areas adjacent to select icons on the National Mall such as Union Square, the Washington Monument, the Lincoln Memorial, the D.C. War Memorial and the Thomas Jefferson Memorial. These alternatives are outlined in depth in the NPS Newsletter 3 document that was made available for public comment (NPS, 2007c). A summary of the four alternatives is offered below.

No-action alternative

The no-action alternative focuses on continuing current management and identifying plans and actions already moving forward. Under this alternative plan, standard maintenance activities will be continued in areas of landscape, public access and wayfinding, interpretation/education, events, visitor facilities, commercial visitor services and recreation. Actions reflecting current management direction include improving the conditions of Union Square and the Mall, and repairing the Tidal Basin wall.

Alternative A

Alternative A focuses on the historic landscape with its memorials and planned views and education. The emphasis of this alternative is especially on the balance between contemporary uses and the planned historic views, character, and visions of the L'Enfant and McMillan plans. Accordingly, educational programmes will be improved by focusing on the historic planned open spaces of the nation's capital as well as the history of the landscape and the various memorials while improving landscape, public access and wayfinding, events management, facilities and commercial visitor services. Some of the proposed actions include redesigning the civic square and reflecting pool in Union Square that highlight the site's history, limiting event timelines, temporary facilities and staging in the Mall, reconstructing the lake in the Constitution Gardens to improve water quality and be self-sustaining for plants, and regulating any event stage, roof, or walls not to obstruct the view to the White House.

Alternative B

Alternative B focuses on a welcoming national civic space for public gatherings, events and high-use levels. Under the alternative plan B, efforts will be made to meet the needs of large groups, pedestrians and event participants. Areas would be redesigned to provide sustainable sites for demonstrations and events, while maintaining a high-quality design and the highest facility maintenance standards to create a sense of place, which would reinforce the civic, historic and symbolic role of the National Mall to the nation. Consequently, educational programmes would add the history of the National Mall, its importance as a First Amendment demonstration site and the evolving nature of ceremonial, celebratory, cultural and visitor uses. Infrastructure improvements would be provided to disperse use. Commercial visitor services such as food service types would be improved. More performance venues would be identified and entertainment sponsorship opportunities as well as music/performances near food service locations would be encouraged. Some of the proposed actions highlight removal of the reflecting pool in Union Square to provide multipurpose visitor facilities, paving walks and additional areas for events in the Mall, and development of event staging areas in Washington Monument.

Alternative C

Alternative C focuses on urban open space, urban ecology, recreation and healthy lifestyles. Alternative C will meet evolving recreational needs by providing a beautiful, enjoyable and ecologically sustainable open space capable of adapting to changing recreational patterns of diverse local and national users. Under this alternative plan, recreational activities will be expanded by redesigning some areas and roads to create more recreation-friendly areas, diversifying recreational opportunities, adding playgrounds, and improving field conditions and providing more rentals. Highlights of proposed actions include redesign of the reflecting pool to be very shallow and drainable for events or frozen for ice skating in Union Square, elimination of some walks between 7th and 9th Streets to increase recreation space on centre lawn panels in the Mall, adding a playground near the carousel in the Mall and the Constitution Gardens, providing separate bike and walking trails in Potomac Riverfront and Tidal Basin Area, and rebuilding the paddleboat facilities and adding rental boats such as kayaks and rowboats.

Conclusion

Importantly, the alternatives for the future of the National Mall do not represent either-or options. Public, stakeholder and internal feedback continuing in 2008 will be used to select the most appropriate options from each of the alternatives. A preferred alternative will then be constructed and once again be made available for public and stakeholder comment. The final report will be used to leverage government funding in order to bring the ideas to fruition.

The National Mall Plan represents an exhaustive multi-year endeavour to appropriately and thoroughly articulate management goals as informed by the historic vision, current status and future needs of this park space. Enjoyed and appreciated by national and international visitors alike, the mandate is to ensure that the symbolic spaces, monuments and memorials can be maintained and sustained for generations to come.

Part VI
Conclusion

21 The future of the national park concept

C. Michael Hall and Warwick Frost

> Among all of the debates affecting America's national parks, the most enduring – and most intense – is where to draw the line between preservation and use.
>
> (Runte 1990: 1)

The concept of a national park has changed significantly over time. Originally used in the nineteenth century to refer to protected areas that should be conserved for the consumption of outstanding scenery and natural beauty for the benefit of visitors, the term is now often used as a 'catch-all' by members of the public to refer to protected areas of national environmental and cultural significance. This situation is not helped by the fact that many national park agencies manage more than just national parks as well as the number of different types of protected areas. The IUCN has six different categories of protected area with national parks usually being a component of a wider system of protected areas. However, some jurisdictions have even more categories, For example, in Austria there are 12 different types of protected areas and 11 in Germany (Mose and Wixlbaumer 2007). In addition, 'one cannot deny a certain image-hierachy between the different categories. In contrast to the prestigious and financially lucrative Category II (national park), the Category V (protected landscape) receives only little attention' (Mose and Wixlbaumer 2007: 5). This is despite the significant spatial area occupied by protected landscapes (see Chapter 1), and especially in Europe where it is by far the most substantial protected area category in terms of area occupied. According to Mose and Wixlbaumer (2007: 5) there are several reasons for the dominance of 'national parks' over other categories of protected areas in the public and policy imagination:

- The outstanding image of national parks as the 'premium category of the protected areas';
- The stringent legal and spatial planning rules underlying the national parks (for example statutues instead of regulations, zoning of activities);
- The supraregional competence of a governmental administrative body;

- The differently weighted overriding management objectives.

Furthermore, and perhaps not surprisingly, many people often do not recognize the differences that exist between the different types of protected areas or, in some cases, the concept is alien to their world-view (e.g. Kalternborn 1994, Neumann 1998, Proctor 1998, Stoll-Kleemann 2001, Trakolis 2001a, 2001b, Hall and Piggin 2002, Kächele and Dabbert 2002, Kalternborn et al. 2002, McLean and Stræde 2003, Middleton 2003, Phillips 2003, Mascarenhas and Scarce 2004, Hovardas and Poirazidis 2007, Schenk et al. 2007).

As Hall and Frost (see Chapter 1) discussed, the IUCN (1994) defines a national park as a 'Natural area of land and/or sea, designated to: (a) protect the ecological integrity of one or more ecosystems for present and future generations; (b) exclude exploitation or occupation inimical to the purposes of designation of the area; and (c) provide a foundation for spiritual, scientific, educational, recreational and visitor opportunities, all of which must be environmentally and culturally compatible'. According to the IUCN, ownership and management should normally be by the highest competent authority of the nation having jurisdiction over it. However, they may also be vested in another level of government, council of indigenous people, foundation or other legally established body that has dedicated the area to long-term conservation.

Although the IUCN categories and definitions are internationally recognized, different national jurisdictions use different legal definitions of national parks. This means that in many countries' national park systems the term 'national park' actually applies to a number of different IUCN categories of protected area, rather than being strictly interpreted. Mose and Wixlbaumer (2007: 6) go so far as to claim that 'it is conceded that an accurate classification of the different national protected areas has its limits and possibly has to happen even arbitrarily'. Furthermore, the application and interpretation of the management categories of protected areas in Europe are different from those of other continents. From a landscape-ecological perspective, ecosystems have different degrees of 'nativeness', which is reflected in the degree of hemmeroby – intensity of cultural effects on vegetation – which is, in turn related to the relative intensity of cultural effects on vegetation as interpreted in IUCN protected area categories. 'This is reflected in the continentally diverse distribution of the strict nature reserves and wilderness areas of Category I on the one hand and the protected landscapes of Category V on the other hand. Thus, the large national parks in the United States and Canada, as opposed to European nature parks, have in many cases a dominant ahemerob [not affected by cultural effects] portion of the surface areas' (Mose and Wixlbaumer 2007: 6).

To further complicate matters some jurisdictions also incorporate areas of national cultural significance within national park systems. Although Kaltenborn et al. (2002) argue that this more or less unique complexity is actually the essence of national park identity, the challenge still remains of how to make the national park concept work and remain relevant, given the challenges its faces.

As outlined in a number of chapters in this volume the reason for the various approaches to designating national parks and national park systems lies in the historical development of the national park concept and in changing values and ideas associated with conservation and heritage. Nevertheless, two elements tend to lie at the core of the majority of national park legislation throughout the world. First, that parks should be protected areas to enable environmental and scenic conservation. Second, that they should be accessible for public recreation, including tourism and enjoyment. These elements have been a part of the creation of national parks and national park systems since before the First World War and lie at the core of national park establishment in the New World and in Europe. The potential contradictions between these two elements have often served as a major problem for park management. As Runte (1997a: xi–xii) noted with respect to the American experience of national parks, 'many are tempted to celebrate national parks as the ideal expression of landscape democracy, despite evidence reaffirming that many parks have also been compromised or mismanaged', particularly with respect to the 'extraordinary growth in traffic and visitation'. In the case of Yellowstone, Runte pointed out:

> The country that invented national parks held just thirty million people. As late as World War I, Yellowstone's annual visitation rarely exceeded 50,000. Moreover, the large majority came by train and stagecoach, part of a community of travelers bound to responsibility by limited access, poorer roads, and rustic accommodations.
>
> The nation about to carry Yellowstone into another millennium has ten times the population of 1872. Park visitation, both domestic and foreign, now exceeds three million every year.
>
> (Runte 1997a: xii)

However, population growth is only one element in the changing face of parks and potential future challenge. As noted above, and throughout this volume, tourism is integral to the diffusion of the national park concept and its adoption and local interpretation in so many jurisdictions. Yet that adoption outside of its US model has also led to substantial debate as to its effectiveness as a model for integrating conservation and tourism goals. Therefore, the history of the social construction of landscape and nature is critical to understanding the development of the national park concept and its future direction. 'National parks are historically and culturally contingent representations of a particular nature aesthetic. Parks are landscapes of consumption, upon which are projected ideas of culture and nature and of where (literally) to draw the boundary between then' (Neumann 1998: 14). For example, in the case of Norway, Arnesen (1998) argues that landscapes are part of the identity and 'memory-bank' of social groups, and can function as a tool for public sector management by which a national park regime is an attempt to preserve a small number of the landscapes within an area. According to Arnesen the definition of the 'original' Norwegian landscape found in the national-romantic painting tradition in Norway that dates back about 100 years is one reason why national parks policy in Norway is heavily biased towards area protection in remote mountain regions. As a result, Norwegian national park policy has diverted attention from what is happening to landscapes outside the parks.

The importance of the role of parks in relation to the social construction of landscape, nature, culture and national identity has been a central theme of this book. Each location has its own particular social construction of the national park concept, which has emerged as the result of the interplay between local and international, often an American or at least Western, interpretation of the national park ideal.

Although the concept of national parks is often regarded as an American idea, the first call for the conservation of a natural area as a national task actually came from the English poet William Wordsworth at the turn of the nineteenth century. European influences have long been extremely significant in the development of national park systems, but have had different emphases to that of the US. Although the first national park was established by the United States Congress in Yellowstone in 1872, there were important precursors of environmental conservation in the British, Dutch and German colonial empires that underlay not only American thinking regarding national parks, but the environmental conservation systems of postcolonial states (Anderson and Grove 1987a, Grove 1995). In fact, conservation via the use of protected areas has always been a concept characterized by international innovation and diffusion (McCormick 1992).

Until the late twentieth century it was often assumed that European and colonial attempts to respond to environmental change derived exclusively from metropolitan and northern models and attitudes. However, there is substantial evidence to suggest that many of the seeds of modern conservation thought and practice developed as part of the European encounter with the tropics and with local classifications and interpretations of the natural world from the sixteenth to the mid-nineteenth centuries. As colonial expansion proceeded, the environmental experiences of those living at the colonial periphery played a continuing and dynamic part in the construction of new European and American ideas of nature and in the growing awareness of the damaging impacts of European economic and social activity on the environments and peoples of the newly colonized territories (Hall and Tucker 2004). In fact, the absolutist nature of much colonial rule allowed the introduction of interventionist forms of land management and conservation that, at the time, would have been extremely difficult to impose in Europe or in the American states (Grove 1995). For example, what has been described as the most comprehensive form of conservation legislation in the British Colonies in the nineteenth century – the *Cape Colony's Forest and Herbage*

Preservation Act – was passed in 1859 and remained as a statute until 1910 (Anderson and Grove 1987b).

Environmental conservation in the developing world during the colonial period tended to have a strong economic orientation and demonstrated marked parallels and interrelationships with the progressive conservation movement of the United States that saw the development of the United States Forest System. Indeed, the support of President Theodore Roosevelt for forest conservation in the United States can be favourably compared with his support for measures to manage game in the African colonies. Nevertheless, despite his support for game reserves it should be noted that in his most widely published safari to East Africa, which lasted between April 1909 and March 1910, Roosevelt travelled with over 200 trackers, skinners, porters and gun bearers and shot, preserved and shipped to Washington DC more than 3,000 specimens of African game (Grove 1995).

The significance of economic utilitarian conservation ideas for national park system development in the colonial territories cannot be overestimated. Whereas the development of the American national park system received much support from the romantic idealism of the likes of the Sierra Club and John Muir, no equivalent figures existed in the developing countries. Instead, the desire to protect the aesthetic attraction of some landscapes and wildlife came to be realized from economic concerns rather than preservationist idealism, with the economic interests being expressed through management of game hunting and its by-products such as ivory, as well as the desire to continue the attraction of international and domestic colonial tourists who came to shoot game or who accompanied those who shot game. Nevertheless, the colonial game and forestry reserves served as the basis for the creation of many national parks during the postcolonial era.

In the case of much of Africa, conservation policies, as in many of the New World park systems, were aimed at protecting valuable wildlife from the perceived destructive forces of humans and valueless wildlife. Wildlife conservationists and colonial officials believed that, for wildlife in the East Africa Protectorate to be adequately and effectively protected, nature conservation areas had to be established and boundaries demarcated, which separated wildlife from human activities that included not only the activities of colonial and European visitors, but also the activities of the indigenous inhabitants of the areas set aside as national parks. This was a clear repeat of the American understanding of national parks as places where no humans lived permanently – even if, it was later realised, that a) many of the parks were not unoccupied lands and b) the indigenous peoples who lived on the land now occupied by parks had actually contributed to the landscape that was valued by colonial societies in the first place.

In Kenya, as elsewhere in Africa, Asia and the Americas, wildlife management policies and programmes for much of the twentieth century were a consequence of conservation and administrative officials' Western experience and environmental values (Akama 1996). The general perception among pioneer naturalists in the colonies was that most human land use practices were incompatible with the principles of nature conservation in general, and wildlife protection in particular. Most often, government officials and conservationists classified African modes of natural resource use, such as subsistence hunting, pastoralism and shifting cultivation, as at best 'unprogressive' and at worst 'barbaric' and to be eliminated. In the case of the East Africa Protectorate local people were prohibited from entering the park and utilizing resources that they had traditionally had access to including pasture, wildlife, water and fuelwood. Thus, whereas wildlife safari tourism, an entirely European recreational phenomenon, was allowed in the protected game parks, subsistence hunting by indigenous people was banned and was, officially, classified as 'poaching' (Grove 1995)

Upon independence, post-colonial nations inherited colonial systems of environmental conservation and the political-economic structures that had put those systems in place. Kenya, which became independent in 1963, for example, retained the colonial national park system given the economic importance of safari tourism (Akama 1996). However, nature-based tourism proved to be an economic justification for the creation of national park systems in many developing countries post-independence in Africa and Asia as well as in Central and South America where the development of ecotourism provided for the establishment of national parks and reserves to conserve rare and endangered species. In addition to the use of national parks as a direct tool for economic development through nature-based tourism, justification for the establishment of national park and reserve systems has also occurred on environmental grounds in relation, not only to the conservation of individual species, but also ecological systems. Nevertheless, the nature of environmental management in national park systems has been dramatically transformed from the 1960s to the present day.

In the 1960s and early 1970s many national park systems in the developing world adopted park management models from North America and other developed countries that encouraged the removal of indigenous communities from within park boundaries. This had two immediate results. First, it reduced the level of community support for national parks by cutting access for communities to traditional natural resources. This therefore typically led to poaching and other illegal activities that often damaged the resources that were intended to be protected. Secondly, in many instances the environment that was being protected within national parks had often developed as a result of complex interaction with human activity, especially in terms of fire and subsistence agriculture. When such activity was stopped the capacity of the environment to support the previous diversity of species was often reduced. A further complication in terms of the relationships between local communities and the creation of national parks was that often, local communities were not able to benefit from tourism to parks because of management and business structures that were often geared towards external interests. Such a situation sometimes led to charges that the creation of national parks were often the result of the ecological imperialism of the West in trying to conserve environments in the developing world without paying attention to the broader development needs of local communities (Hall 1994).

By the late 1970s the inappropriateness of applying Western national park models in many of the environments of the developed world gradually began to be recognized. In Central America the concept of eco-tourism began to develop which saw conservation being beneficial, not just to the tourist who wanted to see exotic megafauna, but that tourism and conservation area management needed to involve local communities so that they could benefit from the creation of national parks and reserves. Costa Rica, for example, now has one of the highest rates of protected land areas in the world, including both national parks and private reserves, with eco-tourism being an important part of economic development.

Community-based national park and reserve management strategies also began to be developed elsewhere in the world. In southern Africa, particularly in Botswana, Namibia and South Africa, national park management is increasingly integrated with local economic and social development strategies in an effort to reduce poaching, utilize traditional environmental knowledge in park management and improve biodiversity. However, perhaps one of the most interesting aspects of changes in approaches towards the involvement of local people in parks is that such a strategy has now returned to the developed world and to many of the New World parks in particular. In Australia, Canada, New Zealand and the United States there is now far greater involvement of indigenous peoples in park management and, in some cases, even symbolic returns of land ownership (see Chapters 10 and 19). The national park concept has therefore been subject to periodic and substantial change, and it is likely that it will change yet again, something that Runte (1990) has recognized with respect to the changing nature of management of the American national parks:

> The National Park Service organic Act of 1916 itself left every methodology for management deliberately vague, calling simply for protection of scenery and wildlife 'in such manner and by such means as will leave them unimpaired for the enjoyment of future generations.' But just what was meant by 'unimpaired'? In effect, a definition that imprecise extended protection to park resources only by implication. It remained for each generation of Americans to bring its own perspective to the issue, invariably, if only subtly, imposing another viewpoint on existing philosophies of park management and use.
> (Runte 1990: 1–2)

The various chapters in this volume have highlighted how national parks are rarely mono-functionally oriented, but instead fulfil a multitude of tasks simultaneously. In their search for a new model for park development in

Europe in the twenty-first century, Mose and Wixlbaumer (2007: 4) note some of the elements of the 'traditional nature paradigm' that may be applied to protected areas:

- *regulatory functions*: preservation and maintenance of biodiversity;
- *habitat function*: regional and supraregional welfare effects;
- *support function*: gene pool as well as natural disaster-prevention;
- *development function*: sustainable regional development;
- *information function*: environmental education and training.

To which, with respect to national parks, we can also add:

- *cultural function*: the preservation and maintenance of culturally significant landscapes and places, often related to national identity but also indigenous identities;
- *tourism and recreation function*: the availability of the area for visitation.

and an emerging

- *local landscape and place function*: the preservation and maintenance of the relationships of local people to their landscapes embodied within their national park and sense of place.

Such a situation and series of demands being placed upon many national park spaces highlights that, 'The multi-functional orientation of large protected areas bears in equal measure enormous challenges and substantial conflict-potential. The multitude of diverse expectations for these protected areas has great potential to create conflict about the very purpose of beneficial use of the area' (Mose and Wixlbaumer 2007: 5). Undoubtedly, this is true, but it also highlights the fact, emphasized by a number of chapters in this book, that the spaces of national parks have always been socially constructed by different interests as part of an ongoing process of social, economic and environmental change and therefore it should not be a surprise that they therefore potentially represent contested spaces.

As this book has demonstrated, national parks have experienced considerable change with regard to their objectives and tasks. While originally meant to function as reserves for scenic landscapes and as sanctuaries and resorts, they later became connected with the idea of ecological and species conservation. More recently, people have become reintegrated back into the park idea, although arguably, resistance to this notion is still strongest in those countries where indigenous peoples were removed from the earliest parks. Nevertheless, the human dimension, as expressed via indigenous and local connections is becoming more important globally. For example, in the development of new national park systems in countries such as Lithuania, Kaltenborn *et al.* (2002) argue that new models of collaborative, adaptive management will need to consider multiple values and goals and be able to function within rapidly changing political and administrative contexts.

According to Mose (2007), many protected areas are undergoing change again where national parks and other large protected areas are increasingly considered to function also as instruments for regional development, particularly in peripheral rural areas faced with significant economic problems (Saarinen 2007). For example, in the case of the establishment of national parks in Scotland, the *National Parks (Scotland) Act 2000* sets out four aims for the Scottish National Parks, one of which is the promotion of sustainable economic and social development in the parks. According to Lloyd *et al.* (2002), positive and assertive planning and management policies are entirely appropriate in order to promote the provision of high quality and innovative industrial and commercial premises in the Scottish National Parks so as to meet the objectives relating to the sustainable economic and social development of the designated areas:

> ... this new understanding of protected areas has been described as a significant shift of paradigm regarding the underlying concept of nature conservation. In rather sharp contrast to traditional concepts, focusing mainly on the conservation objective, and often trying to restrict human activities in protected areas to a very high degree, new approaches are aiming at consistent integration of conservation and development functions making protected areas real 'living landscapes'.
>
> (Mose 2007: xv)

This 'shift of paradigm' is significant for tourism, given its economic role in peripheral area development and its capacity to financially support nature conservation efforts. But in one sense, such a response is nothing new. In fact the supposedly 'traditional' concept of nature conservation was not itself socially constructed until well into the twentieth century, although it did have nineteenth-century precursors with respect to landscape and scenic conservation. Instead, the national park concept is now arguably returning to its regional development and utilitarian roots, with which tourism is intrinsically connected, albeit now with greater knowledge of ecology and a desire to conserve environments rather than just scenes.

However, a greater understanding of the national park concept and hence its future, including its viability, will arguably depend as much on social scientific understandings of its environment as its natural scientific. While many conservationists may believe that they best understand the park concept, the reality is that it is as much embedded in local and national cultures as it is in American legislation or an IUCN definition – in fact, probably more so. As in any ecological system the national park concept is highly adaptive so as to respond to system dynamics. In fact, this is one of the reasons why it has lasted so long and has been embraced by so many people and continues to be extended in new directions. For example, NASA's 2008 Phoenix mission to Mars was undertaking research within a 'national park' region that mission scientists had been preserving for science (NASA 2008a). Shortly after landing on the planet, one of the scientific team stated with respect to the landing site, 'We've dedicated this polygon as the first national park system on Mars – a "keep out" zone until we figure out how best to use this natural Martian resource' (NASA 2008b). Such developments mirror national park management issues on Earth with respect to park use:

> The most interesting and unique regions on Mars that might merit conservation and preservation are by definition the ones where we might wish to send robots and human explorers to explore and exploit. This creates the same paradox as we face on Earth with sites of scientific or natural beauty – how do we preserve such sites while at the same time allowing them to be explored and studied?
> (Cockell and Horneck 2004: 291)

Clearly, an important and highly neglected task is to understand how the national park concept is understood by different people in different settings, including in a cross-cultural and cross-national context, so that management challenges may be improved. However, trying to impose a single definition will not work and will likely have negative consequences for trying to encourage improved conservation practice and a sustainable future. A single definition may make for good accounting practice, but it also makes for bad management and implementation practice for those affected by national parks. Instead, the capacity of the national park meme to adapt to new environments while maintaining its core idea of both conservation and use, however paradoxical it may seem at times, should continue to be encouraged so that rather than be seen as 'worthless' we can continue to understand national parks as 'the best idea America ever had' (Runte 1997a: xvii).

Bibliography

ABSA Group Economic Research (2002) *Game ranch profitability in Southern Africa*, Sandton: SA Financial Sector Forum.

Acha, M.O. (2003) 'Wirikuta: The Wixarika/Huichol sacred natural site in the Chihuahuan desert, San Luis Potosi, Mexico', in D. Harmon and A. Putney (eds) *The Full Value of Parks: From Economics to the Intangible*, Lanham: Rowman and Littlefield, pp. 295–307.

Adams, M. (2004) 'Negotiating nature: Collaboration and conflict between Aboriginal and conservation interests in New South Wales, Australia', *Australian Journal of Environmental Education*, 20(1): 3–11.

Adams, M. (2005) 'Beyond Yellowstone? Conservation and indigenous rights in Australia and Sweden', in G. Cant, A. Goodall and J. Inns (eds) *Discourses and Silences: Indigenous Peoples, Risks and Resistances*, Christchurch: Department of Geography, University of Canterbury, pp. 127–38.

Adams, W.M. and Mulligan, M. (eds) (2003) *Decolonizing Nature: Strategies for Conservation in a Post-colonial Era*, London: Earthscan.

African Initiatives (2003) 'Whose land is it anyway? The cost of conservation', Briefing Paper. African Initiatives and Ujamaa-Community Resource Trust. Online. Available: http://africaninitiatives.gn.apc.org/articles/conserve.htm (accessed 17 November 2005).

Akama, J.S. (1996) *Wildlife Conservation in Kenya: A Political-Ecological Analysis of Nairobi and Tsavo Regions*, Washington, DC: African Development Foundation.

Alcorn, J.B. (2001) *Good Governance, Indigenous Peoples and Biodiversity Conservation: Recommendations for Enhancing Results Across Sectors*. Biodiversity Support Program. WWF. Online. Available: www.worldwildlife.org/bsp/publications/ (accessed 17 November 2005).

Alderson, L. and Marsh, J.S. (1979) 'J.B. Harkin, National Parks and roads', *Parks News* 15: 9–16.

Améen, L. (1903) 'Luleå-Ofotenbanan', in *Svenska Turistföreningens årsskrift 1903*, Stockholm: Wahlström & Widstrand.

Amend, S. and Amend, T. (eds) (1995) *National Parks Without People? The South American Experience*, Quito: IUCN.

Anderson, B. (1983) *Imagined communities: reflections on the origins and spread of nationalism*, London and New York: Verso, 2006 edn.

Anderson, E. (2000) *Victoria's National Parks: a centenary history*, Melbourne: Parks Victoria.

Anderson, D. and Grove, R. (eds) (1987a) *Conservation in Africa: People, Policies and Practice*, Cambridge: Cambridge University Press.

Anderson, D. and Grove, R. (1987b) 'The scramble for Eden: Past, present and future in African conservation', in D. Anderson and R. Grove (eds) *Conservation in Africa: People policies and practice*, Cambridge: Cambridge University Press, pp. 1–12.

Andersson-Skog, L. (1993) 'Såsom allmänna inrättningar till gagnet, men affärsföretag till namnet', *SJ, järnvägspolitiken icg den ekonomiska omvandlingen efter 1920*, Umea: Umeå Studies in Economic History, Umeå University.

Anrick, C-J. (1935) *STF 1885–1935. En krönika om Svenska turistföreningen vid 50-årsjubileet d. 27 februari 1935*, Stockholm: Centraltryckeriet.

Appleton, J. (1984) *A Sort of National Property: The Discovery of the Lake District*, London: Victoria & Albert Museum.

Ariansen, P. (1992) *Miljöfilosofi*, Lund: Nya Doxa.

Arnesen, T. (1998) 'Landscapes lost', *Landscape Research*, 23(1): 39–50.

Attwell, C.A.M. and Cotterill, F.P.D. (2000) 'Postmodernism and African conservation science', *Biodiversity and Conservation*, 9: 559–77.

Atwood, M. (1972) *Survival: A Thematic Guide to Canadian Literature*, Toronto: Anansi.

Australian Academy of Science Committee on National Parks (Western Australian Sub-Committee) (1963) *National Parks and Nature Reserves in Western Australia*, Perth: The Standing Committee on Conservation of the Royal Society of Western Australia, Australian Academy of Science and the National Parks Board of Western Australia.

Australian Conservation Foundation (ACF) (1969) *The Future of the Great Barrier Reef*, Melbourne: ACF.

Australian National Audit Office (ANAO) (1998) *Commonwealth Management of the Great Barrier Reef: GBRMPA*, Canberra: ANAO.

Backhaus, N. (2003) ' "Non-place Jungle" – The Construction of Authenticity in National Parks of Malaysia', *Indonesia and the Malay World* 31(89): 151–60.

Badè, W.F. (1924) *The Life and Letters of John Muir*, 2 vols, Boston: Houghton-Mifflin.

Banff Heritage Tourism Corporation (2004) *Banff Bow Valley Heritage Tourism Strategy*, Banff: Banff Heritage Tourism Corporation.

Banfield, E.J. (1913) *The Confessions of a Beachcomber: Scenes and Incidents in the Career of an Unprofessional Beachcomber in Queensland*, London: Fisher Unwin.

Barnes, M. and Wells, G. (1985) 'Myles Dunphy father of conservation dies', *National Parks Journal*, 29(1): 7.

Barrett, C. (1930) 'The Great Barrier Reef and its Isles: The Wonder and Mystery of Australia's World-famous Geographical Feature', *The National Geographic Magazine*, 58(3): 354–84.

Barrett, C. (1943) *Australia's Coral Realm: Wonders of Sea, Reef and Shore*, Melbourne: Robertson & Mullens.

Bauer, K.R. (2007) 'Protecting indigenous spiritual values,' *Peace Review: A Journal of Social Justice*, 19: 343–9.

Beeton, S. (2005) *Film-induced tourism*, Clevedon: Channel View.

Bell, C. and Sanders, N. (1980) *A Time to Care: Tasmania's Endangered Wilderness*, Blackman's Bay: C. Bell.

Beltran, J. (ed) (2000) *Indigenous and Traditional Peoples and Protected Areas: Principles, Guidelines and Case Studies*, Best Practice Protected Area Guidelines Series No. 4. Gland & Cambridge: IUCN & WWF International.

Bennett, I. (1971) *The Great Barrier Reef*, Sydney: Lansdowne Press.
Bennett, J.A. (2000) *Pacific Forest: A History of Resource Control and Contest in Solomon Islands, c. 1800–1997*, Cambridge and Leiden: The White Horse Press and Brill Academic Publishers.
Benzaken, D., Smith, G. and Williams, R. (1997) 'A Long Way Together: The Recognition of Indigenous Interests in the Management of the Great Barrier Reef World Heritage', in D. Wachenfeld, J. Oliver and K. Davis (eds), *State of the Great Barrier Reef World Heritage Workshop*, Townsville: GBRMPA.
Berger, D.J. (1996) 'The challenge of integrating Maasai tradition with tourism', in M.F. Price and V.L. Smith (eds) *People and Tourism in Fragile Environments*, Chichester: Wiley, pp. 175–97.
Bergin, A. (1993) *Aboriginal and Torres Strait Islander Interests in the Great Barrier Reef Marine Park*, Townsville: GBRMPA.
Berry, G. and Beard, G. (1980) *The Lake District: A Century of Conservation*, Edinburgh: Bartholomew.
Berzins, B. (1998) 'Before the Sharing: Aborigines and tourism in the Northern Territory to the 1970s', *Journal of Northern Territory History*, 9: 69–80.
Billinge, M. (1996) 'A time and place for everything: An essay on recreation, re-creation and the Victorians', *Journal of Historical Geography*, 22(4): 443–59.
Birckhead, J. and Smith, L. (1991) 'Introduction: conservation and country – a re-assessment', in J. Birckhead, T. De Lacy and L. Smith (eds) *Aboriginal Involvement in Parks and Protected Areas*, Canberra: Aboriginal Studies Press.
Black, A. and Breckwoldt, R. (1977) 'Evolution of systems of national park policymaking in Australia', in D. Mercer (ed) *Leisure and Recreation in Australia*, Melbourne: Sorrett Publishing, 190–9.
Bladh, G. and Sandell, K. (2002) *Biosfäromräden i Sverige? Ett koncept och en kontext*, Institutionen för samhällsvetenskap, Turism & fritid. Arbetsrapport 2003:15.
Blangy, S. and Martin, J.L. (2002a) 'Biodiversity, cultural heritage and ecotourism – disruption and recovery in Haida Gwaii (Canada)', *Policy Matters*, 10: 96–97.
Blangy, S. and Martin, J.L. (2002b) 'Integration of biodiversity in cultural heritage in the development of ecotourism. A case study from Haida Gwaii (Queen Charlotte Islands), B.C., Canada', in F. de Castri and V. Balaji (eds) *Tourism, Biodiversity and Information*, Leiden, Netherlands: Backhuys, pp. 105–16.
Blower, J.H., van der Zon, A.P.M. and Mulyana, Y. (1977) *Proposed Komodo National Park Management Plan, 1978–1982*, Bogor: UNDP/FAO.
Blunden, J. and Curry, N. (1990) *A People's Charter? 40 Years of the 1949 National Parks and Access to the Countryside Act*, London: HMSO.
Bolton, G. (1981) *Spoils and Spoilers: Australians Make their Environment 1788–1980*, Sydney: Allen & Unwin.
Bonyhady, T. (2000) *The Colonial Earth*, Melbourne: Miegunyah Press [and Melbourne University Press].
Boomgaard, P. (1999) 'Oriental nature, its friends and its enemies', *Environment and History*, 5(3): 257–92.
Borchers, H. (2008) 'Dragon Tourism Revisited: the Sustainability of Tourism Development in Komodo National Park', in M. Hitchcock, V.T. King and M. Parnwell (eds) *Tourism in Southeast Asia Revisited*, Copenhagen: NIAS.
Borrini-Feyerabend, G. (1999) 'Collaborative management of protected areas', in S. Stolton and N. Dudley (eds) *Partnerships for Protection: New Strategies for Planning and Management for Protected Areas*, London: Earthscan, pp. 224–34.

Borrini-Feyerabend, G., Kothari, A. and Oviedo, G. (2004) *Indigenous and Local Communities and Protected Areas: Towards Equity and Enhanced Conservation: Guidance on Policy and Practice for Co-managed Protected Areas and Community Conserved Areas*. Best Practice Protected Area Guidelines Series No. 11. Gland & Cambridge: IUCN.

Bowen, J. (1994) 'The Great Barrier Reef: Towards Conservation and Management', in S. Dovers (ed), *Australian Environmental History: Essays and Cases*, Melbourne: Oxford University Press.

Bowen, J. and Bowen, M. (2002) *The Great Barrier Reef: History, Science, Heritage*, Cambridge: Cambridge University Press.

Boyd, S.W. (1995) 'Sustainability and Canada's National Parks: Suitability for policy, planning, and management', unpublished PhD thesis, University of Western Ontario.

Boyd, S.W. (2004) 'National parks: Wilderness and culture', in A. Lew, C.M. Hall and A. Williams (eds) *A Companion to Tourism*, Oxford: Blackwell (pp. 473–83).

Boyd, S.W. (2006) 'The TALC model and its application to national parks: A Canadian example', in R.W. Butler (ed) *The Tourism Area Life Cycle* Vol. 1, Clevedon: Channelview (pp. 119–138).

Boyd, S.W. and Butler, R.W. (2000) 'Tourism and National Parks: the Origin of the Concept', in R.W. Butler and S.W. Boyd (eds) *Tourism and National Parks: Issues and Implications*, Chichester: Wiley, pp. 13–27.

Brechin, S.R., Wilshusen, P.R., Fortwangler, C.L. and West, P. (2003) *Contested Nature–Promoting International Biodiversity with Social Justice in the Twenty-first Century*, Albany: State University of New York Press.

Breedon, S. (1994) *Uluṟu: Looking after Uluṟu-Kata Tjuṯa the Aṉangu Way*, Sydney: Simon and Schuster.

Brett, M.R. (1989) *The Pilanesberg: Jewel of Bophuthatswana*, Sandton: Southern Books.

Broberg, G. and Johannisson, K. (1986) ' "Styr som örnen din färd till fjällen": några glimtar i den tidiga turismens idéhistoria', in *STF 100 år. Svenska Turistföreningens årsskrift 1986*. Uppsala.

Brockington, D. (2001) *Fortress Conservation: The Preservation of the Mkomazi Game Reserve, Tanzania*, Oxford: James Currey.

Brockington, D. (2002) *Fortress Conservation: The Preservation of the Mkomazi Game Reserve, Tanzania*, Oxford: International African Institute in association with James Currey.

Brooks, S. (2005) 'Images of "wild Africa": nature tourism and the (re)creation of Hluhluwe Game Reserve, 1930–1945', *Journal of Historical Geography*, 31: 220–40.

Brosius, J.P. (2004) 'Indigenous peoples and protected areas at the World Parks Congress,' *Conservation Biology*, 18(3): 609–12.

Brown, R.C. (1969) 'The doctrine of usefulness: Natural resources and national park policy in Canada, 1887–1914', in J.G. Nelson and R.C. Scace (eds) *The Canadian National Parks: Today and Tomorrow*, Vol.1, Calgary: Department of Geography, University of Calgary.

Brown, T.J. (1999) 'Antecedents of culturally significant tourist behaviour,' *Annals of Tourism Research*, 26(3): 676–700.

Bryan, B.A. (2002) 'Reserve selection for nature conservation in South Australia: Past, present and future', *Australian Geographical Studies*, 40(2): 196–209.

Buchholtz, C.W. (1983) *Rocky Mountains National Park: A History*, Boulder: Colorado Associated University Press.
Buchholtz, D. (2005) 'Cultural Politics or Critical Public History? Battling on the Little Big Horn', *Journal of Tourism and Cultural Change*, 3: 18–35.
Buggey, S. (1999) *An Approach to Aboriginal Cultural Landscapes*, Ottawa: Parks Canada.
Bunce, O.B. (1872) 'Our great National Park', in W.C. Bryant (ed) *Picturesque America*, Vol.1., New York: Appleton and Company.
Burnam Burnam (1988) *Burnam Burnam's Aboriginal Australia: A Traveller's Guide*, Sydney: Angus and Robertson.
Burnham, P. (2000) *Indian Country, God's Country: American Indians and National Parks*, Washington DC: Island Press.
Büscher, B. and Dressler, W. (2007) 'Linking neoprotection and environmental governance: on the rapidly increasing tensions between actors in the environment-development nexus', *Conservation and Society*, 5(4): 586–611.
Bushell, R. and Eagles, P.F.J. (eds) (2007) *Tourism and Protected Areas: Benefits Beyond Boundaries. The Vth IUCN World Parks Congress*, Wallingford: CABI.
Butler, D. (1980) 'Salute and farewell to Marie Byles', *The Sydney Bushwalker*, January: 2–5.
Butler, R.W. and Hinch, T. (eds) (2007) *Tourism and Indigenous Peoples: Issues and Implications*, Oxford: Elsevier.
Byles, M.B. (1938) 'The need for wildernesses', *Bushland*, 1(2): 6–7.
Caffyn, A. and Prosser, B. (1998) 'A Review of Policies for "Quiet Areas" in the National Parks of England and Wales', *Leisure Studies* 17(4): 269–291.
Caillois. R. (1988) *Man and the Sacred*, Glencoe: Free Press.
Calway, D. (2001) *The Economic Impact of Tourism in Cumbria: STEAM 2000*, Windermere: Cumbria Tourist Board.
Carino, J. (2004) 'Indigenous voices at the table: Restoring local decision-making on protected areas', *Cultural Survival Quarterly* 28(1).
Carlquist, L. (1986) 'Resetjänst: Hågkomster från aktiva decennier', in: *STF 100 år. Svenska Turistföreningens årsskrift 1986*, Uppsala.
Carothers, P., Vaske, J.J. and Donnelly, M.P. (2001) 'Social values vs interpersonal conflict among hikers and mountain bikers', *Leisure Sciences*, (23): 47–61.
Carr, A. (2001) ' "Locating" culture: Visitor experiences of significant landscapes' in *IAA Conference 2001*, Melbourne: Interpretation Australia Association, pp. 40–5.
Carr, A. (2004) 'Mountain places, cultural spaces – interpretation and sustainable visitor management of culturally significant landscapes: a case study of Aoraki/Mount Cook National Park,' *Journal of Sustainable Tourism*, 12(5): 432–59.
Carr, A. (2007) 'Maori nature tourism businesses: Connecting with the land', in R. Butler and T. Hinch (eds) *Tourism and Indigenous Peoples: Issues and Implications*, Oxford: Elsevier, pp. 113–127.
Carr, A. (2008) 'Cultural landscape values as a heritage tourism resource,' in B. Prideaux, D.J. Timothy and K. Chon (eds) *Cultural and Heritage Tourism in Asia and the Pacific*, London: Routledge, pp. 35–48.
Carron, L.T. (1985) *A History of Forestry in Australia*, Canberra: Australian National University Press.
Carruthers, J. (1994) 'Dissecting the myth: Paul Kruger and the Kruger National Park', *Journal of Southern African Studies*, 20(2): 263–83.

Carruthers, J. (1995a) *The Kruger National Park: A Social and Political History*, Pietermaritzburg: University of Natal Press.

Carruthers, J. (1995b) *Game Protection in the Transvaal, 1846 to 1926*, Pretoria: Archives Year Book.

Carruthers, J. (1997) 'Nationalism and national parks: Comparative examples from the post-imperial experience', in T. Griffiths and L. Robin (eds) *Ecology and Empire: Comparative History of Settler Societies*, Edinburgh: Keele University Press.

Carruthers, J. (2001) *Wildlife and Warfare: The Life of James Stevenson-Hamilton*, Pietermaritzburg: Natal University Press.

Carruthers, J. (2006) 'Mapungubwe: An historical and contemporary analysis of a World Heritage cultural landscape', *Koedoe*, 41(1): 1–14.

Carruthers, J. (2007) 'South Africa: A world in one country: Land restitution in national parks and protected areas', *Conservation and Society*, 5(3): 292–306.

Cary, G., Lindenmayer, D. and Dovers, S. (eds) (2003) *Australia Burning: Fire Ecology, Policy and Management Issues*. Melbourne: CSIRO.

Casado, S. (2000) 'Ciencia y política en los orígenes de la conservación de la naturaleza en España', *Scripta Vetera* 78. Online: www.ub.es/geocrit/sv-78.htm.

Castree, N. (2003) 'Commodifying what nature?' *Progress in Human Geography*, 27 (3): 273–97.

Centre for Development and Enterprise (2005) *Land Reform in South Africa: A 21st century perspective*, Johannesburg: CDE.

Chape, S., Blyth, S., Fish, L., Fox, P. and Spalding, M. (compilers) (2003) *2003 United Nations List of Protected Areas*, Gland and Cambridge: IUCN and UNEP-WCMC.

Chen, J.H. and Qin, Y.Z. (2006) 'On the innovation of the new tourism routes in Wuyishan', *Fujian Forestry Science and Technology*, 33(2): 197–200.

Cheung, S. (2008) 'Wetland Tourism in Hong Kong: From Birdwatcher to Mass Ecotourist', in Cochrane, J. (ed) *Asian Tourism: Growth and Change*, London: Elsevier, pp. 259–67.

Child, B. (2004) *Parks in Transition: Biodiversity, Rural Development and the Bottom Line*, London: Earthscan.

Chinese Construction Ministry, Wuyishan (1998), *Report to IUCN*.

Chronis, A. (2005) 'Coconstructing heritage at the Gettysburg storyscape', *Annals of Tourism Research*, 32: 386–406.

Clay, J.W., Alcorn, J.B. and Butler, J.R. (2000) *Indigenous Peoples, Forestry Management and Biodiversity Conservation: An Analytical Study for the World Bank Forestry Policy Implementation Review and Strategy Development Framework*, WWF Report for World Bank.

Coates, P. (2004) 'Emerging from the Wilderness (or, from Redwoods to Bananas): Recent Environmental History in the United States and the Rest of the Americas', *Environment and History*, 10(4): 407–38.

Cochrane, J. (2003) 'Ecotourism, Conservation and Sustainability: A Case Study of Bromo Tengger Semeru National Park, Indonesia', unpublished PhD thesis, University of Hull.

Cochrane, J. (2007) 'Ecotourism and biodiversity conservation in Asia: Institutional challenges and opportunities', in J. Higham (ed) *Critical Issues in Ecotourism: Understanding a Complex Tourism Phenomenon*, Oxford: Elsevier, pp. 297–307.

Cockell, C. and Horneck, G. (2004) 'A planetary park system for Mars', *Space Policy* 20: 291–5.

Cohen, M.P. (1984) *The Pathless Way: John Muir and American Wilderness*, Madison: University of Wisconsin Press.
Colchester, M. (1996) 'Beyond "participation": Indigenous peoples, biological diversity conservation and protected areas management', *Unasylva*, 47 (186): 33–9. Online. Available: www.fao.org/docrep/w1033e/w1033e08.htm (accessed 29 January 2008).
Colchester, M. (1997) 'Salvaging nature: Indigenous peoples and protected areas', in K.B. Ghimire and M.P. Pimbert (eds) *Social Change and Conservation: Environmental Politics and Impacts of National Parks and Protected Areas*, London: Earthscan.
Colchester, M. (2001) 'This park is no longer your land,' *The UNESCO Courier*, July/August.
Colchester, M. (2003a) 'The politics of parks: Indigenous peoples assert their rights against mining, markets and tourism', *Multinational Monitor*, 24(11): 19–21.
Colchester, M. (2003b) *Salvaging Nature: Indigenous Peoples, Protected Areas and Biodiversity Conservation*, World Rainforest Movement and Forest Peoples Programme.
Colchester, M. and Erni, C. (eds) (2000) *Indigenous Peoples and Protected Areas in South and Southeast Asia*, Copenhagen: IWGIA.
Cole, D. (1996) 'Magazines for Early Motorists', *Society for Commercial Archaeology Journal* (Spring): 6–11.
Coleman, N. (1990) *Australia's Great Barrier Reef*, Brookvale: National Book Distributors.
Colfelt, D. (2004) *100 Magic Miles of the Great Barrier Reef: The Whitsunday Islands*, Berry: Winward Publications.
Collins, R.M. (1897) 'The South-Eastern Highlands of Queensland', *Proceedings and Transactions of the Queensland Branch of the Royal Geographical Society of Australasia*, XII, 20–5.
Collison Black, R.D. and Kinekamp, R. (eds) (1972) *Papers and Correspondence of William Stanley Jevons*. Vol. 1, London: Macmillan.
Committee of Inquiry into the National Estate (1974) *Report of the National Estate*, Canberra: AGPS.
Common, M.S. and Norton, T.W. (1992) 'Biodiversity: Its conservation in Australia', *Ambio* 21(3): 258–65.
Commonwealth of Australia (2000) *Uluru-Kata Tjuta National Park: Plan of Management*, Uluru-Kata Tjuta National Park and Parks Australia.
Commonwealth of Australia (2007) *Conserving Australia: Australia's national parks, conservation reserves and marine protected areas*, Canberra: Senate Standing Committee on Environment, Communications, Information Technology and the Arts.
Commonwealth of Australia (2008) *Uluru-Kata Tjuta National Park*. Online available at www.environment.gov.au/parks/Uluru (accessed 14 March 2008).
Conway, H. (1991) *People's Parks: The Design and Development of Victorian Parks in Britain*, Cambridge: Cambridge University Press.
Conwentz, H. (1904) 'Om skydd åt det naturliga landskapet jämte dess växt- och djurvärld, särskildt i Sverige', in *Ymer*, Svenska Sällskapet för Antropologi och Geografi, Stockholm: Centraltryckeriet.
Coombes, B. and Hill, S. (2005) ' "Nawhenua, na Tuhoe. Ko D.o.C. te partner" Prospects for comanagement of Te Urewera National Park', *Society and Natural Resources*, 18(2): 135–52.

Cosgrove, D. (1998) *Social Formation and Symbolic Landscape*, Madison: University of Wisconsin Press.
Couldry, N. (2000) *The place of media power: pilgrims and witnesses of the media age*, London: Routledge.
Couldry, N. (2001) 'The hidden injuries of media power', *Journal of Consumer Culture*, 1(2): 155–177.
Cox, T.R. (1983) 'The "worthless lands" thesis: another perspective', *Journal of Forest History* 27(3): 144–5.
Craig, D. (1993) 'Environmental law and Aboriginal rights: Legal framework for Aboriginal joint management of Australian national parks', in J. Birckhead, T. De Lacy and L. Smith (eds) *Aboriginal Involvement in Parks and Protected Areas*, Canberra: Aboriginal Studies Press.
Craik, J. (1987) 'A Crown of Thorns in Paradise: Tourism on Queensland's Great Barrier Reef', in M. Bouquet and M. Winter (eds) *Who From Their Labours Rest? Conflict and Practice in Rural Tourism*, Aldershot: Avebury, pp. 135–58.
Craik, J. (1991) *Resorting to Tourism, Cultural Policies for Tourism Development in Australia*, Sydney: Allen and Unwin.
Craik, W. (1990) *Management of Recreational Fishing in the Great Barrier Reef Marine Park*, Townsville: GBRMPA.
Cranz, G. (1982) *The politics of park design: a history of urban parks in America*, Cambridge, MA: Massachusetts Institute of Technology Press.
Creative Research (2002) *Cumbria Tourism Survey 2002: Report of Findings*, London: Creative Research.
Cresswell, C. and MacLaren, F. (2000) 'Tourism and national parks in emerging tourism countries', in R.W. Butler and S.W. Boyd (eds) *Tourism and National Parks: Issues and Implications*, pp. 283–99, Chichester: Wiley.
Cribb, R. (1995) 'International tourism in Java', *South East Asia Research*, 3(2): 193–204.
Cronin, L. (2000) *Australia's National Parks*, Annandale: Envirobook.
Crouch D. and Lubbren N. (eds) (2003) *Visual Culture and Tourism*, Oxford: Berg.
Crouch D., Jackson R. and Thompson F. (eds) (2005) *The Media and the Tourist Imagination: Convergent cultures*, London: Routledge.
Crowley, K. (1999) 'Lake Pedder's loss and failed restoration: Ecological politics meets liberal democracy in Tasmania', *Australian Journal of Political Science*, 34(3): 409–24.
Culler. J. (1981a) 'Semiotics of tourism', *American Journal of Semiotics*, 1: 127–40.
Cumming, David H.M. (1999) *Study on the development of Transboundary Natural Resource Management Areas in Southern Africa – Environmental Context: Natural Resources, Land Use, and Conservation*, Washington D.C.: Biodiversity Support Program.
Curry N. (1994) *Countryside Recreation, Access and Land Use Planning*, London: Spon.
Curtis, E. (1988) *The turning years: a Tamborine Mountain history*, Tamborine: author.
Cutright, P.R. (1985) *Theodore Roosevelt: the making of a conservationist*, Urbana: University of Illinois Press.
Daly, M., Dehne, A., Leffman, D. and Scott, C. (1999) *Australia: The Rough Guide (Fourth Edition)*, London: Rough Guides.
Dann G. (1996a) 'The People of Tourist Brochures', in T. Selwyn (ed) *The Tourist Image. Myths and Myth Making in Tourism*, Chichester: Wiley.

Dann G. (1996b) *The Language of Tourism: A Sociolinguistic Interpretation*, Wallingford: CABI.

Darrell, J., Hind, D., Maurice, O., Royce, K. and Tiplady, P. (2004) 'The Planning and Management of Tourism in the Lake District', in D. Hind and J. Mitchell (eds) *Sustainable Tourism in the English Lake District*, Sunderland: Business Education, pp. 101–48.

Davidson, J. and Spearritt, P. (2000) *Holiday Business: Tourism in Australia Since 1870*, Melbourne: Melbourne University Press.

Davies, R. (2000) 'Madikwe Game Reserve: A partnership in conservation', in H.H.T. Prins, J.G. Grootenhuis and T.T. Dolan (eds), *Wildlife Conservation by Sustainable Use*, Boston: Kluwer.

Dearden, P. and Rollins, R. (eds) (1993) *Parks and Protected Areas in Canada: Planning and Management*, Toronto: Oxford University Press.

DECC (Department of Environment & Climate Change) (2007) Aboriginal co-management of parks. Online: www.nationalparks.nsw.gov.au/npws.nsf/Content/Aboriginal+co-management+of+parks (accessed 7 February 2008).

Department for Communities and Local Government (2007) Special Grants Programme.

Department of Environment and Heritage (DEH) (2006) *Review of the Great Barrier Reef Marine Park Act 1975*, Canberra: Government of Australia.

Destination Management Partnership Annual Report (2007) *Visit Peak District and Derbyshire*, Chesterfield: Destination Management Partnership.

De Villiers, B. (1999) *Land Claims and National Parks: The Makuleke Experience*, Pretoria: HSRC.

Diamond, J. (1986) 'The Golden Age That Never Was', *Discover Magazine* (Dec): 71–9.

Diamond, J. and Liddle, J. (2005) *Management of Regeneration*, London: Routledge.

Digance, J. (2003) 'Pilgrimage at contested sites,' *Annals of Tourism Research*, 30(1): 143–59.

Dower, J. (1945) *National Parks in England and Wales*, Cmnd 6378, London: HMSO.

Down to Earth (2002) 'Gunung Merapi National Park', in *Down to Earth Bulletin* 55.

Dunlap, T.R. (1999) *Nature and the English Diaspora: environment and history in the United States, Canada, Australia, and New Zealand*, Cambridge: Cambridge University Press.

Durkheim. E. (1995) *The Elementary Forms of Religious Life*, London: Allen and Unwin.

Dustin, D.L. and Schneider, I.E. (2001) 'Collaborative conflict resolution at Devils Tower National Monument,' *Parks & Recreation*, 36: 80–5.

Dustin, D., Schneider, I., McAvoy, L. and Frakt, A. (2002) 'Cross-cultural claims on Devils Tower National Monument: A case study,' *Leisure Sciences*, 24: 1–10.

Dutton, C.E. (1886) Dutton to Steel, February 27, 1886, Steel Letters, Box 1, Item 195, Museum Collection, Crater Lake National Park.

Dutton, S. and Archer, F. (2004) 'Transfrontier parks in South Africa,' *Cultural Survival Quarterly*, 28(1).

Eade, F. (1987) 'Land in Many Hands', *National Parks Today*, Winter Edition.

Eagles, P. (2007) 'Global trends affecting tourism in protected areas', in R. Bushell and P.J. Eagles (eds) *Tourism and Protected Areas: Benefits Beyond Boundaries*, Wallingford: CABI.

Eagles, P.F.J. and McCool, S.F. (2002) *Tourism in National Parks and Protected Areas: Planning and Management*, Wallingford: CABI.

East, K. (1991) 'Joint management of Canada's northern national parks', in P.C. West and S.R. Brechin (eds) *Resident Peoples and National Parks: Social Dilemmas and Strategies in International Conservation*, Tucson: University of Arizona Press, pp. 333–45.

Edensor, T. (1998) *Tourists at the Taj: Performance and Meaning at a Symbolic Site*, London and New York: Routledge.

Edwards, R. (1991) *Fit for the Future: Report of the National Parks Review Panel*, CCP 334, Cheltenham: Countryside Commission.

Eghenter, C. (2004) 'Social science research as a tool for conservation: the case of Kayan Mentarang National Park (Indonesia)', in *IUCN Policy Matters* 13, 224–33.

Eidsvik, H.K. (1983) 'Parks Canada, conservation and tourism: a review of the seventies – a preview of the eighties', in P.E. Murphy (ed) *Tourism in Canada: Selected issues and options*, Victoria: Department of Geography, University of Victoria, pp. 241–69.

Ell, G. (2000) 'Sandra Lee: Fresh perspectives on conservation,' *Forest & Bird*, 296: 4.

English, A.J. and Lee, E. (2003) 'Managing the intangible', in D. Harmon and A.D. Putney (eds) *The Full Value of Parks: From Economics to the Intangible*, Lanham: Rowman & Littlefield, pp. 43–55.

Eutlain, R.W. (comp.) (1999) *Does the Frontier Experience Make America Exceptional?* Boston: Bedford/St. Martins.

Evans, S. (1999) *The Green Republic: a Conservation History of Costa Rica*, Austin: University of Texas Press.

Ewert, A. and Stewart, B. (2004) 'Philosophical perspectives on natural resources: Examining the past to understand the future', in M.J. Manfredo, J.J. Vaske, B.R. Bruyere and Y. Brown (eds) *Society and Natural Resources: A Summary of Knowledge*, Jefferson: Modern Litho.

FAO (Food and Agriculture Organization of the United Nations) (1980) *Bromo-Tengger-Gunung Semeru Proposed National Park, Management Plan 1981–85*, Bogor: Field Report of the UNDP/FAO National Park Development Project, INS/78/061.

FAO (1982) *National Conservation Plan for Indonesia, Introduction, Evaluation Methods and Overview of National Nature Richness*, Bogor: Field Report of UNDP/FAO National Park Development Project INS/78/061.

Federation of Maori Authorities (2003) *Conservation Estate Draft Discussion Paper*. Online. Available: http://www.foma.co.nz/static/pdf/conservation%20estate%20research.pdf (accessed 20 January 2008).

Fernández, J. and Pradas Regal, R. (1996) *Los Parques Nacionales Españoles*, Colección Parques Nacionales: Madrid.

Fearnhead, P. and Mabunda, D. (2003) 'Towards sustainability', in Hall-Martin, A. and Carruthers, J. (eds), *South African National Parks: A Celebration*, Johannesburg: Horst Klemm.

Finlay, H., Armstrong, M. and Wheeler, T. (1998) *Islands of Australia's Great Barrier Reef (3rd Edition)*, Melbourne: Lonely Planet.

Fiske, J., Hodge, B. and Turner, G. (1987) *Myths of Oz: Reading Australian Popular Culture*, Sydney: Allen & Unwin.

Fitter, R. and Scott, P. (1978) *The Penitent Butchers: the Fauna Preservation Society 1903–1978*, London: Collins.

Fitzgerald, F. (1982) *From the Dreaming to 1915: A History of Queensland*, Brisbane: Queensland University Press.

FNNPE (Federation of Nature and National Parks of Europe) (1993) *Loving Them to Death? Sustainable Tourism in Europe's Nature and National Parks*, Grafenau: FNNPE.

Foley, P. and Martin, S. (2000) 'Perceptions of community led regeneration: community and central government viewpoints', *Journal of Regional Studies* 34(8): 783–7.

Forest Peoples Programme (FPP) (2003) WPC Recommendation 24: Indigenous peoples and protected areas. World Parks Congress 2003. Online. Available: www.forestpeoples.org/documents/conservation/bases/wpc_base.shtml (accessed 29 November 2005).

Forsslund, K-E. (1914) *Som gäst hos fjällfoket*, Uppsala: Almqvist & Wiksell AB.

Forsslund, K-E. (1915) *Fridlysta Vildmarker: Skildringar och historier från Sveriges nationalparker*, Stockholm: Wahlström & Widstrand.

Foster, D. (1997) *Gurig National Park: The First Ten Years of Joint Management*, Canberra: Australian Institute of Aboriginal and Torres Strait Islander Studies.

Fox, S. (1981) *John Muir and his Legacy: the American Conservation Movement*, Boston: Little, Brown and Company.

Foxall, G.R. (1993) 'Consumer behaviour as an evolutionary process', *European Journal of Marketing*, 27(8): 46–57.

Francis, R. (1982) 'Changing images of the west', *Journal of Canadian Studies*, 17(3): 5–19.

Fredman, P., Hörnsten Friberg, L. and Emmelin, L. (2007) 'Increased visitation from national park designation', *Current Issues in Tourism*, 10(1): 87–95.

Fredman, P. and Heberlein, T. (2005) 'Visits to the Swedish mountains: Constraints and motivations', *Scandinavian Journal of Hospitality and Tourism Research*, 3: 177–92.

Fredman, P. and Hörnsten, L. (2004) 'Perceived crowding and visitor satisfaction in Fulufjäll National Park, Sweden', in I. Camarda, M.J. Manfredo, F. Mulas, and T.L. Teel (eds) *Global Challenges of Parks and Protected Area Management*, Sassari: Carlo Delfino Editore.

Frost, W. (1997) 'Farmers, government and the environment: the settlement of Australia's "Wet Frontier", 1870–1920', *Australian Economic History Review*, 37 (1): 19–38.

Frost, W. (2002) 'Did they really hate trees? Attitudes of farmers, tourists and naturalists towards nature in the rainforests of eastern Australia', *Environment and History*, 8(1): 3–19.

Frost, W. (2004) 'Tourism, rainforests and worthless lands: the origins of National Parks in Queensland', *Tourism Geographies*, 6 (4): 493–507.

Frost, W (2005) 'Making an edgier interpretation of the Gold Rushes: contrasting perspectives from Australia and New Zealand', *International Journal of Heritage Studies*, 11(3): 235–50.

Frost, W. (2006) '*Braveheart*-ed *Ned Kelly*: historic films, heritage tourism and destination image', *Tourism management*, 27(2): 247–54.

Frost, W. and Roehl, W. (2008) 'Zoos, aquaria and tourism: extending the research agenda', *Tourism Review International*, 13(3).

Fry, W. and White, J.R. (1930) *Big Trees*, Palo Alto: Stanford University Press.

Furze, B., Lacy, T. de and Birckhead, J. (1996) *Culture, conservation and biodiversity: the social dimension of linking local development and conservation through protected areas*, Chichester: John Wiley & Sons.

Gale, D. and Buultjens, J. (2007) 'Mt Warning visitation and indigenous concerns:

Visitors' perceptions', in J. Buultjens and D. Fuller (eds), *Striving for Sustainability: Case Studies in Indigenous Tourism*, Lismore: Southern Cross University Press, pp. 247–90.

Garlick. S. (2002) 'Revealing the unseen: Tourism, art and photography', *Cultural Studies* (16)2: 289–305.

Genosko. G. (2003) 'The bureaucratic beyond: Roger Caillois and the negation of the sacred in Hollywood cinema', *Economy and Society*, 32(1): 74–89.

Ghimire, K.B. and Pimbert, M.P. (eds) (1997) *Social Change and Conservation: Environmental Politics and Impact of National Parks and Protected Areas*, London: UNRISD and Earthscan.

Gillett, K. (1980) *The Australian Great Barrier Reef in Colour*, Sydney: Reed.

Glick, D. and Alexander, B. (2000) 'Development by default, not design: Yellowstone National Park and the Greater Yellowstone Ecosystem', in G.E. Machlis and D.R. Field (eds), *National Parks and rural development: practice and policy in the United States*, Washington DC: Island, pp. 181–205.

Goldstein, W. (1979a) 'National Parks – South Australia', *Parks and Wildlife*, 2(3–4): 123–9.

Goldstein, W. (1979b) 'National Parks – Victoria', *Parks and Wildlife*, 2(3–4): 117–22.

Goldstein, W. (1979c) 'National Parks – Queensland', *Parks and Wildlife*, 2(3–4): 130–40.

Goldstein, W. (1979d) 'National Parks – Tasmania' *Parks and Wildlife*, 2(3–4): 141–8.

Goldstein, W. (1979e) 'National Parks – New South Wales', *Parks and Wildlife*, 2(3–4): 93–107.

Gössling, S. and Hultman, J. (eds) (2006) *Ecotourism in Scandinavia: Lessons in Theory and Practice*, Wallingford: CABI.

Götmark, F. and Nilsson, C. (1992) 'Criteria used for protection of natural areas in Sweden 1909–1986', *Conservation Biology*, 6(2): 220–31.

Government Caucus Committee (1979) *Review of the Administrative Structure of National Parks and Reserves Administered by the Department of Lands and Survey*, Wellington: Government Caucus Committee.

Grant, W.E. (1994) 'The inalienable land: American wilderness as sacred symbol', *The Journal of American Culture* 17(1): 79–86.

Great Barrier Reef Marine Park Authority (GBRMPA) (1994) *The Great Barrier Reef: Keeping it Great – A 25 Year Strategic Plan for the Great Barrier Reef World Heritage Area*, Townsville: GBRMPA.

Great Barrier Reef Marine Park Authority (GBRMPA) (1999) *Great Barrier Reef: Reference File*, Townsville: GBRMPA.

Green, G. and Lal, P. (1991) *Charging Users of the Great Barrier Reef Marine Park*, Townsville: GBRMPA.

Green, P. (1996) 'Indigenous communities', in L. Woodward (ed) *World Heritage Managers Workshop Ravenshoe April 11–13 1996: Papers and Proceeding*, Canberra: Department of Environment, Sport and Territories, pp. 69–70.

Greene, L. (1987) *Historic Resource Study, Yosemite National Park*, Denver: National Park Service.

Grieg-Gran, M. and Mulliken, T. (2004) *The Commercial Record of Community-based Sustainable Use Initiatives*, London: IIED.

Griffiths, T. (1996) *Hunters and collectors: the antiquarian imagination in Australia*, Cambridge: Cambridge University Press.

Groom, A. (1949) *One Mountain After Another*, Sydney: Angus & Robertson.

Grove, R.H. (1995) *Green Imperialism: Colonial Expansion, Tropical Island Edens and the Origins of Environmentalism*, Cambridge: Cambridge University Press.
Hadfield, M. (1985) *A History of British Gardening*, Harmondsworth: Penguin.
Hadley, E.J. (1956) 'John Muir's Views of Nature and their Consequences', unpublished PhD thesis, University of Wisconsin.
Hägerstrand, T. (1991) 'Tillkomsten av nationalparker I Sverige. En idés väg från "andskap" till landskap', in *Svenska Geografisk Årsbok 1991*, Lund.
Haines, A.L. (1972) 'Foreword', in N.P. Langford, *The discovery of Yellowstone Park: journal of the Washburn Expedition to the Yellowstone and Firehole Rivers in the year 1870*, Lincoln: University of Nebraska Press, pp. vi–xxi.
Hall, C.M. (1985) 'Outdoor recreation and national identity: a comparative study of Australia and Canada', *Journal of Canadian Culture*, 2(2): 25–39.
Hall, C.M. (1987) 'John Muir in New Zealand', *New Zealand Geographer* 43(2): 99–103.
Hall, C.M. (1988a) 'John Muir: the grandfather of national parks', *Australian Science Magazine* 26(1): 44–7.
Hall, C.M. (1988b) 'Wilderness in New Zealand', *Alternatives: Perspectives on Science, Technology and the Environment*, 15(3): 40–6.
Hall, C.M. (1989) 'The worthless lands hypothesis and Australia's national parks and reserves', in K. Frawley and N. Semple (eds) *Australia's Ever Changing Forests*, Canberra: Australian Defence Force Academy, pp. 441–56.
Hall, C.M. (1992) *Wasteland to World Heritage: Preserving Australia's Heritage*, Melbourne: Melbourne University Press.
Hall, C.M. (1993) 'John Muir's travels in Australasia 1903–1904: their significance for environmental and conservation thought', in S. Miller (ed) *John Muir: Life and Work*, Albuquerque: University of New Mexico Press, pp. 286–308.
Hall, C.M. (1994) 'Ecotourism in Australia, New Zealand and the South Pacific: appropriate tourism or a new form of ecological imperialism?', in E.A. Cater and G.A. Bowman (eds) *Ecotourism: A Sustainable Option?*, Chichester/London: Wiley/Royal Geographical Society, pp. 137–58.
Hall, C.M. (1996) 'Tourism and the Maori of Aotearoa (New Zealand)', in R.W. Butler and T. Hinch (eds) *Tourism and Indigenous Peoples*, London: International Thompson Business Press, pp. 155–70.
Hall, C.M. (2000) 'Tourism, national parks and Aboriginal people', in R.W. Butler and S.W. Boyd (eds) *Tourism and National Parks: issues and implications*, Chichester: Wiley, pp. 51–71.
Hall, C.M. (2000) 'Tourism and the Establishment of National Parks in Australia', in R.W. Butler and S.W. Boyd (eds) *Tourism and National Parks: Issues and Implications*, Chichester: Wiley, pp. 29–38.
Hall, C.M. (2002a) 'The changing cultural geography of the frontier: National parks and wilderness as frontier remnant', in S. Krakover and Y. Gradus (eds) *Tourism in Frontier Areas*, Lanham: Lexington Books, pp. 283–98.
Hall, C.M. (2002b) 'Tourism in capital cities', *Tourism: An International Interdisciplinary Journal*, 50(3): 235–248.
Hall, C.M. (2007) 'Changing geographies of Australia's wilderness heritage', in R. Jones and B.J. Shaw (eds) *Loving a Sunburned Country? Geographies of Australian Heritages*, Aldershot: Ashgate.
Hall, C.M. and Higham, J. (2000) 'Wilderness management in the forests of New Zealand: Historical development and contemporary issues in environmental

management', in X. Font and J. Tribe (eds) *Forest Tourism and Recreation: Case Studies in Environmental Management*, Wallingford: CABI, pp. 143–60.
Hall, C.M. and Page, S. (2006) *The Geography of Tourism and Recreation*, 3rd edn, London: Routledge.
Hall, C.M. and Piggin, R. (2002) 'Tourism business knowledge of World Heritage sites: A New Zealand case study', *International Journal of Tourism Research*, 4(5): 401–11.
Hall, C.M. and Shultis, J. (1991) 'Railways, Tourism and Worthless Lands: The Establishment of National Parks in Australia, Canada, New Zealand and the United States', *Australian Canadian Studies*, 8(2): 57–74.
Hall, C.M. and Tucker, H. (eds) (2004) *Tourism and postcolonialism: contested discourses, identities and representations*, London and New York: Routledge.
Hall-Martin, A. and Van der Merwe, J. (2003) 'Developing a national park system', in A. Hall-Martin and J. Carruthers (eds) *South African National Parks: A Celebration*, Johannesburg: Horst Klemm.
Hall-Martin, A., Braack, L. and Magome, H. (2003) 'Peace parks or transfrontier conservation areas', in A. Hall-Martin and J. Carruthers (eds) *South African National Parks: A Celebration*, Johannesburg: Horst Klemm.
Hamberg, A. (1896) 'Berättelse om en resa i Sarjekfjällen sommaren 1895', in *Svenska Turistföreningens årsskrift 1896*, Stockholm: Wahlström & Widstrand.
Hamilton, A. (1984) 'Spoon-Feeding the Lizards: Culture and Conflict in Central Australia', *Meanjin*, 43(3): 363–78.
Hampton, H. (1971) *How the US Cavalry saved our National Parks*, Bloomington: Indiana University Press.
Hampton, H. (1981) 'Opposition to national parks', *Journal of Forest History*, 25(1): 36–45.
Hanson, J.R. and D. Moore (1999) 'Applied anthropology at Devils Tower National Monument', *Plains Anthropologist* 44(170): 53–60.
Harkin, M. (2002) 'Towering conflicts: Bear Lodge/Devils Tower and the climbing moratorium,' *International Journal of Environmental, Cultural, Economic & Social Sustainability*, 2(3): 181–91.
Harmon, D. (2003) 'Intangible values of protected areas,' *Policy Matters* (IUCN), 12: 53–63.
Harmon, D. and Putney, A.D. (eds) (2003) *The Full Value of Parks: From Economics to the Intangible*, Lanham: Rowman & Littlefield.
Harney, B. (1963) *To Ayers Rock and Beyond*, Adelaide: Rigby.
Harris, C. (1974) *The National Parks and Reserves of South Australia*. Adelaide: Department of Geography, University of Adelaide.
Harris, W.W. (1974) 'Three parks: An analysis of the origins and evolution of the national parks movement', unpublished thesis, University of Canterbury, Christchurch.
Harroy, J-P. (1972) *World National Parks: Progress and Opportunities*, Brussels: Hayez.
Hart, E. (1983) *The Selling of Canada: The CPR and the Beginnings of Canadian Tourism*, Banff: Altitude Publishing.
Harvey. D. (1989) *The condition of postmodernity*, Oxford: Basil Blackwell.
Havard, W.L. (1934) 'The romance of Jenolan Caves', *Royal Australian Historical Society Journal and Proceedings*, 20(1): 18–65.
Hay-Edie, T. (2003) 'The cultural values of protected areas', in D. Harmon and

A.D. Putney (eds) *The Full Value of Parks: From Economics to the Intangible*, Lanham: Rowman & Littlefield, pp. 91–102.

Haynes, R. (1998) *Seeking the Centre: The Australian Desert in Literature, Art and Film*, Cambridge: Cambridge University Press.

Häyrynen, M. (1998) 'Visual symbolism and aesthetic constructions: National landscapes in the making of Finland', in P.D. Murphy (ed) *Literature of Nature: An International Sourcebook*, Chicago: Fitzroy Dearborn Publishers, pp. 415–22.

Hayward, P. (2004) 'Tourism, Contact and Cultural Commoditisation: A Case Study of Local Musical Entertainment on the Whitsunday Islands, Queensland Australia from the 1930s to 1990s', in D. Harrison (ed), *Pacific Islands Tourism*, New York: Cognizant, pp. 125–39.

Head, L. (2000) 'Identity, heritage and tourism', in *Cultural Landscapes and Environmental Change*, London: Hodder Headline, pp. 136–54.

Healy, R.G. (2006) 'The Commons Problem and Canada's Niagara Falls', *Annals of Tourism Research*, 33(2): 525–544.

Heberlein, T.A., Fredman, P. and Vuorio, T. (2002) 'Current tourism patterns in the Swedish mountain region', *Mountain Research and Development*, 22(2): 142–9.

Heckscher, E. (1907) *Till belysning av järnvägarnas betydelse för Sveriges ekonomiska utveckling*, Stockholm: Centraltryckeriet.

Hernández Pacheco, E. (1933) *La Comisaría de Parques Nacionales y la protección de la naturaleza en España*, Organismo Autónomo Parques Nacionales: Madrid, 2000 edn.

Higham, J. (1997) 'Wilderness Recreation Motivations Held by International Visitors to New Zealand', in W. Nuryanti (ed) *Tourism and Heritage Management*, Yogyakarta: Gadjah Mada University Press, pp. 327–39.

Hill, B. (1994) *The Rock: Travelling to Uluru*, Sydney: Allen & Unwin.

Hill, R. (2004) 'Governance, Indigenous peoples and community conservation', in *Global Trends in Protected Areas: A Report from the 5th World Parks Congress 2003*, Cairns: Rainforest CRC, pp. 19–26.

Hind, D. (2004) 'The Context of Tourism in the Lake District', in D. Hind and J. Mitchell (eds) *Sustainable Tourism in the English Lake District*, Sunderland: Business Education Publishers, pp. 49–74.

Hinds, W.T. (1979) 'The cesspool hypothesis versus natural areas for research in the United States', *Environmental Conservation*, 6(1): 13–20.

Hines, T.S. (2002) 'The Imperial Mall: The city beautiful movement and the Washington Plan of 1901–1902', in R. Longstreth (ed), *The Mall in Washington, 1971–1991*, Washington: Yale University Press, pp.79–100.

Hitchcock, M. (1993) 'Dragon tourism in Komodo, Eastern Indonesia', in M. Hitchcock, V.T. King and M. Parnwell (eds) *Tourism in South-East Asia*, London: Routledge, pp. 303–16.

Hitchcock, M. and Putra, D. (2008) 'Old tourists and new tourists: management challenges for Bali's tourism industry', in J. Cochrane (ed) *Asian Tourism: Growth and Change*, London: Elsevier, pp. 209–20.

Hobhouse, A. (Chairman) (1947) *Report of the National Park Committee (England and Wales)*, Cmnd 6628, London: HMSO.

Hobsbawm, E. (1994) *Age of Extremes: the short twentieth century 1914–1991*, London: Abacus.

Hoggart, K., Buller, H. and Black, R. (1995) *Rural Europe: Identity and Change*, London: Arnold.

Holden, P. (2007) 'Conservation and human rights – the case of the Khomani San (bushmen) and the Kgalagadi Transfrontier Park, South Africa,' *Policy Matters*, 15: 57–68.

Hollinsworth, D. (1998) 'Aboriginality as Cultural Marker in Post-colonial Australia', in D. Day (ed), *Australian Identities*, Melbourne: Australian Scholarly Publishing, pp. 186–97.

Honey, M. (1999) *Ecotourism and Sustainable Development: Who Owns Paradise?* Washington DC: Island.

Hooker, J.D. (1886) 'Hooker to Muir, March 19 1886', in R.H. Limbaugh and K.E. Lewis (eds) (1986) *Microfilm edition of the John Muir Papers*, Stockton, CA: University of Pacific, reel 19.

Hopley, D. (1989) *The Great Barrier Reef: Ecology and Management*, Melbourne: Longman Cheshire.

Horne, J. (2005) *The pursuit of wonder: how Australia's landscape was explored, nature discovered and tourism unleashed*, Melbourne: Miegunyah.

Hough, J. (1991) 'The Grand Canyon National Park and the Havasupai people: Cooperation and conflict', in P.C. West and S.R. Brechin (eds) *Resident Peoples and National Parks: Social Dilemmas and Strategies in International Conservation*, Tucson: University of Arizona Press, pp. 215–30.

Hovardas, T. and Poirazidis, K. (2007) 'Environmental policy beliefs of stakeholders in protected area management', *Environmental Management* 39(4): 515–25.

Howard, P. (2003) *Heritage: management, interpretation, identity*, London and New York: Continuum.

Hundloe, T., Neumann, R. and Halliburton, M. (1988) *Great Barrier Reef Tourism*, Brisbane: Institute of Applied Environmental Research, Griffith University.

Hunt, J.D. (1986) *Garden and Grove: The Italian Renaissance Garden in the English Imagination, 1600–1750*, Philadelphia: University of Pennsylvania Press.

Hutton, D. and Connors, L. (1999) *A History of the Australian Environment Movement*, Cambridge: Cambridge University Press.

Ifopo, F.M. (2007) 'World heritage on our doorstep,' *Te Karaka*, 35: 24–27.

Igoe, J.J. (2004) *Conservation and Globalization: A Study of National Parks and Indigenous Communities from East Africa to South Dakota*, Belmont: Thomson/Wadsworth.

Indian and Northern Affairs Canada (2006) Rich west coast heritage enhances park trails. Online. Available: www.ainc-inac.gc.ca/bc/fnbc/sucsty/suscom/ecoptr/rchwst_e.html/ (accessed 20 January 2008).

Indrianto, A. (2008) 'Interpreting the past: creating the Surabaya heritage trail', in J. Cochrane (ed) *Asian Tourism: Growth and Change*, London: Elsevier, pp. 357–68.

Ise, J. (1961) *Our National Park Policy*. Washington DC: Resources for the Future.

IUCN (1994) *Guidelines for Protected Area Management Categories*, Gland: IUCN.

IUCN (2004) *Participatory Conservation: Paradigm Shifts in International Policy*, Gland & Cambridge: IUCN.

IWGIA (1998) *Indigenous Peoples and Biodiversity Conservation in Latin America*, Copenhagen: IWGIA.

Jaireth, H. and D. Smyth. (2003) *Innovative Governance: Indigenous Peoples, Local Communities and Protected Areas*, New Delhi: Ane Books.

Jakarta Post (2004) 'Community-based national park to curb illegal logging in Sumatra', reprinted in *Indonesia Nature Conservation Bulletin*, 7(20a).

James, S. (2007) 'Constructing the Climb: Visitor decision making at Uluṟu', *Geographical Research*, 45(4): 398–407.
Jenkins. O. (2003) 'Photography and travel brochures: The circle of representation', *Tourism Geographies* 5(3), 305–28.
Jepson, P. and Whittaker, R.J. (2002) 'Histories of protected areas: internationalisation of conservation values and their adoption in the Netherlands Indies (Indonesia)', *Environment and History*, 8(2): 129–72.
Johannisson, K. (1984) 'Det sköna i det vilda: en aspekt på naturen som mänsklig resurs', in T. Frängsmyr (ed) *Paradiset och vildmarken: studier kring synen på naturen och naturresurserna*, Stockholm: LiberFörlag.
Johns, E., Kornhauser, E.M. and Sayers, A. (1998) *New Worlds from Old: 19th Century Australian and American Landscapes*, Canberra: National Gallery of Australia.
Johnson, R.U. (1905) 'Personal impressions of John Muir', *Outlook*, 80: 303–4.
Johnston, A.M. (2006) *Is the Sacred for Sale? Tourism and Indigenous Peoples*, London: Earthscan.
Johnston, J. (2006) 'Cooperative management with Aboriginal people in Canada's national parks', in M. Lockwood, G. Worboys and A. Kothari (eds) *Managing Protected Areas: A Global Guide*, London: Earthscan.
Jokinen. E. and McKie. D. (1997) 'The Disorientated Tourist: The Figuration of the Tourist in Contemporary Cultural Critique', in C. Rojek and J. Urry (eds) *Touring Cultures*, London: Routledge
Jones, B.T.B. (1999) 'Policy Lessons from the Evolution of a Community-Based Approach to Wildlife Management, Kunene Region, Namibia', *Journal of International Development*, 11: 295–304.
Journal of Forest History (1984) 'Readers respond to "Worthless Lands" forum', *Journal of Forest History*, 28(1).
Kächele. H. and Dabbert. S. (2002) 'An economic approach for a better understanding of conflicts between farmers and nature conservationists: an application of the decision support system MODAM to the Lower Odra Valley National Park', *Agricultural Systems* 74(2): 241–55.
Kajala, L., Almik, A., Dahl, R., Diksaite, L, Erkkonen, J., Fredman, P., Jensen, F. Søndergaard, Karoles, K., Sievänen, T., Skov-Petersen, H., Vistad, O. and Wallsten, P. (2007) *Visitor Monitoring in Nature Areas: A manual based on experiences from the Nordic and Baltic countries*. Copenhagen: Nordic Council of Ministers.
Kaliandra Sejati Foundation (2007) Project proposal submitted (successfully) to the Tourism & Biodiversity Fund of the Netherlands Committee of IUCN.
Kalland, A and Asquith, P.J. (1997) 'Japanese perceptions of nature: ideals and illusions', in P.J. Asquith and A. Kalland (eds) *Japanese images of nature: cultural perspectives*, London: Curzon, pp. 1–35.
Kaltenborn, B.P. (1994) 'Recreational use of Jotunheimen national park: some implications for management and planning', *Norsk Geografisk Tidsskrift – Norwegian Journal of Geography*, 48(4): 137–49.
Kaltenborn, B.P., Vistad, O.I. and Stanaitis. S. (2002) 'National parks in Lithuania: old environment in a new democracy', *Norsk Geografisk Tidsskrift – Norwegian Journal of Geography*, 56(1): 32–40.
Kathirithamby-Wells, J. (2005) *Nature and nation: forests and development in Peninsular Malaysia*, Singapore: Singapore University Press.
Keller, R. and Turek, M. (1998) *American Indians and National Parks*, Tucson: University of Arizona Press.

Kellert, S.R. (1995) 'Concepts of nature east and west', in M.E. Soulé and G. Lease (eds) *Reinventing Nature? Responses to Postmodern Deconstruction*, Washington D.C.: Island.

Kelly, P. (2001) *100 Years: The Australian Story*, Sydney: Allen & Unwin.

Kelly, G. and Doherty, M. (2003) 'Healthy, wealthy and wise: A systemic approach to environmental and social health', *Environmental Health* 3(2): 11–25.

Kemf, E. (1993) 'In search of a home: People living in or near protected areas', in E. Kemf (ed) *Indigenous Peoples and Protected Areas: The Law of Mother Earth*, London: Earthscan, pp. 3–11.

Kimes, W.F. and M.B. (1986) *John Muir: A Reading Bibliography*, Fresno: Panorama West.

King, B.E.M. (1997) *Creating Island Resorts*, London: Routledge.

Kleymeyer, C.D. (1994) 'Cultural traditions and community-based conservation', in D. Western and R.M. Wright (eds) *Natural Connections: Perspectives in Community-based Conservation*, Washington DC: Island, pp. 323–46.

Kline, M.B. (1970) *Beyond the Land Itself: Views of Nature in Canada and the United States*, Cambridge, MA: Harvard University Press.

Knobel, R. (1979) 'The economic and cultural values of South African National Parks', in I. Player (ed), *Voices of the Wilderness*, Johannesburg: Jonathan Ball.

Knudsen, D.C. and Greer, C.E. (2008) 'Heritage tourism, heritage landscapes and wilderness preservation; the case of National Park Thy', *Journal of Heritage Tourism*, 3(1): 18–35.

Kompas (2003) 'TN Wasur membuat masyarakat Merauke miskin (Wasur NP is impoverishing the people of Merauke)', *Indonesian Nature Conservation Bulletin* 6 (39b).

Kothari, A. (2006a) 'Collaboratively managed protected areas', in M. Lockwood, G. Worboys and A. Kothari (eds) *Managing Protected Areas: A Global Guide*, London: Earthscan, pp. 528–48.

Kothari, A. (2006b) 'Community conserved areas', in M. Lockwood, G. Worboys and A. Kothari (eds) *Managing Protected Areas: A Global Guide*, London: Earthscan, pp. 549–72.

Kress. G. and Van. Leeuwen. T. (1996) *Reading Images: The Grammar of Visual Design*, London: Routledge.

LaLande, J. (2003) 'The "Forest Ranger" in Popular Fiction', *Forest History Today* (Spring/Fall): 2–28.

Langford, N.P. (1905) *The discovery of Yellowstone Park: journal of the Washburn Expedition to the Yellowstone and Firehole Rivers in the year 1870*, Lincoln: University of Nebraska Press, reprinted 1972.

Lappalainen, M. (2000) *Finland's National Parks: Seas of Blue, Seas of Green*, Vantaa: Metsähallitus.

Larsen, P.B. (2000) *Co-managing Protected Areas with Indigenous Peoples: A Global Overview for IUCN/WCPA and WWF*, WWF International.

Larsen, P.B. and Oviedo, G. (2005) 'Protected areas and indigenous peoples: The Durban contributions to reconciliation and equity', in J.A. McNeely (ed) *Friends for Life: New Partners in Support of Protected Areas*, Gland & Cambridge: IUCN, pp. 113–28.

Lasdun, S. (1992) *The English Park: Royal, Private, and Public*, New York: Vendome.

Lash, S. and Urry, J. (1994) *Economies of Signs and Space*, London: Sage.

Lasimbang, J. (2004) 'National parks: Indigenous resource management principles in protected areas and indigenous peoples of Asia', *Cultural Survival Quarterly*, 28(1).
Lawrence, D., Kenchington, R. and Woodley, S. (2002) *The Great Barrier Reef: Finding the Right Balance*, Melbourne: Melbourne University Press.
Layton, R. and Titchen, S. (1995) 'Uluṟu: An outstanding Australian Aboriginal cultural landscape', in B. von Droste, H. Plachter and M. Rossler (eds), *Cultural Landscapes of Universal Value*, Stuttgart and New York: Gustav Fischer Verlag Jena with UNESCO.
LDNPA (Lake District National Park Authority) (2000) *Education Service: Footpath Erosion Factsheet*, Kendal: LDNPA.
LDNPA (2004a) *Lake District National Park Management Plan*, Kendal: LDNPA.
LDNPA (2004b) *A Social and Economic Profile of the Lake District National Park*, Kendal: LDNPA.
LDNPA (2005) *Promoting Sustainable Tourism*, Kendal: LDNPA.
LDNPA (2006) *The State of Tourism*, Kendal: LDNPA.
LeConte, J. (1886) LeConte to Steel, January 5, 1886, Steel Letters, Box 1, Item 210, Museum Collection, Crater Lake National Park.
Lee, E. (2000) 'Cultural connections to the land – a Canadian example,' *Parks*, 10(2): 3–12.
Lehtinen, A.A. (2006) *Postcolonialism, Multitude, and the Politics of Nature: On the Changing Geographies of the European North*, Oxford: Oxford University Press.
Lennon, J. (2006) 'Cultural heritage management', in M. Lockwood, G. Worboys and A. Kothari (eds) *Managing Protected Areas: A Global Guide, London:* Earthscan, pp. 448–73.
Libby, R. (2000) Yellowstone, Edens and Local Place. Public Lecture, Institute for Advanced Studies, University of Western Australia, www.ias.uwa.edu.au/ Previous_Conferences_and_Symposia/programs_2000/papers_from_the_july-_workshop/libby_robin.
Light, D. (2007) 'Dracula tourism in Romania: cultural identity and the state', *Annals of Tourism Research*, 34(3): 746–65.
Liljekvist, F. (1906) 'Turisthärbärget i Abisko', in *Svenska Turistföreningens årsskrift 1906*, Stockholm: Wahlström & Widstrand.
Lindenmayer, D. and Burgman, M. (2005) *Practical Conservation Biology*, Melbourne: CSIRO.
Lindenmayer, D.B. and Fischer, J. (2006) *Habitat Fragmentation and Landscape Change: An Ecological and Conservation Synthesis*, Washington D.C.: Island.
Lindhagen, A. (1996) *Forest Recreation in Sweden. Four Case Studies Using Quantitative and Qualitative Methods.* Sveriges lantbruksuniversitet, Institutionen för skoglig landskapsvård, rapport 64.
Linge, G. (2000) 'Ensuring the full freedom of religion on public lands: Devils Tower and the protection of Indian sacred sites,' *Boston College Environmental Affairs Law Review*, 27: 307–39.
Lloyd, G. and Illsley, B. (2002) 'National parks in Scotland: Balancing environment and economy', *European Planning Studies*, 10(5): 665–70.
Lloyd, G., McCarthy, J. and Illsley, B. (2004) 'Commercial and industrial developments in national parks in England and Wales: Lessons for the Scottish agenda', *Journal of Environmental Policy & Planning*, 6(3 and 4): 289–304.
Lockwood, M. (2006) 'Values and benefits', in M. Lockwood, G.L. Worboys and

A. Kothari (eds) *Managing Protected Areas: A Global Guide*, London: Earthscan, pp. 101–15.
Lockwood, M. and Kothari, A. (2006) 'Social context', in M. Lockwood, G. Worboys and A. Kothari (eds) *Managing Protected Areas: A Global Guide*, London: Earthscan, pp. 41–72.
Lockwood, M., Worboys, G. and Kothari, A. (eds) (2006) *Managing Protected Areas: A Global Guide*, London: Earthscan.
Longstreth, R. (2002) 'Introduction: Change and continuity on the Mall, 1791 – 1991', in R. Longstreth (ed), *The Mall in Washington, 1971–1991*, Washington: Yale University Press, pp.11–18.
Lothian, W.F. (1977) *A History of Canada's National Parks*, Ottawa: Parks Canada.
Loukaki, A. (1997) 'Whose Genius Loci?: Contrasting Interpretations of the Sacred Rock of the Athenian Acropolis', *Annals of the Association of American Geographers* 87(2), 306–29.
Lövgren, O. (1992) 'Varför är det så lätt att älska naturen? Om glappet mellan ideal och vardag', in L. Lundgren (ed) *Livsstil och miljö*, Lund.
Lowenthal, D. (1998) *The heritage crusade and the spoils of history*, Cambridge: Cambridge University Press.
Lowry, W.R. (1998) *Preserving Public Lands for the Future: The Politics of Intergenerational Goods*, Washington D.C.: Georgetown University Press.
Lubetkin, M.J. (2006) *Jay Cooke's Gamble: the Northern Pacific Railroad, the Sioux and the Panic of 1873*, Norman: University of Oklahoma Press.
Lucas, P.H.C. (1970) *Conserving New Zealand's Heritage: Report on a Study Tour of National Park and Allied Areas in Canada and the United States*, Wellington: Government Printer.
Luria, S. (2005) *Capital Speculations*, Lebanon, NH: University of New Hampshire Press.
MacCannell, D. (1976) *The Tourist: A New Theory of the Leisure Class*, New York: Schocken.
MacEwan, M. and MacEwan, A. (1982) *National Parks: Conservation or Cosmetics?*, London: Allen & Unwin.
MacEwan, M. and MacEwan, A. (1987) *Greenprints for the Countryside? The Story of Britain's National Parks*, London: Allen & Unwin.
Machlis, G.E. and Field, D.R. (2000) *National Parks and Rural Development. Practice and Policy in the United States*, Washington, D.C.: Island.
MacKay, F. (2002) *Addressing Past Wrongs – Indigenous Peoples and Protected Areas: The Right to Restitution of Lands and Resources*, UK: Forest Peoples Programme. Online. www.forestpeoples.org/publications/addressing_past_wrongs_eng.shtml (accessed 29 November 2005).
MacKay, F. (2007) 'Indigenous peoples, protected areas and the right to restitution – the jurisprudence of the Inter-American Court of Human Rights,' *Policy Matters*, 15: 209–22.
MacKay, F. and Caruso, E. (2004) 'Indigenous lands or national parks?', *Cultural Survival Quarterly* 28(1).
MacKenzie, J.M. (1988) *The Empire of Nature: hunting, conservation and British imperialism*, Manchester: Manchester University Press.
Magome, D.T. and Collinson, R.F.H. (1998) 'From protest to pride: A case study of Pilanesberg National Park, South Africa', World Bank/WBI's CBNRM Initiative. http://srdis.ciesin.columbia.edu/cases/south_africa-003.html.

Magome, M. (2003) 'Preface', in Hall-Martin, A. and Carruthers, J. (eds) *South African National Parks: A Celebration*, Johannesburg: Horst Klemm.

Magome, H. and Murombedzi, J. (2003) 'Sharing South African National Parks: Community land and conservation in a democratic South Africa', in W.M. Adams and M. Mulligan (eds) *Decolonizing Nature: Strategies for Conservation in a Postcolonial Era*, London: Earthscan, pp. 108–34.

Mair, J. and Delafons, J. (2001) 'The policy origins of Britain's National Parks: the Addison Committee 1929–1931', *Planning Perspectives*, 16: 293–309.

The Management Committee of Wuyi Resort (1995) Master Plan of Wuyishan Tourism Resorts.

The Management Committee of Wuyi Resort (2006) Summary of the achievements of Wuyishan National Resorts, 2006 (internal document).

The Management Committee of Wuyishan Scenic Park (2001) Report to the Hujian Provincial Planning Bureau on the progress of development of Wuyishan Scenic Park.

The Management Committee of Wuyishan Scenic Park (2004) The outline of the Eleventh Year Plan of Wuyishan Scenic Park.

Manuera, E., Te Heuheu, T. and Prime, K. (1992) 'The conservation estate, the tangata whenua', in J. Birckhead, T. de Lacy and L. Smith (eds) *Aboriginal Involvement in Parks and Protected Areas*, Canberra: Aboriginal Studies Press, pp. 327–37.

Marcus, J. (1997) 'The Journey Out to the Centre: The Cultural Appropriation of Ayers Rock', in G. Cowlishaw and B. Morris (eds) *Race Matters: Indigenous Australians and 'Our' Society*, Canberra: Aboriginal Studies Press.

Marius (pseud) (1897) *Järnvägen Gellivare-Ofoten och de militära intressena*. Stockholm: Central-tryckeriet.

Mark, A.F. (2001) 'Symposium: Managing protected natural areas for conservation, ecotourism, and indigenous peoples,' *Journal of the Royal Society of New Zealand*, 31(4): 811–12.

Marsh, G.P. (1864) *Man and Nature; or, Physical Geography as Modified by Human Action*, Cambridge, MA: Belknap Press of Harvard University Press, 1965 edn.

Marsh, J.G. (1983) 'Canada's national parks and tourism: A problematic relationship', in P.E. Murphy (ed) *Tourism in Canada: Selected Issues and Options*. Victoria: Department of Geography, University of Victoria.

Marsh, J.G. (1985) 'The Rocky and Selkirk Mountains and the Swiss Connection 1885–1914', *Annals of Tourism Research* 12: 417–433.

Marsh, J.S. (1982) 'The evolution of recreation in Glacier National Park, 1880 to the present', in G. Wall and J.S. Marsh (eds) *Recreational Land Use: Perceptions on its evolution in Canada*, Ottawa: Carleton University Press.

Marshall, J. and Walton, J. (1981) *The Lake Counties from 1830 to the Mid-20th Century*, Manchester: Manchester University Press, pp.62–76.

Marty, S. (1984) *A Grand and Fabulous Notion: The First Century of Canada's Parks*, Toronto: New Canada Publications.

Mascarenhas, M. and Scarce, R. (2004) 'The intention was good: legitimacy, consensus-based decision-making, and the case of forest planning in British Columbia, Canada', *Society and Natural Resources* 17: 17–38.

Mason, V. (2007) President of Bali Bird Club and organiser of bird-walks, personal communication, Bali, August 14.

Mbenga, E. (2003) 'A study of visitor wildlife viewing preferences and experiences in Madikwe Game Reserve, South Africa', unpublished thesis, University of Natal.

McAllister, E. (1894) 'Report of the Board of Directors', *Sierra Club Bulletin* 1(4).
McAvoy, L. (2002) 'American Indians, place meanings and the old/new west', *Journal of Leisure Research*, 34(4): 383–96.
McAvoy, L., McDonald, D. and Carlson, M. (2003) 'American Indian/First Nation place attachment to park lands: The case of the Nuu-chah-nulth of British Columbia', *Journal of Park and Recreation Administration*, 21(2): 84–104.
McCarthy, J., Lloyd, G. and Illsley, B. (2002) 'National parks in Scotland: Balancing environment and economy', *European Planning Studies*, 10(5): 665–70.
McCarthy, M.A. and Lindenmayer, D.B. (1999) 'Conservation of the greater glider (Petauroides volans) in remnant native vegetation within exotic plantation forest', *Animal Conservation* 2: 203–9.
McCormick, J. (1992) *The Global Environmental Movement*, London: Belhaven.
McCrone, D., Morris, A. and Kiely, R. (1995) *Scotland – the brand: the making of Scottish heritage*, Edinburgh: Edinburgh University Press.
McDonald, G. (1987) 'Land Use Planning at the Economic Periphery', unpublished paper presented at Geography and Public Policy, the Institute of Australian Geographers 22nd Conference, Australian Defence Force Academy, Canberra.
McFarlane, C. (2006) 'Knowledge, Learning and Development: a post-rationalist approach', *Progress in Development Studies* 6(4): 287–305.
McGrath, A. (1991) 'Travels to a Distant Past: The Mythology of the Outback', *Australian Cultural History: Travellers, Journeys, Tourists*, 10: 113–25.
McIntyre, N., Jenkins, J. and Booth, K. (2001) 'Global influences on access: The changing face of access to public lands in New Zealand,' *Journal of Sustainable Tourism*, 9(5): 434–50.
McKercher, B. and du Cros, H. (1998) 'I Climbed to the Top of Ayers Rock but Still Couldn't See Uluṟu: The Challenge of Reinventing a Tourist Destination', in B. Faulkner, C. Tidswell and D. Weaver (eds) *Proceedings of the 8th Annual Australian Tourism and Hospitality Conference*, Canberra: Bureau of Tourism Research.
McLean, F. (1998) 'Museums and the construction of national identity: a review', *International Journal of Heritage Studies*, 3: 244–52.
McLean J. and Stræde S. (2003) 'Conservation, relocation, and the paradigms of park and people management – A case study of Padampur Villages and the Royal Chitwan National Park, Nepal', *Society and Natural Resources*, 16(6): 509–26.
McVetty, D. and Deakin, M. (2007) 'Optimising the outcomes of tourism in co-managed protected heritage areas: The cases of Aulavik National Park and Gwaii Haanas National Park Reserve/Haida Heritage site', George Wright Society Biennial Conference, Online. Available online: http://nsgi.uri.edu/washu/washuw99003/18-McVetty_and_Deakin.pdf (accessed 20 January 2008).
Millenium Ecosystem Assessment (2003) *Ecosystems and human well-being: A framework for assessment*, Washington DC: Island Press.
Medford Mail (1895) September 27, 1895, in Steel Scrapbook 2:2, Museum Collection, Crater Lake National Park.
Meinig, D.W. (1979) 'Symbolic landscapes', in D.W. Meinig (ed) *The interpretation of ordinary landscapes: geographical essays*, New York: Oxford University Press, pp. 164–92.
Mels, T. (1999) *Wild Landscapes. The Cultural Nature of Swedish National Parks*, Lund: Department of Social and Economic Geography, Lund University.
Mendel, L.C. (2002) 'The consequences for wilderness conservation in the

development of the national park system in Tasmania, Australia', *Australian Geographical Studies* 40(1): 71–83.
Mendel, L.C. and Kirkpatrick, J.B. (2002) 'Historical progress of biodiversity conservation in the protected-area system of Tasmania, Australia', *Conservation Biology*, 16(6): 1520–9.
Mendoza, J.G. (1998) 'The persistence of romantic ideas and the origins of Natural Park policy in Spain', *Finisterra* XXXIII, 65: 51–63.
Mercer, D. (1998) 'The uneasy relationship between tourism and native peoples: The Australian experience', in W.F. Theobald (ed) *Global Tourism*, 2nd edn, London: Butterworth Heinemann, pp. 98–128.
Mercer, D. and Petersen, J. (1986) 'The revocation of national parks and equivalent reserves', *Search*, 17(5–6): 134–40.
Meredith, P. (1999) *Myles and Milo*, Sydney: Allen & Unwin.
Meyrowitz. J. (1992) *No Sense of Place*, New York: Routledge.
Middleton, B. (2003) 'Ecology and objective-based management: Case study of the Keoladeo National Park, Bharatpur, Rajasthan', in V.K. Saberwal and M. Rangajan (eds) *Battles Over Nature: Science and the Politics of Conservation*, Delhi: Permanent Black, 86–116.
Miles, J.C. (1995) *Guardians of the Parks: A History of the National Parks and Conservation Association*, Washington D.C.: Taylor and Francis.
Miller, P. (1967) *Nature's Nation*, Cambridge, MA: Harvard University Press.
Miller, L.R., Dickenson, J.E. and Pearlman-Houghie, D.J. (2001) 'Quiet enjoyment in the National Parks of England and Wales: public understanding of the term and its influence on attitudes towards recreational activities', *Leisure Studies*, 20: 19–40.
Minchin, L. (2007) 'Reef Facing Extinction', Melbourne: *The Age*, 30 January, p. 1.
Mitchell, B. (2007) 'Private protected areas', unpublished paper presented at a summit on the IUCN categories, Andalusia, Spain, 7–11 May 2007.
Mitchell, N., Slaiby, B. and Benedict, M. (2002) 'Local community leadership: Building partnerships for conservation in North America,' *Parks*, 12(2): 55–66.
Moosa, M.V. and Morobe, M. (2003), in Hall-Martin, A. and Carruthers, J. (eds) *South African National Parks: A Celebration*, Johannesburg: Horst Klemm.
Moran, D. Tresidder, E. and McVittie, A. (2006) 'Estimating the recreational value of mountain biking sites in Scotland using count data models', *Tourism Economics*, 12(1) 123–35.
Morgan, N.J. and Pritchard, A. (1999) *Tourism promotion and power: creating images, creating identities*, Chichester: Wiley.
Morley. D. (2001) 'Belongings: Place, space and identity in a mediated world', *Cultural Studies*, 4(4): 425–48.
Morrison, J. (1993) *Protected Areas and Aboriginal Interests in Canada – Discussion Paper*, Toronto: World Wildlife Fund Canada.
Morrison, J. (1997) 'Protected areas, conservationists and Aboriginal interests in Canada', in K. Ghimire and P.M. Pimbert (eds) *Social Change and Conservation: Environmental Politics and Impacts of National Parks and Protected Areas*, London: UNRISD and Earthscan, pp. 270–96.
Morse, J., King, J. and Bartlett, B. (2005) *Walking to the Future . . . Together: A Shared Vision for Tourism in Kakadu National Park*, Canberra: Department of the Environment and Heritage. Online. Available: www.deh.gov.au/parks/publications/kakadu/tourism-vision/index.html (accessed 20 January 2008).

Mose, I. (2007) 'Foreword', in Mose, I. (ed) *Protected Areas and Regional Development in Europe: Toward a New Model for the 21st Century*, Aldershot: Ashgate, xv–xvii.

Mose, I. and Wixlbaumer, N. (2007) 'A new paradigm for protected areas in Europe?' in Mose, I. (ed) *Protected Areas and Regional Development in Europe: Toward a New Model for the 21st Century*, Aldershot: Ashgate, 3–20.

Mosley. J.G. (1963) 'Aspects of the Geography of Recreation in Tasmania', unpublished Ph.D. Thesis, Australian National University.

Mosley, J.G. (1978) 'A history of the wilderness reserve idea in Australia', in J.G. Mosley (ed) *Australia's Wilderness: Conservation Progress and Plans, Proceedings of the First National Wilderness Conference*, Melbourne: Australian Conservation Foundation, 27–33.

Mowforth, M. and Munt, I. (2003) *Tourism and Sustainability: Development and New Tourism in the Third World*, London: Routledge.

Muir, J. (1890a) 'The Treasures of Yosemite', *Century* 40 (August), 483–500.

Muir, J. (1890b) 'Features of the Proposed Yosemite National Park', *Century* 40 (September): 656–67.

Muir, J. (1892) Muir to Steel, October 2, 1892, Steel Library, Box 1, Item 164.v. Museum Collection, Crater Lake National Park.

Muir, J. (1894) *The Mountains of California*, Boston: Houghton-Mifflin.

Muir, J. (1897) 'The National Parks and Forest Reservations', *Harpers Weekly* 16(2111) (June 5): 566.

Muir, J. (1901a) 'Hunting Big Redwoods', *Atlantic Monthly* 88 (September), 304–320.

Muir, J. (1901b) *Our National Parks*, Boston: Houghton-Mifflin.

Muir, J. (1902) Muir to Steel, February 19, 1902, Steel Scrapbook 22(2) p. 46, Museum Collection, Crater Lake National Park.

Muir, J. (1910) 'The Hetch-Hetchy Valley: A national question', *American Forestry* 16(5): 263–69.

Muir, J. (1914) *The Yosemite*, Boston: Houghton Mifflin.

Muir, J. *et al.* (1905) 'Statement concerning the proposed recession of Yosemite Valley and Mariposa Big Tree Grove by the State of California to the United States', *Sierra Club Bulletin* 5(3): 242–50.

Mulongoy, K.J. and Chape, S. (2004) *Protected Areas and Biodiversity: An Overview of Key Issues*, Nairobi: UNEP World Conservation Monitoring Centre.

Mulvaney, K. (1999) 'Management strategies and the component of indigenous sacred places,' *The George Wright Forum*, 16(4): 37–49.

Mundraby, D. (2005) 'Does nature conservation preserve the rights of indigenous rainforest people?' *Message Stick: The Newsletter of the North Queensland Land Council*, 6–7.

Murphy, P.E. (1985) *Tourism: A Community Approach*, New York: Methuen.

Murtha, M. (1996) 'British Columbia parks' partnership with Aboriginal People', *Trends* 33 (4): 40–5.

Mylne, L. (2005) *Frommer's Portable Australia's Great Barrier Reef*, 3rd edn, Hoboken: Wiley.

Nance, C. (1986) 'Perceptions of the natural environment', in C. Nance and D.L. Speight (eds) *A land transformed: environmental change in South Australia*, Melbourne: Longman Cheshire, 200–25.

NASA (2008a) Phoenix makes first trench in Science Preserve, NASA News Release, June 17, www.jpl.nasa.gov/news/news.cfm?release=2008-111c.

NASA (2008b) NASA's Phoenix spacecraft commanded to unstow arm. Mission

News May 29, www.nasa.gov/mission_pages/phoenix/news/phoenix-20080528.html.
Nash, R. (1967) *Wilderness and the American Mind*, New Haven: Yale University Press.
Nash, R. (1969) 'Wilderness and man in North America', in J.G. Nelson and R.C. Scace (eds) *The Canadian national parks: today and tomorrow*, Vol 1, Calgary: Dept of Geography, University of Calgary, 66–93.
Nash, R. (1970) 'The American invention of National Parks', *American Quarterly*, 22(3): 726–35.
National Coalition to Save Our Mall, Inc. (2008) 'A monument to democracy: History of the Mall'. Online. Available: www.savethemall.org/mall/resource-hist01.html (accessed 21 January 2008).
National Development Planning Agency (1993) *Biodiversity Action Plan for Indonesia*, Jakarta: Ministry of National Development Planning.
National Park Service (NPS) (1990) 'National Park Service: The First 75 Years'. Online. Available: www.nps.gov/history/history/online_books/sontag/sontagt.htm (accessed 17 March 2008).
National Park Service (NPS) (2007a) *A History of the National Mall and Pennsylvania Avenue National Historic Park*, Washington: National Park Service.
National Park Service (NPS) (2007b) 'Enriching your American experience: The National Mall Plan'. Online. Available: www.nps.gov/nationalmallplan/ (accessed 17 December 2007).
National Park Service (NPS) (2007c) 'Making choices for the future of the National Mall and Pennsylvania Avenue National Historic Park'. Online. Available: www.nps.gov/nationalmallplan/Newsletters.html (accessed 17 December 2007).
National Park Service (NPS) (2007d) 'National Mall & Memorial Parks: District of Columbia'. Online. Available: www.nps.gov/nama/ (accessed 17 December 2007).
National Park Service (NPS) (2008) 'Washington Monument'. Online. Available: www.4uth.gov.ua/usa/english/travel/npsname/index373.htm (accessed 10 January 2008).
Nature Conservancy (1996) *Traditional Peoples and Biodiversity Conservation in Large Tropical Landscapes*, The Nature Conservancy.
Naturvårdsverket (1989) *Nationalparksplan för Sverige*, Stockholm: Swedish Environmental Protection Agency.
Naturvårdsverket (2004) *Protect, Preserve, Present: A Programme for Better Use and Management of Protected Areas, 2005–2015*, Swedish Environmental Protection Agency Report 5483. Stockholm: Swedish Environmental Protection Agency.
Naturvårdsverket (2007) *Nationalparksplan för Sverige. Utkast och remissversion*, Stockholm: Swedish Environmental Protection Agency.
Neave, S. (1991) *Medieval Parks of East Yorkshire*, Hull: University of Hull and the Hutton Press.
Negi, C.S. and Nautiyal, S. (2003) 'Indigenous peoples, biological diversity and protected area management – policy framework towards resolving conflicts', *International Journal of Sustainable Development and World Ecology*, 10(2): 169–79.
Nelson, J.G. (1973) 'Canada's national parks: Past, present and future', *Canadian Geographical Journal*, 86: 69–89.
Nelson, J.G. (1982) 'Canada's national parks, Past, present and future', in G. Wall and

J.S. Marsh (eds) *Recreational Land Use: Perceptions on its evolution in Canada*, Ottawa: Carleton University Press, pp. 41–61.

Nelson, J.G. (2000) 'Tourism and national parks in North America: An overview' in R.W. Butler and S.W. Boyd (eds) *Tourism and National Parks: Issues and Implications*, Chichester: Wiley, pp. 303–12.

Nelson, J. and Hossack, L. (eds) (2003) *Indigenous Peoples and Protected Areas in Africa: From Principles to Practice*, Moreton-in Marsh: Forest Peoples Programme.

NEPA (National Environmental Protection Agency) (2004), A summary table of Chinese Natural Reserves. Online. Available: www.zhb.gov.cn/natru/zyb/zrbhq/200511/t20051130_72109.htm.

Neufeld, D. (2002) 'The commemoration of northern Aboriginal peoples by the Canadian government', *The George Wright Forum*, 19(3): 22–33.

Neufeld, D. (2005) 'Writing our histories into the land: First Nations initiatives', in R. Cooley, K. MacKay and M. Peniuk (eds) *Parks and Protected Areas: Dynamic Landscape or Museum?* Winnipeg: University of Manitoba Press, pp. 10–11.

Neumann, R.P. (1995) 'Ways of seeing Africa: Colonial recasting of African society and landscape in Serengeti National Park', *Ecumene* 2(2): 149–69.

Neumann, R.P. (1996) 'Dukes, earls, and ersatz Edens: aristocratic nature preservationists in colonial Africa', *Environment and Planning D: Society and Space*, 14: 79–98.

Neumann, R.P. (1998) *Struggles over Livelihood and Nature Preservation in Africa*, San Francisco: University of California Press.

New Zealand 1887, *Parliamentary Debates, 1887*, 57: 399.

New Zealand 1894, *Parliamentary Debates, 1894*, 86: 579.

Nicholls, G. (2000) 'Risk and adventure education', *Journal of Risk Research*, 3(2): 121–134.

Nicholson, N. (1963) *Portrait of the Lakes*, London: Robert Hale.

Nicol, J.I. (1969) 'The national parks movement in Canada', in J.G. Nelson and R.C. Scace (eds) *The Canadian National Parks: Today and Tomorrow*, Vol.1, Calgary: Department of Geography, University of Calgary.

Nichol, J.I. (1970) 'The national parks movement in Canada', in J.G. Nelson (ed) *Canadian Parks in Perspective*, Montreal: Harvest House, pp. 19–34.

Nicolson, M.H. (1959) *Mountain Gloom and Mountain Glory: The Development of the Aesthetics of the Infinite*, Ithaca: Cornell University Press.

Nilsson, P.Å. (1999) Fjällturismens historia: En studie av utvecklingen i Åredalen. Inst. f. Turismvetenskap Rapport No. 1999:1, Östersund: Mitthögskolan,.

Nilsson Dahlström, Å. (2003) *Negotiating Wilderness in a Cultural Landscape: Predators and Saami Reindeer Herding in the Laponian World Heritage Area*, Acta Universitatis Upsaliensis, Uppsala Studies in Cultural Anthropology no. 32, Uppsala.

Northwest Regional Development Agency (NWDA) (2005) *Lake District Economic Futures: Policy Statement*, Warrington: NWDA.

Notzke, C. (1994) 'Native people and protected areas', in *Aboriginal Peoples and Natural Resources in Canada*, York: Captus University Press.

Notzke, C. (2006) *The Stranger, the Native and the Land: Perspectives on Indigenous Tourism*, York: Captus University Press.

Novak, B. (1980) *Nature and Culture: American Landscape Painting, 1825–1875*, New York: Oxford University Press.

Nyíri, P. (2006) *Scenic spots: Chinese tourism, the state, and cultural authority*, Seattle and London: University of Washington Press.
Ödmann, E., Bucht, E. and Nordström, M. (1982) *Vildmarken och välfärden*, Stockholm: Liber Förlag.
Oelschlaeger, M. (1991) *The Idea of Wilderness: From Prehistory to the Age of Ecology*, New York: Yale University Press.
Office of the Deputy Prime Minister (ODPM) (2004) *Planning Policy Statement 7: Sustainable Development in Rural Areas*, London: ODPM. Online. Available: (www.opdm.gov.uk/index.asp?id=1143823).
Olmsted, F.L. (1852) *Walks and talks of an American farmer in England*, Ann Arbor: University of Michigan Press, reprint no date.
Olmsted, F.L. (1858) 'Letter to Parke Godwin, 1 August 1858', in C.E. Beveridge and D. Schuyler (eds), *The Papers of Frederick Law Olmsted: Vol 3 Creating Central Park 1857–1861*, Baltimore: Johns Hopkins University Press.
Olmsted, F.L. (1865) 'Preliminary report upon the Yosemite and Big Tree Grove', in V.P. Ranney (ed), *The Papers of Frederick Law Olmsted: Vol. 5 The California Frontier 1863–1865*, Baltimore: Johns Hopkins University Press, 1990, pp. 488–516.
O'Neill, C. and Walton, J. (2004) 'Tourism and the Lake District: Social and Cultural Histories', in D. Hind and J. Mitchell (eds) *Sustainable Tourism in the English Lake District*, Sunderland: Business Education Publishers, pp. 19–47.
Oredsson, S. (1969) *Järnvägarna och det allmänna: Svensk järnvägspolitik fram till 1890*. Bibliotheca Historica Lundensis XXIV, Lund: Gleerups.
Oregon Department of Geology and Mineral Industries (2003) *Oregon's Geothermal Energy Potential in Spotlight*, Portland: Oregon Department of Geology and Mineral Industries.
Orsi, R.J. (1985) 'Wilderness saint and "robber baron": The anomalous partnership of John Muir and the Southern Pacific Company for preservation of Yosemite National Park', *Pacific Historian*, 29(2/3), 136–56.
Ortuño Medina, F. (1980) *Los Parques Nacionales de las Islas Canarias*, Madrid: MAP.
Orwell, G. (1950) *Shooting an Elephant and Other Essays*, London: Secker & Warburg.
Osborne, M.A. (1994) *Nature, the Exotic and the Science of French Colonialism*, Bloomington and Indianapolis: Indiana University Press.
Oviedo, G. and Brown, J. (1999) 'Building alliances with indigenous peoples to establish and manage protected areas', in S. Stolton and N. Dudley (eds) *Partnerships for Protection: New Challenges for Planning and Management of Protected Areas*, London: Earthscan, pp. 99–108.
Oviedo, G., Maffi, L. and Larsen, P.B. (2000) *Indigenous and Traditional Peoples of the World and Ecoregion Conservation: An Integrated Approach to Conserving the World's Biological and Cultural Diversity*, WWF and Terralingua.
Pabst, M. (2002) *Transfrontier Peace Parks in Southern Africa*, Stuttgart: SAFRI.
Pannell, S. (2006) *Reconciling Nature and Culture in a Global Context? Lessons from the World Heritage List*, Cairns: Rainforest CRC.
Parker, G. and Ravenscroft, N. (2000) 'Tourism, national parks and private lands', in R.W. Butler and S.W. Boyd (eds) *Tourism and National Parks: Issues and Implications*, Chichester: Wiley, pp. 95–106.
Parks Australia (2005) *Uluṟu-Kata Tjuṯa National Park: Visitor guide and maps*, Director of National Parks, Australia.

338 Bibliography

Parks Canada (1971) *National Parks System Manual*, Ottawa: Department of Indian and Northern Affairs, Ottawa.
Parks Canada (1985) *National Parks Managing Planning Process Manual*, Ottawa: Department of the Environment.
Parks Canada (1994) *Parks Canada – Guiding Principles and Operational Policies*, Ottawa: Ministry of Supply and Services Canada.
Parks Canada (1997) *Banff National Park Management Plan*, Ottawa: Government of Canada.
Parks Canada (2000) *Unimpaired for Future Generations? Protecting Ecological Integrity with Canada's National Parks*, 2 vols, Ottawa: Report of the Panel on the Ecological Integrity of Canada's National Parks.
Parks Canada (2002) Press Release: The Government of Canada announces action plan to protect Canada's Natural Heritage, October 3rd.
Parks Canada (2007a) 'An approach to Aboriginal cultural landscapes'. Online. Available: www.pc.gc.ca/docs/r/pca-acl/index_e.asp (accessed 20 January 2008).
Parks Canada (2007b) 'Gwaii Haanas National Park Reserve and Haida Heritage Site. Agreements'. Online. Available: www.pc.gc.ca/pn-np/gwaiihaanas/plan/plan2_E.asp (accessed 20 January 2008).
Parks Canada (2007c) 'Parks Canada Attendance 2002–03 to 2006–07', Online. Available: www.pc.gc.ca/docs/pc/attend/table1_e.asp.
Peckett, M.K. (1998) 'Narrowing the road: Co-management with Anishnabe at the Riding Mountain National Park (Winnipeg, Manitoba)', *7th annual conference of the International Association for the Study of Common Property*, Online. Available: http://dlc.dlib.indiana.edu/archive/0000000129/ (accessed 20 January 2008).
Penczer, P.R. (2007) *The Washington National Mall*, Arlington, VA: Oneonta Press.
Perry, A.H.T. (1929) *National and Other Parks*, Cape Town: n.p.
Pettigrew, C. and Lyons, M. (1979) 'Royal National Park – a history', *Parks and Wildlife*, 2(3–4): 15–30.
Pfister, R.E. (2000) 'Mountain culture as a tourism resource: Aboriginal views on the privileges of storytelling', in P.M. Godde, M.F. Price and F.M. Zimmermann (eds) *Tourism and Development in Mountain Regions*, Wallingford: CABI, pp. 115–36.
Phillips, A. (2003) 'Turning ideas on their head – the new paradigm for protected areas,' *The George Wright Forum*, 20(2): 8–32.
Phillips, A. (2004) 'The IUCN management categories: speaking a common language about protected areas'. Online. Available: www.chinabiodiversity.com/protected-area/iucn-categories2-en.htm (accessed 7 April 2008).
Phillips, G. 'Tourism figures', email, 10 April 2008.
Picard, M. (1996) *Bali: Cultural Tourism and Touristic Culture*, Singapore: Archipelago Press.
Picard, D. and Robinson, M. (2006) *Festivals, tourism and social change: remaking worlds*, Clevedon: Channel View.
Pigram, J.J. and Jenkins, J.M. (2006) *Outdoor Recreation Management*, 2nd edn, Oxford: Routledge.
Pinchot, G. (1900) *A Primer of Forestry, Part II-Practical Forestry*, USDA-Bureau of Forestry Bulletin No. 24, Washington, DC: Government Printing Office.
Pinchot, G. (1902a) Pinchot to Steel, February 18, 1902, SL, Box 2, Item 20A. Museum Collection, Crater Lake National Park.
Pinchot, G. (1902b) Pinchot to Steel, May 15, 1902, SL, Box 2, Item 20D. Museum Collection, Crater Lake National Park.

Pinchot, G. (1947) *Breaking New Ground*, New York: Harcourt Brace and Co.
Poirer, R.A. (2007) 'Ecotourism and indigenous rights in Australia,' *Peace Review: A Journal of Social Justice*, 19: 351–8.
Portland Oregonian (1892) December 28, 1892 in Steel Scrapbook 9:1, Mazamas Library, Portland.
Pouliquen-Young, O. (1997) 'Evolution of the system of protected areas in Western Australia', *Environmental Conservation*, 24: 168–81.
Powell, J. (1976) *Environmental Management in Australia 1788–1914*, Melbourne: Oxford University Press.
Power, T. (2002) 'Joint management at Uluṟu-Kata Tjuṯa National Park,' *Environment and Planning Law Journal*, 19(4): 284–301.
Prato, T. and Fagre, D. (2005) *National Parks and Protected Areas: Approaches for Balancing Social, Economic and Ecological Values*, Oxford: Blackwell.
Pressey, R.L. (1992) 'Nature conservation in rangelands: Lessons from research on reserve selection in New South Wales', *The Rangeland Journal* 14(2): 214–26.
Pressey, R.L. and Tully, S.L. (1994) 'The cost of ad hoc reservation: A case study in western New South Wales', *Austral Ecology* 19(4): 375–84.
Pretes, M. (2003) 'Tourism and nationalism', *Annals of Tourism Research*, 30(1): 125–42.
Price, M.F. and Willebrand, T. (2006) 'Editorial', *The International Journal of Biodiversity Science & Management*, 2(4): 303–4.
Prineas, P. (1976/1977) 'The story of the park proposal', *National Parks Journal*, December/January: 9–11.
Prineas, P. and Gold, H. (1983) *Wild Places: Wilderness in New South Wales*, Sydney: Kalianna Press.
Pritchard, A. and Morgan, N.J. (2001) 'Culture, identity and tourism representation: marketing Cymru or Wales?' *Tourism Management*, 22: 167–79.
Proctor, J.D. (1998) 'Environmental values and popular conflict over environmental management: a comparative analysis of public comments on the Clinton Forest Plan', *Environmental Management* 22: 347–58.
Prosper, L. (2007) 'Wherin lies the heritage value? Rethinking the heritage value of cultural landscapes from an Aboriginal perspective,' *The George Wright Forum*, 24(2): 117–24.
Pulsford, J.S. (1993) *Historical Nutrient Usage in Coastal Queensland River Catchments Adjacent to the Great Barrier Reef Marine Park* – Research Publication No. 40, Townsville: GBRMPA.
Purnomo, A. and Lee, R. (2005) 'The Winds of Change – Recent Progress towards Conserving Indonesian Biodiversity', in *Indonesia Nature Conservation Letter*, Issue 8–3, January 31.
Pye-Smith, C. and Hall, C. (1997) *The Countryside We Want: A manifesto for the year 2000*, Bideford: Green Books.
Pyne, S.J. (2004) *Tending Fires: Coping with America's Wildland Fires*, Washington, DC: Island.
Qiu, B.X. (2007) 'Implement the scientific model for development and promote a harmony development of Scenic Parks', *Presentation at the National Meeting on Implementation of National Regulation on Scenic Park*, Beijing. 1 December.
Queensland (1906) *Official records of the debates of the Legislative Council and Legislative Assembly*, Vol. 98.

Rattray, G. (1986) *To Everything its Season: MalaMala, the Story of a Game Reserve*, Johannesburg: Jonathan Ball.

Ray, C. (1998) 'Culture, intellectual property and territorial rural development', *Sociologia Ruralis*, 38(1): 4–21.

Redford, K.H., and Painter, M. (2006) *Natural Alliances between Conservationists and Indigenous Peoples*, WCS Working Paper No. 25, New York: Wildlife Conservation Society.

Reeves, B. (1994) 'Ninaistakis – the Nitsitapii's sacred mountain: Traditional Native religious activities and land use/tourism conflicts', in D.L. Carmichael (ed) *Sacred Sites, Sacred Places*, London: Routledge, pp. 265–95.

Reid, D. (1979) *Our Own Country, Canada: Being an account of the national aspirations of the principal land artists in Montreal and Toronto, 1860–1890*, Ottawa: National Museums of Canada.

Reid, H. (2006) 'Culture, conservation and co-management: Lessons from Australia and South Africa,' *Policy Matters: IUCN Commission on Environmental, Economic & Social Policy*, 14: 255–68.

Relph. E. (1985) *Regional Landscapes and Humanistic Geography*, London: Croom Helm.

Ringdahl, B. (2003) 'A political ecological analysis of the Pilanesberg National Park and the Lebatlane Tribal Reserve, South Africa', unpublished thesis, Lund University.

Riseth, J.Å. (2005) 'Nature protection and the colonial legacy – Sámi reindeer management versus urban recreation: the case of Junkerdal-Balvatn, northern Norway', in T. Peil and M. Jones (ed) *Landscape, Law and Justice*, Oslo: Novus Forlag.

Riseth, J.A. (2007) 'An indigenous perspective on national parks and Sami reindeer management in Norway', *Geographical Research*, 45(2): 177–85.

Robinson, A. and Sumardja, E. (1990) 'Indonesia', in C.W. Allin (ed) *International Handbook of National Parks and Nature Reserves*, Westport: Greenwood, pp. 197–213.

Robinson. M. (2002) 'Between and beyond the pages: Literature-tourism relationships', in M. Robinson and H.C. Andersen (eds) *Literature and Tourism: Essays in the Reading and Writing of Tourism*, London: Thomson.

Roche, M.M. (1987) 'A time and a place for National Parks', *New Zealand Geographer* 43(2): 104–7.

Rojek. C. (1995) *Decentring Leisure: Rethinking Leisure Theory*, London: Sage.

Rollinson, W. (1967) *A History of Man in the Lake District*, London: J.M. Dent & Sons.

Ross, S. (2000) *What Gardens Mean*, Chicago: University of Chicago Press.

Rothman, H.K. (1989) *Preserving different pasts: the American National Monuments*, Urbana: University of Chicago Press.

Runte, A. (1972) 'Yellowstone: it's useless, so why not make it a park?', *National Parks and Conservation Magazine*, March: 4–7.

Runte, A. (1973) 'Worthless' lands – Our national parks: The enigmatic past and uncertain future of America's scenic wonderlands', *American West*, May 10: 4–11.

Runte, A. (1974a) 'Pragmatic alliance: Western railroads and the national parks', *National Parks & Conservation Magazine: The Environmental Journal*, April: 12–21.

Runte, A. (1974b) 'Yosemite Valley Railroad – Highway of history, pathway of promise', *National Parks & Conservation Magazine: The Environmental Journal*, December: 4–9.

Runte, A. (1977) 'The national park idea: Origins and paradox of the American experience', *Journal of Forest History*, 21(2): 64–75.
Runte, A. (1979) *National Parks: the American Experience*, Lincoln: University of Nebraska Press.
Runte, A. (1983) 'Reply to Sellars', *Journal of Forest History*, 27(3): 135–41.
Runte, A. (1984) *Trains of Discovery: Western Railroads and the National Parks*, Flagstaff: Northland.
Runte, A. (1987) *National parks: the American experience*, 2nd edn, Lincoln: University of Nebraska Press.
Runte, A. (1990) *Yosemite: the Embattled Wilderness*, Lincoln and London: University of Nebraska Press.
Runte, A. (1991) *Public Lands, Public Heritage: The National Forest Idea*, Niwot, CO: Roberts Rinehart Publishers in cooperation with the Buffalo Bill Historical Center.
Runte, A. (1997a) *National Parks: The American Experience*, 3rd edn, Lincoln: University of Nebraska Press.
Runte, A. (1997b) Preface to the second edition, in *National Parks: The American Experience*, 3rd edn, Lincoln: University of Nebraska Press. Online. Available: www.nps.gov/history/history/online_books/runte1/preface2.htm.
Runte, A. (2006) *Allies of the Earth: Railroads and the Soul of Preservation*, Kirksville: Truman State University Press.
Ruppert, D. (1994) 'Redefining relationships: American Indians and National Parks', *Practicing Anthropology*, 16(3): 10–13.
Ryan, C. (ed) (2007) *Battlefield tourism: history, place and interpretation*, Oxford: Elsevier.
Ryan, P.J. (1985) 'John Muir and tall trees in Australia', *The Pacific Historian* 29(2–3): 125–35.
Ryan, C. and Aicken, M. (eds) (2005) *Indigenous Tourism: The Commodification and Management of Culture*, Oxford: Elsevier.
Rydén, B.E. (2002) 'Sarek nationalpark – Nordenskiölds idé blev världsarv', in *Till fjälls. Svenska Fjällklubben 75 år*, Svenska Fjällklubbens Årsbok 2001–2002, Årgång 72–73, Stockholm.
Saarinen, J. (2005) 'Tourism in the northern wilderness: Wilderness discourses and the development of nature-based tourism in northern Finland', in C.M. Hall and S. Boyd (eds) *Nature-based Tourism in Peripheral Areas: Development or Disaster?* Clevedon: Channel View.
Saarinen, J. (2007) 'Protected areas and regional development issues in northern peripheries: nature protection, traditional economies and tourism in the Urho Kekkonen National Park, Finland', in I. Mose (ed) *Protected Areas and Regional Development in Europe: Toward a New Model for the 21st Century*, Aldershot: Ashgate, 199–212.
Sahlberg, B., Sehlin, H., Vidén, L. and Wärmark, A. (1993) 'Tourism in Sweden', in H. Aldskogius (ed) *The National Atlas of Sweden: Cultural Life, Recreation and Tourism*, Stockholm: National Atlas of Sweden Publishing.
Salcido, G. (1995) 'Natural protected areas in Mexico', *George Wright Forum*, 12(4): 30–8.
San Francisco Examiner (1895) *San Francisco Examiner*, January 15: 9.
Sandbach, F. (1978) 'The Early Campaign for a National Park in the Lake District', *Transactions of the Institute of British Geographers*, New Series, 3(4): 498–514.

Sandell, K. (1995) 'Access to the "north" – But to what and for whom? Public access in the Swedish countryside and the case of a proposed national park in the Kiruna Mountains', in C.M. Hall and M.E. Johnston (eds) *Polar Tourism*, Chichester: Wiley.

Sandell, K. (2005) 'Access, tourism and democracy: A conceptual framework and the non-establishment of a proposed national park in Sweden', *Scandinavian Journal of Hospitality and Tourism*, 5(1): 63–75.

Sandell, K. (2006) 'The right of public access: Potentials and challenges for ecotourism', in S. Gössling and J. Hultman (eds) *Ecotourism in Scandinavia: Lessons in Theory and Practice*, Wallingford: CABI.

Sandell, K. and Sörlin, S. (eds) (2000) *Friluftshistoria – från 'härdande friluftslif' till ekoturism och miljöpedagogik: Teman i det svenska friluftslivets historia*, Stockholm: Carlssons bokförlag.

Sandwith, T., Shine, C., Hamilton, L. and Sheppard, D. (2001) *Transboundary Protected Areas for Peace and Co-operation*, Gland: IUCN.

Sarono (2002) 'Merapi National Park', letter in *Jakarta Post*, 26th August, in *Indonesian Nature Conservation Newsletter*, 1st September.

Sax, J.L (1976) 'America's national parks: their principles, purposes and prospects', *Natural History*, 85(8): 55–88.

SCETO (Société Centrale pour l'Équipement Touristique Outre-Mer) (1971) *Bali Tourism Study*, draft report to the government of Indonesia, UNDP/IBRD.

Schama, S. (1995) *Landscape and Memory*, London: Harper Collins.

Schenk, A., Hunziker, M. and Kienast, F. (2007) 'Factors influencing the acceptance of nature conservation measures – A qualitative study in Switzerland', *Journal of Environmental Management* 83(1): 66–79.

Scherl, L.M. (2005) 'Protected areas and local and indigenous communities', in J.A. McNeely (ed) *Friends for Life: New Partners in Support of Protected Areas*, Gland & Cambridge: IUCN, pp. 101–12.

Scherl, L.M. and Edwards, S. (2007) 'Tourism, indigenous and local communities and protected areas in developing nations', in R. Bushell and P.F.J. Eagles (eds) *Tourism and Protected Areas: Benefits beyond boundaries*, Wallingford: CABI, pp. 71–88.

Scherl, L.M. Wilson, A., Wild, R., Blockus, J., Franks, P., McNelly, A. and McShane, T. (2004) 'Community conserved areas' in *Can Protected Areas Contribute to Poverty Reduction? Opportunities and Limitations*, Cambridge: IUCN, pp. 35–8.

Schough, K. (2007) *Lake Duortnus, Royal Science, and Nomadic Practices*, Karlstad University Studies 2007: 11. Karlstad.

Schullery, P. and Whittlesey, L. (2003) *Myth and history in the creation of Yellowstone National Park*, Lincoln: University of Nebraska Press.

Scott, J.D. (1969) *We Climb High, A Chronology of the Mazamas 1894–1964*, Portland: Mazamas.

Scott, K.A. (2005) *Yellowstone denied: the life of Gustavus Cheyney Doane*, Norman: University of Oklahoma Press.

Scott, P. (2002) 'This Vast Empire: The iconography of the Mall, 1971–1848', in R. Longstreth (ed) *The Mall in Washington, 1971–1991*, Washington: Yale University Press, pp. 37–60.

Sears, J.F. (1989) *Sacred places: American tourist attractions in the nineteenth century*, New York: Oxford University Press.

Selby. M. (1996) 'Absurdity, phenomenology and place: An existential place

marketing project', in Z. Liu, D. Botterill (eds) *Higher Degrees of Pleasure: Proceedings of the International Conference for Graduate Students of Leisure and Tourism*, Cardiff.
Sehlin, H. (1986a) 'Hur det började', in *STF 100 år. Svenska Turistföreningens årsskrift 1986*, Uppsala.
Sehlin, H. (1986b) 'Svenska Turistföreningens första år i fjällen', in *Till fjälls. Svenska Fjällklubben*, Svenska Fjällklubbens Årsbok.
Selin, P. (1996) *Inlandsbanan: Idé och historia*, Östersund: Björkås förlag.
Sellars, R.W. (1983) 'National parks: Worthless lands or competing land values?' *Journal of Forest History*, 27(3): 130–4.
Sellars, R.W. (1997) *Preserving Nature in the National Parks: a history*, New Haven: Yale University Press.
Shackley, M. (1998) 'Ninstints (Canada): A deserted Haida village in Gwaii Hanaas National Park Reserve (Queen Charlotte Islands)', in M. Shackley (ed) *Visitor Management: Case Studies from World Heritage Sites*, London: Butterworth-Heinemann, pp. 182–93.
Shaffer, M.S. (2001) *See America First: Tourism and National Identity, 1880–1940*, Washington, DC: Smithsonian Institution Press.
Sharpley, R. (2003) 'Rural Touism and Sustainability: A Critique', in D. Hall, I. Roberts and M. Mitchell (eds) *New Directions in Rural Tourism*, Aldershot: Ashgate, pp. 38–53.
Sharpley, R. (2004) 'The Impacts of Tourism in the Lake District', in D. Hind and J. Mitchell (eds) *Sustainable Tourism in the English Lake District*, Sunderland: Business Education Publishers, pp. 207–42.
Sharpley, R. (2007) *Tourism and Leisure in the Countryside*, 5th edn, Huntingdon: Elm.
Sharpley, R. and Pearce, T. (2007) 'Tourism, Marketing and Sustainable Development in the English National Parks: The Role of National Park Authorities', *Journal of Sustainable Tourism* 15(5): 557–73.
Sheail, J. (1975) 'The Concept of National Parks in Great Britain 1900–1950', *Transactions of the Institute of British Geographers* 66(Nov): 41–56.
Sheail, J. (1995) 'Nature protection, ecologists and the farming context: A UK historical context', *Journal of Rural Studies*, 11(1): 79–88.
Sheldrake. P. (2001) *Spaces for the Sacred: Place, Memory and Identity*, Cambridge: SCM Press.
Shepard, P. (1967) *Man in the Landscape: A Historic View of the Esthetics of Nature*, New York: Knopf.
Shepherd, N. (2002) 'How Ecotourism can go Wrong: The Cases of SeaCanoe and Siam Safari, Thailand', *Current Issues in Tourism* 5 (2 and 3): 309–18.
Shepherd, R. (2002) 'Commodification, culture and tourism', *Tourist Studies*, 2(2): 183–201.
Shoard, M. (1982) 'The lure of the moors', in J.R. Gold and J. Burgess (eds) *Valued Environments*, London: Allen & Unwin, pp. 55–73.
Shoard, M. (1999) *A Right to Roam*, Oxford: Oxford University Press.
Short, J.R. (1991) *Imagined Country: Society, Culture and Environment*, London: Routledge.
Sidaway, R (1990) *Birds and walkers: a review of existing research on access to the countryside and disturbance to birds*, Ramblers Association.
Sierra Club (1896) 'Proceedings of the Meeting of the Sierra Club, November 23, 1895', *Sierra Club Bulletin* 1(6): 271–84.

Sierra Club (1944) *Sierra Club Bulletin* 29 (October): 45–9.
Sievanen, L. (2008) 'Eco-tourism for whom?', in *Inside Indonesia* Issue 91, reproduced in *Indonesia Nature Conservation Bulletin*, 11(02a).
Sievert, J. (2000) *The origins of nature conservation in Italy*, Bern: Peter Lang.
Sillanpää, P. (2002) *The Scandinavian Sporting Tour: A Case Study in Geographical Imagology*, ETOUR V 2002:9, Örnsköldsvik: Åbo Akademi University.
Simonian, L. (1995) *Defending the Land of the Jaguar: a History of Conservation in Mexico*, Austin: University of Texas Press.
Slade, B. (1985–86) 'Royal National Park: the people in a people's park', *Geo: Australia's Geographical Magazine*, 7(4): 64–77.
Smith, B. (1992) *Imaging the Pacific in the Wake of the Cook Voyages*, Melbourne: Melbourne University Press.
Smith, L. (2004) 'The contested landscape of early Yellowstone', *Journal of Cultural Geography* 22(1): 3–27.
Smyth, D. (1997) 'Recognition of Aboriginal Maritime Culture in the Great Barrier Reef Marine Park: An Evaluation', in D. Wachenfeld, J. Oliver and K. Davis (eds) *State of the Great Barrier Reef World Heritage Workshop*, Townsville: GBRMPA.
Smyth, D. (2001) 'Joint management of national parks', in R. Baker, J. Davies and E. Young (eds) *Working on Country – Contemporary Indigenous Management of Australia's Lands and Coastal Regions*, Melbourne: Oxford University Press, pp. 75–91.
Sneed, P.G. (1997) 'National parks and northern homelands: Toward co-management of national parks in Alaska and the Yukon', in S. Stevens (ed) *Conservation Through Cultural Survival: Indigenous People and Protected Areas*, Washington DC: Island, pp. 135–54.
Soemarwoto, O. (1979) 'Inter-relations among population resources, environment and development in the ESCAP region with special reference to Indonesia', paper presented at regional seminar on alternative patterns of development and lifestyles in Asia and the Pacific.
Sofield, T. (2008) 'The role of tourism in transition economies of the Greater Mekong Sub-region', in J. Cochrane (ed) *Asian Tourism: Growth and Change*, pp. 39–53.
Sörlin, S. (1988) *Framtidslandet. Debatten om Norrland och naturresurserna under det industriella genombrottet*. Acta Regiae Societatis Skytteanae Nr 33. Malmö: Carlsson Bokförlag.
Sörlin, S. (1999) 'The articulation of territory: Landscape and the constitution of regional and national identity', *Norsk geografisk Tidskrift*, 53: 103–11.
Sörlin, S. (2000) 'Staden bortom staden – tillblivelsen av en plats', in *Kiruna 100-årsboken*. Del 1. Kiruna: Kiruna kommun.
Spence, M. (1996) 'Dispossessing the wilderness: Yosemite Indians and the national park ideal, 1864–1930', *The Pacific Historical Review*, 65(1): 27–59.
Squire, S. (1988) 'Wordsworth and Lake District Tourism: Romantic Reshaping of the Landscape', *Canadian Geographer* 32(3): 237–47.
Stadling, J. (1900) 'När världens nordligaste järnväg byggdes', in *Svenska Turistföreningens årsskrift 1900*, Stockholm: Wahlström & Widstrand.
Stanley, H.J. (compiler) (1977) *History of Royal National Park*, unpublished manuscript, New South Wales National Parks and Wildlife Service Library.
Stanly, N. (2002) 'Chinese theme parks and national identity', in T. Young and R. Riley (eds) *Theme Park Landscapes: Antecedents and Variations*, Washington DC: Dumbarton Oaks, pp. 269–89.

Star, P. and Lochhead, L. (2002) 'Children of the Burnt Bush: New Zealanders and the Indigenous Remnant, 1880–1930', in E. Pawson and T. Brooking (eds) *Environmental Histories of New Zealand*, Melbourne: Oxford University Press, pp. 119–35.

Starbäck, K. (1909) Naturskydd. In *Skogsvårdsföreningens Folkskrifter*, no. 18. Stockholm: Centraltryckeriet.

Statistics Sweden (2007) *Skyddad natur 31 dec 2006*. www.scb.se.

Steel, W.G. (1886) 'Crater Lake and How to See It', *The West Shore* 12:3 (March): 104–6.

Steel, W.G. (1890) *The Mountains of Oregon*, Portland: David Steel.

Steel, W.G. (1898) 'The Valley of the Stehekin', *The State* 2(1) (July 20).

Steel, W.G. (1907) 'The President's Order', *Steel Points*, 1(2) (January): 73.

Steel, W.G. (1925) 'Crater Lake Yesterday. Today and Tomorrow', *Steel Points Junior* 1(2) (August), n.p.

Steel, W.G. (1930) Quoted September 7, 1930, History Files, Crater Lake National Park.

Stevens, S. (ed) (1997) *Conservation Through Cultural Survival: Indigenous Peoples and Protected Areas*, Washington D.C.: Island.

Stevenson-Hamilton, J. (1905) 'Game preservation in the Transvaal', *Journal of the Society for the Preservation of the Wild Fauna of the Empire*, 2: 20–45.

STF (1903) *Svenska Turistföreningens årsskrift*, Wahlström & Widstrand, Stockholm.

STF (1904) *Svenska Turistföreningens årsskrift*, Wahlström & Widstrand, Stockholm.

STF (1910) *Svenska Turistföreningens årsskrift*, Wahlström & Widstrand, Stockholm.

Stoll-Kleemann, S. (2001) 'Barriers to nature conservation in Germany: a model explaining opposition to protected areas', *Journal of Environmental Psychology* 21: 369–85.

Stolton, S. and Dudley, N. (eds) (1999) *Partnerships for Protection: New Strategies for Planning and Management for Protected Areas*, London: Earthscan.

Stolton, S. and Oviedio, G. (2004) 'Using the categories to support the needs and rights of traditional and indigenous peoples in protected areas', in K. Bishop, N. Dudley, A. Phillips and S. Stolton (eds) *Speaking a Common Language: The Uses and Performance of the IUCN System of Management Categories for Protected Areas*, Cardiff: Cardiff University, IUCN & UNEP.

Stone, D. (1994) *Tanah Air: Indonesia's Biodiversity*, Singapore: Editions Didier Millet.

Streatfield, D.C. (2002) 'The Olmsteds and the landscape of the Mall', in R. Longstreth (ed), *The Mall in Washington, 1971–1991*, Washington: Yale University Press, pp. 117–42.

Strittholt, J.R. and Dellasala, D.A. (2001) 'Importance of roadless areas in biodiversity conservation in forested ecosystems: Case study of the Klamath-Siskiyou ecoregion of the United States', *Conservation Biology* 15(6): 1742–54.

Sumardja, E., Harsono and MacKinnon, J. (1984) *Indonesia's Network of Protected Areas*, in Proceedings of Third World Congress on National Parks, Gland: IUCN, pp. 214–23.

Sundin, B. (2001) 'Det svenska naturskyddets framväxt', in F. Sjöberg (ed) *Vad ska vi med naturen till?* Smedjebacken: Nya Doxa.

Sutton, M. (1999) 'Aboriginal ownership of National Parks and tourism', *Cultural Survival Quarterly* 23(2): 55–6.

Svenonius, F. (1908) 'Lappland som turistland', in O. Bergqvist and F. Svenonius (eds) *Lappland*, Stockholm: Kungliga Hofboktryckeriet.

Swedish Association for the Conservation of Nature (1910) *Sveriges Natur. Svenska Naturskyddsföreningens årsskrift 1910*, Stockholm: Wahlström & Widstrand.

Swedish Environmental Protection Agency (2007) *Utkast till ny nationalparksplan för Sverige. Svar på remiss 2007-04-26 och övriga inkomna synpunkter*. Online. Available: www.naturvardsverket.se/sv/Nedre-meny/Aktuellt/Remisser/ Sammanstallning-av-remissvar/Utkast-till-ny-nationalparksplan/ (accessed 15 November 2007).

Swedish Government Writ (2001/02), *En samlad naturvårdspolitik*, Stockholm: The Government Office.

Swedish Parliament (1904a) *Andra kammarens tillfälliga utskott 1904:21*, Utlåtande i anledning af väckt motion om skrivelse till Kung. Maj.t angående skydd för vårt lands naturminnesmärken.

Swedish Parliament (1904b) *Motion i andra kammaren 1904:194*, Motion av herr K. Starbäck, om skrivelse till Kungl. Maj:t angående skyddsåtgärder för vårt lands natur och minnesmärken.

Swedish Parliament (1907) *Betänkande rörande åtgärder till skydd för vårt lands natur och naturminnesmärken*, Stockholm: Isaac Marcus Boktryckeri-aktiebolag.

Swedish Parliament (1909a) *Proposition 1909:125*, Angående åtgärder till skyddande af naturminnesmärken å kronans mark samt avsättande af vissa nationalparker.

Swedish Parliament (1909b) *Svensk Författningssamling 1909:56*, Lag angående nationalparker.

Sycholt, A. (2002) *A guide to the Drakensberg*, Cape Town: Struik.

Taiepa, T., Lyver, P., Horsley, P., Davis, J., Bragg, M. and Moller, H. (1997) 'Co-management of New Zealand's conservation estate by Maori and Pakeha: A review', *Environmental Conservation*, 24(3): 236–50.

Tamura, T. (1957) *National Parks of Japan*, Tokyo: Tokyo News Service.

Taylor, D. (1999) 'Central Park as a model for social control: Urban parks, social class and leisure behaviour in nineteenth century America', *Journal of Leisure Research*, 31(4): 426–77.

Taylor, K. (2000) 'Culture or nature: Dilemmas of interpretation', *Tourism, Culture & Communication*, 2(2): 69–84.

Taylor, M. (2003) *Public Policy in the Community*, Houndmills: Palgrave Macmillan.

Taylor, B. and Geffen, J. (2003) 'Battling religions in parks and forest reserves: Facing religion in conflicts over protected areas', in D. Harmon and A.D. Putney (eds) *The Full Value of Parks: From Economics to the Intangible*, Lanham: Rowman & Littlefield, pp. 281–93.

Taylor, B. and Geffen, J. (2004) 'Battling religions in parks and forest reserves: Facing religion in conflicts over protected areas', *The George Wright Forum*, 21(2): 56–68.

Te Heuheu, T. (1995) 'A sacred gift: Tongariro National Park, New Zealand', in B. von Droste, H. Plachter and M. Rosser (eds) *Cultural Landscapes of Universal Value: Components of a Global Strategy*, Jena: Gustav Fischer Verlag, pp. 170–3.

Terborgh, J. and Peres, C.A. (2002) 'The problem of people in parks', in J. Terborgh, C. van Schaik, L. Davenport and M. Rao (eds) *Making Parks Work: Strategies for Preserving Tropical Nature*, Washington DC: Islands, pp. 307–19.

Te Runanga o Ngai Tahu (2006) 'Ngai Tahu tourism', in *Annual Report 2006*. Te Runanga o Ngai Tahu, pp. 31–2.

Tempo (2004) Warga tolak perluasan Taman Nasional Halimun (Local people reject the extension of Halimun National Park), article dated March 4, in *Indonesian Nature Conservation Bulletin* 7 (9b).

Thacker, C. (1983) *The wildness pleases: the origins of Romanticism*, London: Croom Helm.
Thaning, O. (1986) 'Medlemmarna', in *STF 100 år. Svenska Turistföreningens årsskrift 1986*. Uppsala.
Thompson, J. (1976) *Origin of the 1952 National Parks Act*, Wellington: Department of Lands and Surveys.
TILCEPA (2006) Open letter to the President of the Republic of Indonesia from the co-chairs of the Inter-commission Theme on Indigenous and Local Communities, Equity and Protected Areas (TILCEPA) of the World Conservation Union, dated 1 March 2006. Online. Available: www.iucn.org/themes/ceesp/Wkg_grp/TILCEPA/TILCEPA%20letter%20Mt%20Merapi%2001.03.06.doc).
Timothy, D. and Boyd, S.W. (2003) *Heritage tourism*, Harlow: Prentice Hall.
Toohey, P. (2001) 'Burke Attacks Uluru Closure', *The Australian*, 3.
Tongson, E. and Dino, M. (2004) 'Indigenous peoples and protected areas: The case of the Sibuyan Mangyan Tagabukid, Philippines', in T.O. McShane and M. P. Wells (eds) *Getting Biodiversity Projects to Work: Towards more Effective Conservation and Development*, New York: Columbia University Press, pp. 181–207.
Tongson, E. and McShane, T. (2006) 'Securing Indigenous rights and biodiversity in Sibuyan Island, Romblon, Philippines', *Policy Matters*, 14: 286–97.
Tongue, T.H. (1902) Tongue to Steel, April 18, 1902, Box 2 Item 21F, Museum Collection, Crater Lake National Park.
Tourism Review Steering Committee (TRSC) with the assistance of the GBRMPA and the Office of National Tourism (1997) *Review of the Marine Tourism Industry in the Great Barrier Reef World Heritage Area*, Canberra: TRSC.
Trakolis, D. (2001a) 'Local people's perceptions of planning and management issues in Prespes Lakes National Park, Greece', *Journal of Environmental Management* 61(3): 227–41.
Trakolis, D. (2001b) 'Perceptions, preferences, and reactions of local inhabitants in Vikos-Aoos National Park, Greece', *Environmental Management* 28(5): 665–76.
Transvaal Province (1918) *Report of the Game Reserves Commission, TP 5–'18*, Pretoria: Transvaal Province.
Tresidder. R. (1999) 'Sacred spaces in a post-industrial society', in D. Crouch (ed) *Leisure/tourism Geographies: Practices and Geographical Knowledge*, London: Routledge.
Tresidder. R. (2001) 'Representations of sacred spaces in a post-industrial society', in M. Cotter, W. Boyd and J. Gardiner (eds) *Heritage Landscapes: Understanding Place and Communities*, Lismore: Southern Cross University.
Tribe, J. (2008) 'Critical Tourism', *Journal of Travel Research*, 46: 245–55.
Turner, A. (1979) 'National Parks in New South Wales, 1879–1979: Participation, Pressure Groups and Policy', Unpublished PhD thesis, Australian National University.
Turner, F.J. (1920) *The Frontier in American History*, New York: Henry Holt.
Tuxill, S. (2006) 'Co-management of a cultural landscape, Arizona, US', in M. Lockwood, G. Worboys and A. Kothari (eds) *Managing Protected Areas: A Global Guide*, London: Earthscan, p. 535.
Tyrrell, I. (1999) *True gardens of the gods: Californian-Australian environmental reform, 1860–1930*, Berkeley: University of California Press.
Uekoetter, F. (2006) *The Green and the Brown: a history of conservation in Nazi Germany*, Cambridge: Cambridge University Press.

UN Commission on Sustainable Development (2002) *Dialogue Paper by Indigenous People*. Addendum No. 3. UN Economic and Social Council. Redturs. Online. Available: www.redturs.org/ (accessed 17 November 2005).

UNESCO (2005) *Local & Indigenous Knowledge of the Natural World: An Overview of Programmes and Projects*, International Workshop on Traditional Knowledge, Panama City, 21–23 September 2005, UNESCO. Online. Available: www.un.org/sea/socdev/unpfii/news_workshop_tk.htm (accessed 17 November 2005).

Union of South Africa (1926) *House of Assembly Debates*, 3rd Session, 5th Parliament.

United States Geological Survey (USGS) (1887) 'Report of Capt. C.E. Dutton, Part 1', *USGS Eighth Annual Report, 1886–1887*, Washington D.C.: Government Printing Office.

Unrau, H.D. (1988) *Administrative History Crater Lake National Park, Oregon*, Denver: National Park Service.

Urry, J. (1990) *The Tourist Gaze: Leisure and Travel in Contemporary Society*, London: Sage.

Utley, R.M. (1983) 'Commentary on the "Worthless Lands" thesis', *Journal of Forest History*, 27(3): 142.

Valentine, P., Birtles, A., Curnoch, M., Arnold, P. and Dunstan, A. (2004) 'Getting Closer to Whales – Passenger Expectations and Experiences and the Management of Swim with Dwarf Minke Whale Interactions in the Great Barrier Reef', *Tourism Management* 25(6): 647–55.

Van der Merwe, P., Saayman, M. and Krugell, W. (2004) 'Factors that determine the price of game', *Koedoe,* 47(2): 105–13.

van Tiggelen, J. (2007), 'The Tropical Time Bomb', *The Age* (Good Weekend magazine), 13 October, pp. 19–26.

Vaux, C. (1865) 'Letter to Frederick Law Olmstead', in V.P. Ranney (ed) *The papers of Frederick Law Olmsted: Volume V The California Frontier 1863–1865*, Baltimore, Johns Hopkins University Press, 1990, pp. 383–90.

Vesterlund, O. (1901) 'I civilisationens utkanter', in *Svenska Turistföreningens årsskrift 1901*. Stockholm: Wahlström & Widstrand.

Wahlberg, S. (1986) 'Turism och naturskydd', in *STF 100 år. Svenska Turistföreningens årsskrift 1986*. Uppsala.

Waitt, G., Figueroa, R. and McGee, L. (2007) 'Fissures in the rock: rethinking pride and shame in the moral terrains of Uluṟu', *Transactions of the Institute of British Geographers*, 32: 248–63.

Wakatu (2008) Wakatu Tourism. Wakatu Incorporation. Online. Available: www.wakatu.org/main/tourism/ (Accessed 12 March 2008).

Wall, G. (1982) 'Outdoor recreation and the Canadian identity', in G. Wall and J. Marsh (eds) *Recreational Land Use: Perspectives on its Evolution in Canada*, Ottawa: University of Carlton Press, pp. 418–34.

Wall, G. (1999) 'Partnerships involving indigenous peoples in the management of heritage sites', in P. Boniface (ed) *Tourism and Cultural Conflicts*, Wallingford: CAB International, pp. 269–86.

Wall-Reinius, S. and Fredman, P. (2007) 'Protected areas as attractions', *Annals of Tourism Research*, 34(4): 839–54.

Walter, F. (1989) 'Attitudes toward the environment in Switzerland, 1880–1914', *Journal of Historical Geography* 15(3): 287–99.

Waterson, M (1994) *The National Trust; the first hundred years*, London: National Trust.

Watts, S. (2003) *Rough Rider in the White House: Theodore Roosevelt and the politics of desire*, Chicago: University of Chicago Press.
WCPA (2000) *Protected Areas – Benefits Beyond Boundaries*, IUCN. Online. Available: www.iucn.org/themes/wcpa/ (accessed 20 January 2008).
WCPA (2005) *WCPA Strategic Plan 2005–2012*, IUCN. Online. Available: www.iucn.org/themes/wcpa/pubs/other.htm (accessed 20 January 2008).
WCPA News (2003) 'The Durban Action Plan-Outcome 5: The rights of indigenous peoples,' *WCPA News*, 91: 18–20.
Wearing, S. and Huyskens, M. (2001) 'Moving on from joint management policy regimes in Australian national parks', *Current Issues in Tourism*, 4(2/3/4): 182–209.
Weaver, D. (2001) *Ecotourism*, Brisbane: Wiley.
Weaver, D.B. (2006) 'Indigenous territories', in *Sustainable Tourism: Theory and Practice*, Oxford: Elsevier, pp. 143–6.
Weaver, D. and Lawton, L. (2002) *Tourism Management*, 2nd edn, Brisbane: Wiley.
Webb, D. and Adams, B. (1998) *The Mt Buffalo Story 1898–1998*, Melbourne: Miegunyah.
Weber, R., Butler, J. and Larson, P. (eds) (2000) *Indigenous Peoples and Conservation Organizations: Experiences in Collaboration*. WWF, Online. Available: www.worldwildlife.org/bsp/publications/ (accessed 17 November 2005).
Weiner, D.R. (1988) *Models of Nature: ecology, conservation, and cultural revolution in Soviet Russia*, Bloomington: Indiana University Press.
Weiner, D.R. (1999) *A little corner of freedom: Russian nature protection from Stalin to Gorbachev*, Berkeley: University of California Press.
Weiskel, T. (1976) *The Romantic Sublime: Studies in the Structure and Psychology of Transcendence*, Baltimore: Johns Hopkins University Press.
Wellings, P. (2007) 'Joint management: Aboriginal involvement in tourism in the Kakadu World Heritage Area', in R. Bushell and P. Eagles (eds) *Tourism and Protected Areas: Benefits Beyond Boundaries. The Vth IUCN World Parks Congress*, Wallingford, CABI, pp. 89–100.
Wells, J. (1996) 'Marketing Indigenous Heritage: A Case Study of the Uluru National Park', in C.M. Hall and S. McArthur (eds) *Heritage Management in Australia and New Zealand: The Human Dimension*, 2nd edn, Oxford: Oxford University Press.
West, P.C. and Brechin, S.R. (eds) (1991) 'National parks, protected areas, and resident peoples: A comparative assessment and integration', in *Resident Peoples and National Parks: Social Dilemmas and Strategies in International Conservation*, Tucson: University of Arizona Press, pp. 363–400.
Western, D. and Wright, R.M. (eds) (1994) *Natural Connections: Perspectives in Community-based Conservation*, Washington DC: Island.
White, R. (1991) *It's Your Misfortune and None of My Own: A New History of the American West*, Norman: University of Oklahoma Press.
Whited, T.L., Engels, J.I., Hoffmann, R.C., Ibsen, H. and Verstegen, W. (2005) *Northern Europe: An Environmental History*, Santa Barbara: ABC-CLIO.
Whittaker, E. (1994) 'Public Discourse on Sacredness: The Transfer of Ayers Rock to Aboriginal Ownership', *American Ethnologist*, 21(2): 310–34.
Whitten, A. and Whitten, J. (1992) *Wild Indonesia*, London: New Holland.
Whitten, T., Soeriaatmadja, R.E. and Afiff, S.A. (1996) *The Ecology of Java and Bali*, Singapore: Periplus Editions.

Whittlesey, L.H. (1995) *Death in Yellowstone: Accidents and Foolhardiness in the First National Park*, Lanham: Roberts Rinehart.

Whittlesey, L.H. (2002) 'Native Americans, the earliest interpreters: What is known about their legends and stories of Yellowstone National Park and the complexities of interpreting them,' *The George Wright Forum*, 19(3): 40–51.

Williams, D.R. (2002) 'Social construction of Arctic wilderness: Place meanings, value pluralism, and globalization', in A. Watson, L. Alessa, and J. Sproull (eds) *Wilderness in the Circumpolar North: Searching for Compatibility in Ecological, Traditional, and Ecotourism Values*, Proceedings RMRS-P26, May 15–16, 2001, Anchorage: Department of Agriculture, Forest Service, Rocky Mountain Research Station.

Wilson, G, and Bryant, R.L. (1997) *Environmental Management: New Directions for the Twenty-First Century*, London: Routledge.

Wiltshier, P. (2007) 'Mentoring in community-based tourism', *Tourism: An Interdisciplinary Journal*, 55(4): 375–90.

Winer, N. (2003) 'Co-management of protected areas, the oil and gas industry and indigenous empowerment – the experience of Bolivia's Kaa Iya del Gran Chaco', *Policy Matters*, 12: 181–91.

Winks, R.W. (1983) 'Upon reading Sellars and Runte', *Journal of Forest History*, 27 (3): 142–3.

Wolfe, L.M. (1938) *John and the Mountains: the Unpublished Journals of John Muir*, Boston: Houghton-Mifflin.

Wordsworth, W. (1810) *Wordsworth's Guide to the Lakes*, Oxford: Oxford University Press, 1970 reprint.

Wright, G.M., Dixon, J.S. and Thompson, B.H. (1933) *Fauna of the National Parks of the United States. A Preliminary Survey of Faunal Relations in National Parks*, Contribution of Wild Life Survey Fauna Series No.1, May, 1932, Washington D.C.: National Park Service.

Wright, J. and Mazel, A. (2007) *Tracks in a Mountain Range*, Johannesburg: Witwatersrand University Press.

Wright, R. (1989) *The bureaucrat's domain: space and the public interest in Victoria 1836–84*, Melbourne: Oxford University Press.

Wright, R.G. and Mattson, D.J. (1996) 'The origin and purpose of national parks and protected areas', in R.G. Wright (ed) *National Parks and protected areas: their role in environmental protection*, Cambridge USA: Blackwell, pp. 3–14.

Wuyishan Municipal Bureau of Statistics (2006) *Statistics Year Book of Wuyishan*.

Wuyishan Tourism Bureau (2006) *The newsletter of Wuyishan Tourism*. October, 23.

WWF (World Wildlife Fund) (2000) *Map of Indigenous and Traditional Peoples in Ecoregions*, Gland: WWF.

WWF (2005) WWF statement of principles on indigenous peoples and conservation. WWF. Policy. Online. Available: www.panda.org/about_wwf/what_we_do/policy/people_environment/indigenous_people/index.cfm (accessed 17 November 2005).

Wyatt, J. (1987) *The Lake District National Park*, Exeter: Webb & Bower.

Xu, H.G. (2006) 'The Limit of Growth of Resource Dependent Tourism Destination', *China Population Resources and Environment*, 5: 18–22.

Xu, S.L. (2003) 'Institutional Reform of Chinese World Natural and Cultural Heritage', *Management World*, 6: 63–73.

Young, T. (2002) 'Virtue and irony in a US national park', in T. Young and R. Riley (eds), *Theme park landscapes: antecedents and variations*, Washington DC: Dumbarton Oaks, pp. 157–81.

Young, T. (2005a) 'Between a Rock and a Hard Place: Backpackers at Uluṟu, Central Australia', in B. West (ed) *Down the Road: Backpacking and Independent Travel*, Perth: API Network.

Young, T. (2005b) 'Going by the Book: Backpacker Travellers in Aboriginal Australia and the Negotiation of Text and Experience', unpublished thesis, University of Newcastle, Australia.

Yunupingu, G. (1997) 'From Bark Petition to Native Title', in G. Yunupingu (ed) *Our Land is Our Life: Land Rights – Past, Present and Future*, Brisbane: University of Queensland Press.

Zachrisson, A., Sandell, K., Fredman, P. and Eckerberg, K. (2006) 'Tourism and protected areas: motives, actors and processes', *International Journal of Biodiversity Science and Management*, 2(4): 350–8.

Zancai, X., Yi, W. and Wei, C. (2007) 'Aesthetic Values in Sustainable Tourism Development: A Case Study in Zhangjiajie National Park of Wuling Yuan, China', unpublished discussion paper.

Zell, L. (2006) *Diving and Snorkelling Great Barrier Reef*, 2nd edn, Melbourne: Lonely Planet.

Zeppel, H. (1998) 'Land and culture: Sustainable tourism and indigenous peoples', in C.M. Hall and A. Lew (eds) *Sustainable Tourism: A Geographical Perspective*, London: Addison Wesley Longman, pp. 60–74.

Zeppel, H. (2006) *Indigenous Ecotourism: Sustainable Development and Management*, Wallingford: CABI.

Zeppel, H. (2007) 'Indigenous ecotourism, conservation and resource rights', in J. Higham (ed) *Critical Issues in Ecotourism: Understanding a Complex Tourism Phenomenon*, Oxford: Elsevier, pp. 308–48.

Zimmer, O. (1998) 'In search of national identity: Alpine landscape and the reconstruction of the Swiss nation', *Comparative Studies in Society and History*, 40(4): 637–65.

Zon, R. (1927) *Forests and Water in the Light of Scientific Investigation*, Washington DC: Government Printing Office.

Index

Abel Tasman National Park (New Zealand) 277
Abruzzo National Park (Italy) 43
Accessibility 116, 120, 158, 168, 174, 175–7, 183, 206, 290, 291
Adirondack Forest Preserve (USA) 27
Aesthetics and conservation 7. 8, 12, 13, 33, 35, 46–7, 49, 58, 59, 61, 62, 91, 145, 152, 154, 185, 187, 188, 196, 252, 270, 271, 285, 290, 293, 303, 305
Africa 37–40, 265, 305–6
Agriculture 22, 43, 44, 47, 53, 54, 58, 59, 60, 126, 157, 187, 212, 222, 246, 252, 306
Algeria 14
Amboseli National Park (Kenya) 40
American culture 30–2
Ancient Troya National Park (Turkey) 73
Angkor Wat 14, 40
Aoraki/Mt Cook National Park (New Zealand) 277
Arkansas Hot Springs (USA) 33
Art 23–4, 26, 47, 65–9, 81–4, 158, 185, 188, 191, 212, 248, 271
Asia 37–8, 40–1, 265
Australia 3, 5, 10, 25, 35, 36–7, 45, 50–61, 72, 114–27, 128–40, 184, 266, 278–81; New South Wales 10, 52–4, 272, 278; [Sydney 3, 6; Port Hacking 3, 4]; Northern Territory 128–40, 272, 278; Queensland 37, 57–8, 114–27; South Australia 25, 55–6, 60 [Adelaide 25]; Tasmania 58–9, 59–60, 69; Victoria 56–7 [Melbourne 25]; Western Australia 54–5, 61 [Perth 54]
Austria 301
Authenticity 125, 130, 170
Automobile 34, 84–5, 86–7, 105

Badlands National Park (USA) 264
Banff Hot Springs Reserve (Canada) 33
Banff National Park (Canada) 102, 110–13
Battle of Little Bighorn National Monument (USA) 76
Belgium 74
Biodiversity 61, 114, 145, 151, 204, 207, 213, 221, 223, 232, 239, 244, 253, 260–2, 272, 307
Biosphere Reserve 9, 197, 200, 239, 262
Birkenhead Park 25
Blue Ridge Parkway 87
Bolivia 267
Braveheart 65
Byles, Marie 10
Byron, Lord 24

Cambodia 40
Canada 33–4, 68–9, 102–14, 276–7; Alberta 105, 106; British Columbia 105, 106, 107, 109; Manitoba 106; New Brunswick 107; Newfoundland and Labrador 107, 109; Northwest Territories 106, 107, 109; Nova Scotia 107; Nunavut 109; Ontario 105, 107, 109; Prince Edward Island 107; Quebec 107; Saskatchewan 106; Yukon 107, 109
Canadian Pacific Railway 33, 68–9, 103–4
Canadian Parks Act 110
Canyon de Chelly National Monument (USA) 264
Castlemaine Diggings National Heritage Park (Australia) 73–4
Catlin, George 23
'Cesspool hypothesis' 62

354 Index

Central Park (USA) 26
Century Magazine 90
China 221, 225–37
Circeo National Park 43, 44, 70
Coastal parks 37, 72, 124
Colonialism 37–41, 71, 74–5, 188, 305
Co-management 111–2, 224, 260–1, 265, 273–81; *see also* national park management
Committee of Inquiry into the National Estate 59
Conservation 6, 8, 9, 12, 13, 14, 15, 44, 59–60, 85, 89, 91, 97, 102, 116, 118, 119, 120–3, 143, 145, 150, 153, 155, 156, 162, 164, 167, 178, 185, 187–8, 195, 197, 199, 211, 213, 220, 226, 234, 239, 243, 267, 305; *see also* preservation
Conservation movement 86, 88, 93, 146, 160, 175, 221
Convention for the Preservation of Animals, Birds and Fish in Africa 38
Cook, Thomas 25
Cooke, Jay 19, 21
Costa Rica 35
Covadonga Mountain National Park (Spain) 148, 149
Crater Lake National Park (USA) 17, 28, 48, 84, 88–101
Crockett and Boone Club 37, 67
Cultural landscape 81–7, 259–81

Darwin, Charles 114
Death Valley National Park (USA) 264
Declaration on the Rights of Indigenous Peoples 261
Democracy 3, 26, 30, 43, 168, 223, 243–4, 303
Denmark 72
Department of Lands and Surveys 10
Devils Tower National Monument (USA) 272
Dominion Forest Reserves and Parks Act 33
Dunphy, Milo 17
Dunphy, Myles 9

Ecology 15, 49–50, 126, 139, 241, 255, 270, 309
Economic development 21–2, 34, 35, 48–9, 90, 100, 102, 145, 163, 178–9, 187, 197, 213, 226, 231, 235–6, 239, 306, 307

Ecotourism 35, 101, 237, 263, 267, 274, 306
Egmont National Park (New Zealand) 58
England and Wales 6–7, 23, 25, 41, 42, 155–66, 167–183; Lake District 155–66; Peak District 167–183
Europe 41–3, 69–71, 301–2
Everglades National Park (USA) 28
Evolution 114

Fauna Preservation Society 37, 39
Federated Mountain Clubs 10
Film 65, 177
Finland 185–6; Punkaharju 185
Forest of Dean (England) 42
Forestry 40, 51, 59, 60, 91, 99, 149, 161, 223
France 41, 82
Frontier 84, 85
'Frontier commons' 85, 86

Gallipoli Peninsula Historic National Park (Turkey) 73
Game reserves 39
Gardens 81–2
General Grant National Park (USA) 28
Geothermal energy 100
Germany 38–9, 41, 43, 301; Schorfheide 43
Glacier National Park (Canada) 106
Glacier National Park (USA) 48, 264
Golden Gate Park (USA) 26
Göring, Hermann 43
Gran Paradiso National Park 43
Grand Canyon National Park (USA) 48, 264
Great Barrier Reef (Australia) 114–5
Great Barrier Reef Marine Park (Australia) 114–27
Great Barrier Reef Marine Park Authority (GBRMPA) 114, 119–21
Great Basin National Park (USA) 88
Great Smoky Mountains National Park (USA) 42
Greece 41, 70
Guyana 267

Haida 276–7
Hall, C.M. 50, 266
Harkin, J.B. 34
Hedges, Cornelius 16–17
Health 36–7, 42, 52
Heritage 15, 32, 41, 63–77, 87, 105, 123,

125, 129, 131, 133, 137–40, 156, 162, 164, 165, 185, 189, 204, 220, 225, 231, 238, 249–50, 261, 266, 270, 271, 278, 287, 303
Heritage tourism 111, 263
Hetch Hetchy Valley (USA) 100
Hot springs 7, 16, 21, 33, 104, 106
Hudson River School of Landscape Painting 66
Humpty Dumpty 3
Hunting 37–8, 42, 305
Hyde Park (England) 25

Iceland 41, 70, 73
India 40
Indigenous peoples 8, 11, 28, 31, 72, 75, 103, 123–5, 128–40, 187, 217, 224, 259–81, 302, 305, 306, 307, 308
Indonesia 38, 211–24
Interest groups 88
International Union for the Conservation of Nature (IUCN) 6, 8–9, 12–13, 14, 32, 38, 44, 120, 156, 200, 214, 225, 242, 244, 245, 256, 260, 261, 263, 268
Ireland 73
Italy 41, 43, 46, 70, 81, 82

Japan 40–1, 44, 72–3
John Forrest National Park (Australia) 54–5
Johnson, Robert Underwood 89–90

Kaa-Iya del Gran Chaco National Park (Bolivia) 267
Kakadu National Park (Australia) 278
Kalahari Gemsbok National Park 244
Kenya 39–40
Kgalagadi Transfrontier Park (South Africa) 267
Kings Canyon National Park 88
King's Park (Australia) 54
Kruger National Park (South Africa) 39, 75, 246–54

Lake District National Park (England) 155–66
Lamington National Park (Australia) 58
Land ownership 42–3
Landscape 63–9, 81–7, 128, 130–1, 137, 138, 139, 144, 145, 150, 154, 155, 156, 158, 161, 163, 165, 171–3, 185, 187, 189, 198, 200, 212–3, 214, 223, 226, 251–2, 259–73, 288, 292–3, 303–4

Langford, Nathaniel 16–17, 22
Latin America 35
Little Bighorn Battlefield National Monument (USA) 75
Loch Lomond National Park (Scotland) 89
Ludwigshöhe National Park (Poland) 42

Maasai 265
Mackinac National Park (USA) 27
Malaysia 46, 71
Mammoth Cave (USA) 82
Mammoth Hot Springs (USA) 21
Management *see* national park management
Maori 266, 267, 271
Marine parks 118–27
Marine reserves 15
Mars 310
Marsh, G.P. 51
Mather, Stephen 34
Mesa Verde National Park (USA) 266
Mexico 35
Military 28–9
Mining 33, 47, 48, 53, 54, 59, 106, 187, Monumentalism 214
Moors 72
Mootwingee Historic Site (Australia) 278
Mount Field National Park (Australia) 59
Mount Kenya National Park (Kenya) 40
Mount Rainier National Park (USA) 28, 48, 84
Muir, John 17, 81, 88–101, 305
Murray River Reserve (Australia) 54
Mussolini, Benito 43, 70

Nairobi National Park (Kenya) 40
Nash, Roderick 30
Natal Game Protection Association 39
National identity 9, 26, 41, 44, 63–77
National Mall and Memorial Parks (National Mall) (USA) 282–97
National monuments 73
National park management 11, 90, 100, 106, 110, 111, 116, 121, 138, 151, 163, 199, 215, 228, 230, 237, 251, 259, 265, 267–9, 273–6, 303, 306; *see also* co-management
National park development models 30–2: New World 32–7; Developing countries 37–41; European 41–3; Totalitarian countries 43–4

356 Index

National parks: definition of 3–15, 22–3, 36, 120, 156, 200, 214, 225, 242, 244–6, 256, 268, 301–3; as national institutions 26–7; renaming 75, 256; spiritual values 270–1
National Parks Act (Canada) 106–7, 108, 109
National Parks Act (New Zealand) 10, 11
National Parks and Access to the Countryside Act 1949 6–7
National Parks and Conservation Association (NPCA) 88
National Pleasure Resorts Act (South Australia) 56
Netherlands 38
Netherlands Indies Society for the Protection of Nature 38
New South Wales National Parks and Wildlife Service 10
New Zealand 10–11, 34–5, 46, 266, 267, 271, 277–8
Niagara Falls 18
Niagara Falls Reservation 27
Northern Pacific Railroad Company 19, 21, 48, 81
Norway 303–4

Olmsted, Frederick Law 25–6
Ordesa National Park (Spain) 148

Pacific Rim National Park (Canada) 5, 107, 276–7
Pallin, Paddy 81
Parks movement 4–5, 81
Peak District National Park (England) 167–183
Photography 83–4
Pilanesberg National Park (South Africa) 254–6
Pinchot, Gifford 97, 98–9
Platt National Park 28
Poland 41, 42
Population growth 303
Post-colonialism 71, 188, 304, 306
Preservation 10, 11, 34, 37, 47, 49, 50, 58–9, 67, 89, 95, 98, 100, 101, 102, 103, 106, 107, 109, 120, 138, 140, 160, 184, 196, 198, 207, 214, 223, 235, 247, 268, 269, 286, 301, 305, 308, 310; *see also* conservation
Pride and Prejudice 177
Protected Area Management Categories 8–9, 12–13, 14, 36, 120, 156, 200, 214, 225, 242, 244–6, 256, 268, 301–3

Public parks 24–6, 91
Punkaharju Forest Park (Finland) 185

Railways 19, 21, 48–9, 53, 55, 56, 96, 98, 184–7, 193–5
Recreation 5, 6, 25, 36, 42, 52, 55–6, 94, 101, 107, 155, 175, 177, 198–9, 266
Redwood (*Sequoia Gigantea*) 18, 66
Renaissance 81–2
Roads 86–7, 96
Robertson, Sir John 3–4
Robben Island World Heritage Site (South Africa) 240
Rocky Mountains National Park (Canada) 33, 104–5
Rocky Mountains National Park (USA) 28, 48
Rocky Mountains School of Landscape Painting 66
Romania 41, 65, 70
Romantic movement 23–4, 68, 145, 185
Roosevelt, Theodore 17, 24, 37, 67
Royal National Park (Australia) 3, 5–6, 35, 52–3
Royal Society of New South Wales 3
Runte, Alfred 45, 46–7, 49, 61, 67, 302
Russia 43, 46
'Rustic architecture' 83

Sami 186, 188, 193, 195, 200
Scenery 10, 21, 47, 59
Scenery Preservation Act (Tasmania) 59
Scotland 42, 65, 71, 89; Dunbar 89
Sequoia National Park (USA) 28, 84
Sierra Club 88, 95–6, 98, 305
Sierra Nevada National Park (Spain) 150
Society for the Preservation of Monuments of Nature in the Netherlands 38
Solomon Islands 46
South Africa 39, 46, 75, 228–55, 267; KwaZulu-Natal 242
South America 265
Spain 41, 143–54; Canary Islands 151
Steel, William Gladstone 88–101
Sublime 24, 81, 82–4, 145, 158
Sully's Hill National Park 28
Sweden 41, 184–96, 197–207
Switzerland 41, 67

Taburiente Crater Park (Spain) 151
Tallgrass Prairie National Preserve (USA) 88
Teide National Park (Spain) 151
Thailand 221
Thermal Springs Districts Act 1881 (New Zealand) 7
Thy National Park (Denmark) 72
Tongariro National Park (New Zealand) 34, 267
Tourism 9, 17–22, 25, 33, 35, 44, 49, 52, 59, 63–77, 90, 93, 94, 100, 101, 102, 104, 107, 111, 115–17, 121–3, 131–7, 151, 155, 157–60, 165, 169, 172, 189–90, 197–8, 218–20, 240, 266
Tower Hill National Park (Australia) 56
Tragedy of the Commons 222
Transport 19, 84–7, 96, 98, 120, 122, 132, 176, 185–6, 192–4, 195, 212, 232, 246, 267, 289, 291; *see also* automobile, railways
Tsavo National Park (Kenya) 40, 252, 265
Turkey 73

Uganda 39
Uluru-Kata Tjuta National Park (Australia) 37, 75, 128–39, 266
United States 3, 16–29, 30–1, 45, 65–8, 73, 75, 81–7, 88–101, 266, 282–97; Alaska 88; Arizona 264; California 18, 23, 26, 67, 87, 88, 264 [San Franciso 26]; Kentucky 82; Montana 21–2, 264; Nevada 88; New York 27; North Dakota 28; Oklahoma 28; South Dakota 264; Washington DC 282–97; Wyoming 3, 22, 81, 272
United States Forest Service 99, 100
United States National Park Service 32, 34, 73

Urban national parks 282–97
Urban parks 3, 6, 25, 216, 282–97

Vancouver Island 277
Vietnam 221

Wallace Monument (Scotland) 65
Washburn Expedition 16–18, 24, 26
Waterton Lakes National Park (Canada) 33
Wicklow Mountains National Park (Ireland) 73
Wilderness 10, 11, 23–4, 30, 33, 49, 52, 59, 68, 72, 103, 107, 120, 124, 161, 175, 189, 195, 211, 212, 213, 216, 217, 222, 224, 239, 243, 250, 256, 259, 268, 276, 302
Wilderness Society (USA) 88
Wilson's Promontory National Park (Australia) 56
Wordsworth, William 23
World Heritage 9, 14, 32, 114–27, 128, 138, 165, 197, 200, 221, 227, 228, 230–1, 239, 240, 242, 270, 277
'Worthless lands hypothesis' 34, 63–77, 266
Wuyishan Scenic Park (China) 226–37

Yellowstone National Park (USA) 3, 7, 16–29, 39, 43, 47–8, 52, 53, 54, 67, 81, 83, 84, 85, 92, 147, 184, 190, 239, 243, 244, 245, 259, 263–4, 268, 269, 303
Yosemite National Park (USA) 23, 28, 47, 66–7, 83, 84, 88–101, 184, 264
Yugoslavia 71

Zhangjiajie National Park (China) 221
Zoning 108–9, 120–1, 154, 216, 301

Þingvellir National Park (Iceland) 73

eBooks – at www.eBookstore.tandf.co.uk

A library at your fingertips!

eBooks are electronic versions of printed books. You can store them on your PC/laptop or browse them online.

They have advantages for anyone needing rapid access to a wide variety of published, copyright information.

eBooks can help your research by enabling you to bookmark chapters, annotate text and use instant searches to find specific words or phrases. Several eBook files would fit on even a small laptop or PDA.

NEW: Save money by eSubscribing: cheap, online access to any eBook for as long as you need it.

Annual subscription packages

We now offer special low-cost bulk subscriptions to packages of eBooks in certain subject areas. These are available to libraries or to individuals.

For more information please contact webmaster.ebooks@tandf.co.uk

We're continually developing the eBook concept, so keep up to date by visiting the website.

www.eBookstore.tandf.co.uk

LIVERPOOL INSTITUTE
OF HIGHER EDUCATION

LIBRARY

WOOLTON ROAD,
LIVERPOOL, L16 8ND

Economic Survey of the Baltic States